Ang Yang, Yin Shan and Lam Thu Bui (Eds.)

Success in Evolutionary Computation

Studies in Computational Intelligence, Volume 92

Editor-in-chief
Prof. Janusz Kacprzyk
Systems Research Institute
Polish Academy of Sciences
ul. Newelska 6
01-447 Warsaw
Poland
E-mail: kacprzyk@ibspan.waw.pl

Further volumes of this series can be found on our
homepage: springer.com

Vol. 68. Cipriano Galindo, Juan-Antonio
Fernández-Madrigal and Javier Gonzalez
*A Multi-Hierarchical Symbolic Model of the Environment
for Improving Mobile Robot Operation*, 2007
ISBN 978-3-540-72688-3

Vol. 69. Falko Dressler and Iacopo Carreras (Eds.)
*Advances in Biologically Inspired Information Systems:
Models, Methods, and Tools*, 2007
ISBN 978-3-540-72692-0

Vol. 70. Javaan Singh Chahl, Lakhmi C. Jain,
Akiko Mizutani and Mika Sato-Ilic (Eds.)
Innovations in Intelligent Machines-1, 2007
ISBN 978-3-540-72695-1

Vol. 71. Norio Baba, Lakhmi C. Jain and Hisashi Handa
(Eds.)
*Advanced Intelligent Paradigms in Computer
Games*, 2007
ISBN 978-3-540-72704-0

Vol. 72. Raymond S.T. Lee and Vincenzo Loia (Eds.)
Computation Intelligence for Agent-based Systems, 2007
ISBN 978-3-540-73175-7

Vol. 73. Petra Perner (Ed.)
Case-Based Reasoning on Images and Signals, 2008
ISBN 978-3-540-73178-8

Vol. 74. Robert Schaefer
Foundation of Global Genetic Optimization, 2007
ISBN 978-3-540-73191-7

Vol. 75. Crina Grosan, Ajith Abraham and Hisao Ishibuchi
(Eds.)
Hybrid Evolutionary Algorithms, 2007
ISBN 978-3-540-73296-9

Vol. 76. Subhas Chandra Mukhopadhyay and Gourab Sen
Gupta (Eds.)
Autonomous Robots and Agents, 2007
ISBN 978-3-540-73423-9

Vol. 77. Barbara Hammer and Pascal Hitzler (Eds.)
Perspectives of Neural-Symbolic Integration, 2007
ISBN 978-3-540-73953-1

Vol. 78. Costin Badica and Marcin Paprzycki (Eds.)
Intelligent and Distributed Computing, 2008
ISBN 978-3-540-74929-5

Vol. 79. Xing Cai and T.-C. Jim Yeh (Eds.)
*Quantitative Information Fusion for Hydrological
Sciences*, 2008
ISBN 978-3-540-75383-4

Vol. 80. Joachim Diederich
Rule Extraction from Support Vector Machines, 2008
ISBN 978-3-540-75389-6

Vol. 81. K. Sridharan
Robotic Exploration and Landmark Determination, 2008
ISBN 978-3-540-75393-3

Vol. 82. Ajith Abraham, Crina Grosan and Witold
Pedrycz (Eds.)
Engineering Evolutionary Intelligent Systems, 2008
ISBN 978-3-540-75395-7

Vol. 83. Bhanu Prasad and S.R.M. Prasanna (Eds.)
*Speech, Audio, Image and Biomedical Signal Processing
using Neural Networks*, 2008
ISBN 978-3-540-75397-1

Vol. 84. Marek R. Ogiela and Ryszard Tadeusiewicz
*Modern Computational Intelligence Methods
for the Interpretation of Medical Images*, 2008
ISBN 978-3-540-75399-5

Vol. 85. Arpad Kelemen, Ajith Abraham and Yulan Liang
(Eds.)
Computational Intelligence in Medical Informatics, 2008
ISBN 978-3-540-75766-5

Vol. 86. Zbigniew Les and Mogdalena Les
Shape Understanding Systems, 2008
ISBN 978-3-540-75768-9

Vol. 87. Yuri Avramenko and Andrzej Kraslawski
Case Based Design, 2008
ISBN 978-3-540-75705-4

Vol. 88. Tina Yu, David Davis, Cem Baydar and Rajkumar
Roy (Eds.)
Evolutionary Computation in Practice, 2008
ISBN 978-3-540-75770-2

Vol. 89. Ito Takayuki, Hattori Hiromitsu, Zhang Minjie
and Matsuo Tokuro (Eds.)
Rational, Robust, Secure, 2008
ISBN 978-3-540-76281-2

Vol. 90. Simone Marinai and Hiromichi Fujisawa (Eds.)
*Machine Learning in Document Analysis
and Recognition*, 2008
ISBN 978-3-540-76279-9

Vol. 91. Horst Bunke, Kandel Abraham and Last Mark (Eds.)
Applied Pattern Recognition, 2008
ISBN 978-3-540-76830-2

Vol. 92. Ang Yang, Yin Shan and Lam Thu Bui (Eds.)
Success in Evolutionary Computation, 2008
ISBN 978-3-540-76285-0

Ang Yang
Yin Shan
Lam Thu Bui
(Eds.)

Success in Evolutionary Computation

With 142 Figures and 68 Tables

 Springer

Ang Yang
CSIRO Land and Water
Clunies Ross Street
Black Mountain
Canberra, ACT 2601
Australia
ayang@ieee.org

Lam Thu Bui
Artificial Life and Adaptive Robotics
Laboratory (ALAR)
School of Information Technology
and Electrical Engineering
University of New South Wales
Northcott Drive
Canberra, ACT 2600
Australia
L.Bui@adfa.edu.au

Yin Shan
Data Analysis and Research
Coordination
Program Review Division
Medicare Australia
Canberra, ACT 2901
Australia
yin.shan@logichoice.com

ISBN 978-3-540-76285-0   e-ISBN 978-3-540-76286-7

Studies in Computational Intelligence ISSN 1860-949X

Library of Congress Control Number: 2007939404

Cover design: Deblik, Berlin, Germany

Printed on acid-free paper

9 8 7 6 5 4 3 2 1

springer.com

Preface

Darwinian evolutionary theory is one of the most important theories in human history for it has equipped us with a valuable tool to understand the amazing world around us. There can be little surprise, therefore, that Evolutionary Computation (EC), inspired by natural evolution, has been so successful in providing high quality solutions in a large number of domains as diverse as engineering and finance.

EC includes a number of techniques, such as Genetic Algorithms, Genetic Programming, Evolution Strategy and Evolutionary Programming, which have been used in a diverse range of highly successful applications. This book brings together some of these EC applications in fields including electronics, telecommunications, health, bioinformatics, supply chain and other engineering domains, to give the audience, including both EC researchers and practitioners, a glimpse of this exciting and rapidly evolving field. We believe that it is also important for the users of EC to be aware of the theoretical basis underlying EC to unleash the potential for powerful tools that will emerge from the developments and advances in theory. Therefore, this book also includes several contributions concerning the latest theoretical progress in EC which are likely to be immediately applicable to solving problems in the real world.

The contributors' responses to this book were positive with 31 full paper submissions. These submissions were strictly peer reviewed by at least two and generally three experts in the field to ensure the quality of the book. We accepted only 16 papers resulting in an acceptance rate of 52%.

We thank Dr. Janusz Kacprzyk for giving us the opportunity to bring together the latest advances in EC application and for his advice through the editing process. We cannot overstate the importance of the advice from the editors from Springer, Dr. Thomas Ditzinger and Heather King. We express our gratitude to Dr. Hussein Abbass for his help, whose assistance is always available. We also thank the Prediction and Reporting Technologies group in the Land and Water Division of the Commonwealth Scientific and Industrial Research Organization, the Data Analysis and Research Coordination Section of Medicare Australia and Artificial Life and Adaptive Robotics Laboratory (ALAR) at the University of New South Wales at ADFA for their support during the editing of this book. We also had great personal support from David Collings, Sun Xuemei and our families without whom this project would not have been possible.

Canberra, Australia   *Ang Yang, Yin Shan, Lam Thu Bui*
August 2007

Contents

Part I Theory

Adaptation of a Success Story in GAs:
Estimation-of-Distribution Algorithms for Tree-based
Optimization Problems
P.A.N. Bosman and E.D. de Jong 3

The Automated Design of Artificial Neural Networks
Using Evolutionary Computation
J.-Y. Jung and J.A. Reggia 19

A Versatile Surrogate-Assisted Memetic Algorithm
for Optimization of Computationally Expensive
Functions and its Engineering Applications
Y. Tenne and S.W. Armfield 43

Data Mining and Intelligent Multi-Agent Technologies
in Medical Informatics
F. Shadabi and D. Sharma 73

Part II Applications

Evolving Trading Rules
Adam Ghandar, Zbigniew Michalewicz, Martin Schmidt,
Thuy-Duong Tò, and Ralf Zurbrugg 95

A Hybrid Genetic Algorithm for the Protein Folding
Problem Using the 2D-HP Lattice Model
Heitor S. Lopes and Marcos P. Scapin 121

Optimal Management of Agricultural Systems
D.G. Mayer, W.A.H. Rossing, P. deVoil, J.C.J. Groot, M.J. McPhee,
and J.W. Oltjen 141

Evolutionary Electronics: Automatic Synthesis of Analog
Circuits by GAs
Esteban Tlelo-Cuautle and Miguel A. Duarte-Villaseñor 165

Intuitive Visualization and Interactive Analysis of Pareto
Sets Applied on Production Engineering Systems
H. Müller, D. Biermann, P. Kersting, T. Michelitsch, C. Begau,
C. Heuel, R. Joliet, J. Kolanski, M. Kröller, C. Moritz, D. Niggemann,
M. Stöber, T. Stönner, J. Varwig, and D. Zhai 189

Privacy Protection with Genetic Algorithms
Agusti Solanas 215

A Revision of Evolutionary Computation Techniques
in Telecommunications and An Application
for The Network Global Planning Problem
Pablo Cortés, Luis Onieva, Jesús Muñuzuri, and Jose Guadix 239

Survivable Network Design with an Evolution Strategy
Volker Nissen and Stefan Gold 263

Evolutionary Computations for Design Optimization
and Test Automation in VLSI Circuits
A.K. Palit, K.K. Duganapalli, K. Zielinski, D. Westphal, D. Popovic,
W. Anheier, and R. Laur 285

Evolving Cooperative Agents in Economy Market Using
Genetic Algorithms
Raymond Chiong 313

Optimizing Multiplicative General Parameter Finite Impulse
Response Filters Using Evolutionary Computation
Jarno Martikainen and Seppo J. Ovaska 327

Applying Genetic Algorithms to Optimize the Cost
of Multiple Sourcing Supply Chain Systems – An Industry
Case Study
Kesheng Wang and Y. Wang 355

Part I

Theory

Adaptation of a Success Story in GAs: Estimation-of-Distribution Algorithms for Tree-based Optimization Problems

Peter A.N. Bosman[1] and Edwin D. de Jong[2]

[1] Centre for Mathematics and Computer Science, P.O. Box 94079, 1090 GB Amsterdam, The Netherlands, `Peter.Bosman@cwi.nl`
[2] Institute of Information and Computing Sciences, Utrecht University, P.O. Box 80.089, 3508 TB Utrecht, The Netherlands, `dejong@cs.uu.nl`

Summary. Fundamental research into Genetic Algorithms (GA) has led to one of the biggest successes in the design of stochastic optimization algorithms: Estimation-of-Distribution Algorithms (EDAs). These principled algorithms identify and exploit structural features of a problems structure during optimization. EDA design has so far been limited to classical solution representations such as binary strings or vectors of real values. In this chapter we adapt the EDA approach for use in optimizing problems with tree representations and thereby attempt to expand the boundaries of successful evolutionary algorithms. To do so, we propose a probability distribution for the space of trees, based on a grammar. To introduce dependencies into the distribution, grammar transformations are performed that facilitate the description of specific subfunctions. The results of performing experiments on two benchmark problems demonstrate the feasibility of the approach.

1 Introduction

Research into evolutionary computation (EC) has shown that effective optimization with evolutionary algorithms (EAs) requires an EA to have an inductive bias that can be fit to the structure of the problem under study quickly and reliably. For a fixed representation of solutions, the use of non-adaptive recombination operators such as classical crossover operators will not lead to efficient optimization unless the bias introduced by the recombination operator fits the problem's structure [28, 29]. Typically, the required resources to optimally solve the optimization problem grow exponentially with the problem size. In contrast, using the optimal crossover operator the required resources to solve the optimization problem optimally typically grow only polynomially with the problem size with the polynomial being of low order [9]. For more efficient optimization, therefore, either the representation has to be adapted to fit the bias of the recombination operator, or the operator has to be adapted to fit the structure of the optimization problem. The

P.A.N. Bosman and E.D. de Jong: *Adaptation of a Success Story in GAs: Estimation-of-Distribution Algorithms for Tree-based Optimization Problems*, Studies in Computational Intelligence (SCI) **92**, 3–18 (2008)
`www.springerlink.com`

latter approach is the most commonly used, that is, problem-specific recombination operators are designed. If we assume to have no prior knowledge of the
problem under study, however, it is impossible to do so. The only alternative
then is to expand the capacity of the model in the recombination operator so
that it can be adapted to the optimization problem.

The possibility to adapt the model to the optimization problem by induc-
ing it from the selected solutions contributes to the attractiveness of EAs.
To adapt the model in a meaningful way, induction must be used on previously evaluated solutions. Because EAs work with a population of solutions,
they are natural candidates for optimization based on model adaptation by
induction since all kinds of statistical methods can be used that work on
collections of samples, such as the estimation of probability distributions.
Estimation-of-distribution algorithms (EDAs) represent one of the most successful types of EA that employ model-adaptation [13, 15, 21]. As such, EDAs
can be seen as an important success in EC. EDAs provide an elegant way
of bringing about an adaptive search bias. The model reflecting the bias is a
probability distribution. Fitting the model to the problem's structure is done
by estimating the probability distribution from the set of selected solutions.
The probability distribution itself is an explicit model for the inductive search
bias. Estimating the probability distribution from data corresponds to tuning
the model for the inductive search bias. Because a lot is known about how
probability distributions can be estimated from data [5, 14] the flexibility of
the inductive search bias of EDAs is potentially large. In addition, the tuning
of the inductive search bias in this fashion also has a rigorous foundation in
the form of the well-established field of probability theory.

Although clearly more computational effort is required to tune the inductive search bias by estimating the probability distribution, the payoff has been
shown to be worth the computational investment (e.g. polynomial scale-up
behavior instead of exponential scale-up behavior and successful applications
to important problems) [1, 4, 8, 17, 18, 20]. It is now interesting to investigate
how this success in EC can be carried over to the domain of solution representations without a fixed length, which are typically trees. Trees are the common
type of representation in genetic programming (GP), a very important area
of evolutionary computation [7, 11].

GP is an important area in EC that offers algorithms to search highly
expressive classes of functions, and has been applied to a diverse range of
problems including circuit design, symbolic regression, and control. Basic GP
employs the subtree-crossover operator, which exchanges randomly selected
subtrees between individuals. Due to the particular structure of trees, subtrees
(rather than e.g. any combination of nodes and arcs) appear a reasonable
choice as the form of the partial solutions that will be exchanged since the
functionality of a subtree is independent of its place within the tree. However,
basic GP does not make informed choices as to which partial solutions will
be exchanged; both the size of the removed subtree and the position where it
will be inserted are chosen randomly.

Several subtree encapsulation methods exist that make more specific choices as to which partial solutions are selected to be further propagated, e.g. GLiB [2], ADFs [12], and ARL [24]. Subtree encapsulation methods have been found to substantially improve GP performance. Yet, the criteria for the selection of partial solutions they employ are still heuristic; typically, either the fitness of the tree in which a subtree occurs is used as an indication of its value, or the partial solution is itself subject to evolution.

In this chapter we explore how the success of EDAs may be used in GP. We investigate whether a more principled approach to the selection of partial solutions and their recombination is possible. If the distribution of high-fitness trees can be estimated, this would directly specify which combinations of elements are to be maintained in the creation of new individuals and thus which combinations of partial solutions may be fruitfully explored. We investigate how the principle of distribution estimation may be employed in the context of tree-based problems. In GAs, the development of EDAs and other linkage learning techniques has yielded a better insight into the design of competent GAs by rendering the assumptions implicitly made by algorithms explicit. Our aim is that the application of distribution estimation techniques to tree-based problems may likewise clarify the design of principled methods for GP. This chapter represents a first step in that direction.

To estimate distributions over the space of trees, a representation must be chosen. In Probabilistic Incremental Program Evolution (PIPE) [25], trees are matched on a fixed-size template, such that the nodes becomes uniquely identifiable variables. The size of the template may increase over time however as trees grow larger. Using this representation, all nodes are treated equally. While this permits the encapsulation of any combination of nodes, it does not exploit the particular non-fixed-size and variable–child–arity structure of trees. The complexity of such an algorithm is determined by the maximally allowed shape for the trees, which must be chosen in advance. In PIPE the distribution is the same for each node, which corresponds to a univariately factorized probability distribution over all nodes. In Extended Compact Genetic Programming (ECGP) [26], trees are represented in the same way as in PIPE. However, the probability distribution that is estimated is a marginal-product factorization that allows the modelling of dependencies between multiple nodes that are located anywhere in the fixed-size template. Moreover, the size of the template is fixed beforehand in ECGP and does not increase over time. In this chapter, a method will be employed that estimates the distribution of trees based on the subtrees that actually occur. The representation specifies a set of rules whose expansion leads to trees. The rules capture local information, thereby offering a potential to exploit the specific structure of trees, while at the same time their use in an EDA offers a potential for generalization that is not provided by using fixed-size templates. Shan et al. [27] use a similar approach (based on stochastic grammars). The main difference is that their approach starts from an explicit description of a specific set of trees and then generalizes the description to represent common subtrees.

In our approach we do the reverse. Our approach starts from a description of all possible trees and then tunes it to more specifically describe the set of trees.

The remainder of this chapter is organized as follows. In Sect. 2 we introduce basic terminology. In Sect. 3 we define a probability distribution over trees. We also show how to draw new samples from the distribution and how the distribution is estimated from data. In Sect. 4 we perform experiments on two benchmark problems and compare the performance of standard GP and an EDA based on the proposed distribution. We conclude this chapter in Sect. 5.

2 Terminology

A grammar G is a vector of l_r production rules $R_j, j \in \{0, 1, \ldots, l_r - 1\}$, that is, $G = (R_0, R_1, \ldots, R_{l_r-1})$.

A production rule is denoted by $R_j : S_k \rightarrow E_j$, where S_k is a symbol that can be replaced with the expansion E_j of the production rule. Let K be the number of available symbols, then $k \in \{0, 1, \ldots, K - 1\}$. We will use only one symbol and allow ourselves to write S instead of S_k.

An expansion E_j of production rule R_j is a tree. We will therefore generally call E_j an expansion tree. The internal nodes of an expansion tree are functions with at least one argument. The leaves are either symbols, constants or input variables. An example of an expansion tree is $E_j = +(\sin(S)), (-(\log(S)), \cos(S)))$ which, in common mathematical notation, represents $\sin(S) + (\log(S) - \cos(S))$.

A sentence is obtained from a grammar if a production rule is chosen to replace a symbol repeatedly until all symbols have disappeared. Sentences can therefore be seen as trees. We denote a sentence by s. We will denote a subtree of a sentence by t and call it a sentence subtree.

A sentence subtree t is said to be matched by an expansion tree E_j, denoted $t \in E_j$ if and only if all nodes of E_j coincide with nodes found in t following the same trails, with the exception of symbol nodes, which may be matched to any non-empty sentence subtree.

The following grammar $G = (R_0, R_1, \ldots, R_{5+n})$ with $l_r = 6 + n$ is an example of a grammar that describes certain n-dimensional real-valued functions:

$$R_0 : S \rightarrow c \text{ (a constant} \in \mathbb{R}) \qquad R_{n+1} : S \rightarrow +(S, S)$$
$$R_1 : S \rightarrow i_0 \text{ (input variable 0)} \qquad R_{n+2} : S \rightarrow \cdot(S, S)$$
$$R_2 : S \rightarrow i_1 \text{ (input variable 1)} \qquad R_{n+3} : S \rightarrow -(S)$$
$$\vdots \qquad \vdots \qquad\qquad R_{n+4} : S \rightarrow \sin(S)$$
$$R_n : S \rightarrow i_{n-1} \text{ (input variable n–1)} \qquad R_{n+5} : S \rightarrow \cos(S)$$

3 Probability Distribution

3.1 Definition

To construct a probability distribution over sentences, we introduce a random variable S that represents a sentence and a random variable T that represents a sentence subtree. Because sentence subtrees are recursive structures we define a probability distribution for sentence subtrees recursively. To do so, we must know where the tree terminates. This information can be obtained by taking the depth of a sentence subtree into account. Let $P^G(T = t|D = d)$ be a probability distribution over all sentence subtrees t that occur at depth d in a sentence. Now, we define the probability distribution over sentences s by:

$$P^G(S = s) = P^G(T = s|D = 0) \tag{1}$$

Since sentence subtrees are constructed using production rules, we can define $P^G(T = t|D = d)$ using the production rules. Since there is only one symbol, we can also focus on the expansion trees. Although depth can be used to model the probability of terminating a sentence at some node in the tree, sentences can be described more precisely if depth is also used to model the probability of occurrence of functions at specific depths. Preliminary experiments indicated that this use of depth information leads to better results.

We define $P_j^E(J = j|D = d)$ to be a discrete conditional probability distribution that models the probability of choosing expansion tree E_j, $j \in \{0, 1, \ldots, l_r\}$ at depth d when constructing a new sentence.

We assume that the values of the constants and the indices of the input variables in an expansion tree are not dependent on the depth. Conforming to this assumption we define $2l_r$ multivariate probability distributions that allow us to model the use of constants and inputs inside production rules other than the standard rules $S \to c$ and $S \to i_k$, $k \in \{0, 1, \ldots, n - 1\}$:

- $P_j^C(\boldsymbol{C}_j)$, $j \in \{0, 1, \ldots, l_r - 1\}$, a probability distribution over all constants in expansion tree E_j, where $\boldsymbol{C}_j = (C_{j0}, C_{j1}, \ldots, C_{j(n_{C_j}-1)})$. Each C_{jk} is a random variable that represents a constant in E_j.
- $P_j^I(\boldsymbol{I}_j)$, $j \in \{0, 1, \ldots, l_r - 1\}$, a probability distribution over all inputs in expansion tree E_j, where $\boldsymbol{I}_j = (I_{j0}, I_{j1}, \ldots, I_{j(n_{I_j}-1)})$. Each I_{jk} is a random variable that represents input variables, i.e. $I_{jk} \in \{0, 1, \ldots, n-1\}$.

The definition of $P_j^I(\boldsymbol{I}_j)$ enforces a single production rule $S \to i$, where i represents all input variables, instead of n production rules $S \to i_j$, $j \in \{0, 1, \ldots, n - 1\}$. This reduces the computational complexity for estimating the probability distribution, especially if n is large. However, it prohibits the introduction of production rules that make use of specific input variables.

We will enforce that any sentence subtree t can be matched by only one expansion tree E_j. The probability distribution over all sentence subtrees at some given depth $D = d$ is then the product of the probability of matching the sentence subtree with some expansion tree E_j and the product of all (recursive) probabilities of the sentence subtrees located at the symbol–leaf nodes in E_j.

Let S_{jk} be the kth symbol in expansion tree E_j, $k \in \{0, 1, \ldots, n_{S_j} - 1\}$. Let $stree(S_{jk}, t)$ be the sentence subtree of sentence subtree t at the same location where S_{jk} is found in the expansion tree E_j that matches t. Let $depth(S_{jk})$ be the depth of S_{jk} in expansion tree E_j. Let $match(t)$ be the index of the matched expansion tree, i.e. $match(t) = j \Leftrightarrow t \in E_j$. We then have:

$$P^G(T = t | D = d) = \quad (2)$$

$$P^E(J = j | D = d) P_j^C(\boldsymbol{C}_j) P_j^I(\boldsymbol{I}_j) \prod_{k=0}^{n_{S_j} - 1} P^G(T = stree(S_{jk}, t) | D = d + depth(S_{jk}))$$

$$\text{where } j = match(t)$$

Summarizing, formulating a probability distribution for sentences requires:

- $P^E(J|D)$; a univariate discrete conditional probability distribution over all possible production rules, conditioned on the depth of occurrence.
- $P_j^C(\boldsymbol{C}_j)$; l_r multivariate probability distributions over n_{C_j} variables.
- $P_j^I(\boldsymbol{I}_j)$; l_r multivariate probability distributions over n_{I_j} variables.

3.2 Estimation from Data

General Greedy Approach

We choose a greedy approach to estimate $P^G(S)$ from a set $\boldsymbol{S}$ of sentences. Greedy approaches have been used in most EDAs so far and have led to good results [3, 13, 16, 19]. The algorithm starts from a given initial grammar. Using transformations, the grammar is made more involved. A transformation results in a set of candidate grammars, each slightly different and each more involved than the current grammar. The goodness of each candidate grammar is then evaluated and the best one is accepted if it is better than the current one. If there is no better grammar, the greedy algorithm terminates.

Transformations

The only transformation that we allow is the substitution of one symbol of an expansion tree with one expansion tree from the base grammar. The base grammar is the grammar that is initially provided. To ensure that after a transformation any sentence subtree can be matched by only one expansion tree we assume that the base grammar has this property. To transform the grammar we only allow to substitute symbols with expansion trees of the base grammar. Here is an example of expanding the base grammar (first column) using a single expansion (second column):

Base grammar	Single expansion	Full expansion
$R_0 : S \rightarrow c$	$R_0 : S \rightarrow c$	$R_0 : S \rightarrow c$
$R_1 : S \rightarrow i$	$R_1 : S \rightarrow i$	$R_1 : S \rightarrow i$
$R_2 : S \rightarrow f(S, S)$	$R_2 : S \rightarrow f(S, S)$	$R_2 : S \rightarrow f(S, S)$
$R_3 : S \rightarrow g(S, S)$	$R_3 : S \rightarrow g(S, S)$	$R_3 : S \rightarrow g(c, S)$
	$R_4 : S \rightarrow g(f(S, S), S)$	$R_4 : S \rightarrow g(i, S)$
		$R_5 : S \rightarrow g(f(S, S), S)$
		$R_6 : S \rightarrow g(g(S, S), S)$

Note that the set of expansion trees can now no longer be matched uniquely to all sentence trees. For instance, sentence $g(f(c, c), c)$ can at the top level now be matched by expansion trees 3 and 4. To ensure only a single match, we could expand every expansion tree from the base grammar into a production rule and subsequently remove the original production rule that has now been expanded (third column in the example above). However, this rapidly increases the number of production rules in the grammar and may introduce additional rules that aren't specifically interesting for modelling the data at hand. To be able to only introduce the rules that are interesting, we equip the symbols with a list of indices that indicate which of the production rules in the base grammar may be matched to that symbol. Once a substitution occurs, the symbol that was instantiated may no longer match with the expansion tree that was inserted into it. For example:

Base grammar	Expanded grammar
$R_0 : S \rightarrow c$	$R_0 : S \rightarrow c$
$R_1 : S \rightarrow i$	$R_1 : S \rightarrow i$
$R_2 : S \rightarrow f(S^{0,1,2,3}, S^{0,1,2,3})$	$R_2 : S \rightarrow f(S^{0,1,2,3}, S^{0,1,2,3})$
$R_3 : S \rightarrow g(S^{0,1,2,3}, S^{0,1,2,3})$	$R_3 : S \rightarrow g(S^{0,1,3}, S^{0,1,2,3})$
	$R_4 : S \rightarrow g(f(S^{0,1,2,3}, S^{0,1,2,3}), S^{0,1,2,3})$

A sentence can now be preprocessed bottom-up in $\mathcal{O}(n)$ time to indicate for each node which expansion tree matches that node, where n is the number of nodes in the tree. The sentence can then be traversed top-down to perform the frequency count for the expansion trees. It should be noted that this approach means that if a symbol list does not contain all indices of the base grammar, then it represents only the set of the indicated rules from the base grammar. In the above example for instance $S^{0,1,3}$ represents $S \rightarrow c$, $S \rightarrow i$ and $S \rightarrow g(S^{0,1,2,3}, S^{0,1,2,3})$. Therefore, the probability associated with this particular symbol is not recursive application of the distribution in (2), but is uniform over the indicated alternatives. This is comparable to the approach of default tables for discrete random variables in which instead of indicating a probability for all possible combinations of values for the random variables, only a subset of them is explicitly indicated. All remaining combinations are assigned an equal probability such that the distribution sums to one over all possible values.

Goodness Measure

The goodness of a grammar is determined using its associated probability distribution. This distribution can be estimated by traversing each sentence and by computing the proportions of occurrence for the expansion trees at each depth. Probability distributions $P_j^C(C_j)$ and $P_j^I(I_j)$ can be estimated after filtering the sentences to obtain for each expansion tree a set of sets of constants and a set of sets of variables (one set of constants or variables for each match of the expansion tree). From these sets, the distributions $P_j^C(C_j)$ and $P_j^I(I_j)$ can be estimated using well-known density estimation techniques (such as proportion estimates for the discrete input variable indices).

Once all parameters have been estimated, the probability-distribution value of each sentence in the data can be computed to obtain the likelihood (or the negative log-likelihood). The goodness measure that we ultimately use to distinguish between probability distributions is the MDL metric [23], which is a form of likelihood penalization, also known as the extended likelihood principle [6]:

$$\text{MDL}\left(P^{\boldsymbol{G}}(S)\right) = -\sum_{i=0}^{|\boldsymbol{\mathcal{S}}|-1} \ln\left(P^{\boldsymbol{G}}(S=\boldsymbol{\mathcal{S}}_i)\right) + \frac{1}{2}\ln(|\boldsymbol{\mathcal{S}}|)|\boldsymbol{\theta}| + \zeta \qquad (3)$$

where $|\boldsymbol{\theta}|$ denotes the number of parameters that need to be estimated in probability distribution $P^{\boldsymbol{G}}(T)$ and ζ denotes the expected additional number of bits required to store the probability distribution. In our case, $|\boldsymbol{\theta}|$ equals:

$$|\boldsymbol{\theta}| = 2d_{\max} + l_r - 4 + \left(\sum_{j=0}^{l_r-1} |\boldsymbol{\theta}| \leftarrow P_j^C(C_j)\right) + \left(\sum_{j=0}^{l_r-1} n^{n_{l_j}} - 1\right) \qquad (4)$$

Since in the latter case, the number of parameters grows exponentially with the number of input variables in an expansion tree, this is likely to be a strong restriction on the number of input variables that the greedy algorithm will allow into any expansion tree. As an alternative, independence can be enforced between input variables. The resulting distribution can then no longer model correlation between multiple input variables in one expansion tree. The number of parameters to be estimated reduces drastically however and this in turn will allow more input variables to be incorporated into the expansion trees. Effectively, we thus enforce $P_j^I(\boldsymbol{I}_j) = \prod_{k=0}^{n_{I_j}-1} P_j^I(I_{jk})$, which changes the last term in the number of parameters $|\boldsymbol{\theta}|$ into $\sum_{j=0}^{l_r-1}(n-1)n_{I_j}$.

The expected additionally required number of bits ζ in our case come from the production rules. The more production rules, the more bits are required. Moreover, longer production rules require more bits. The number of bits required to store a production rule is $\ln(l_{r_{base}})$ times the number of internal nodes in the expansion tree where $l_{r_{base}}$ is the number of production rules in the base grammar. Finally, symbols have lists of production-rule indices. To store one such list, the number of required bits equals $\ln(l_{r_{base}})$ times the length of that list. The value of ζ equals the sum of these two terms over all production rules.

3.3 Sampling

1. Initialize the sentence tree of the new sentence: introduce a single symbol as the root.
2. Set $d = 0$.
3. Visit the symbol at the root.
4. Draw an integer $j \in \{0, 1, \ldots, l_r - 1\}$ randomly from $P^E(J|D = d)$ if the list of production-rule indices associated with the visited symbol is complete, or draw a random integer from the list otherwise.
5. Make a copy of expansion tree E_j.
6. Draw all constants in the copied tree randomly from $P_j^C(\boldsymbol{C}_j)$.
7. Draw all input indices in the copied tree randomly from $P_j^I(\boldsymbol{I}_j)$.
8. Expand the current sentence tree by replacing the symbol that is being visited with the copied tree.
9. Extract the symbols from the copied tree and their relative depths.
10. Recursively visit each symbol in the copied tree after setting the depth d properly (current depth plus relative symbol depth).

4 Experiments

We have performed experiments with our approach to estimating a probability distribution over sentences by using it in an EDA. We applied the resulting EDA as well as a standard GP algorithm that only uses subtree-swapping crossover to two GP–benchmarking problems, namely the Royal Tree Problem by Punch, Zongker and Goodman [22] and the tunable benchmark problem by Korkmaz and Üçoluk [10], which we will refer to as the binary-functions problem. Both problems are non-functional, which means that fitness is defined completely in terms of the structure of the sentence and input variables have an empty domain.

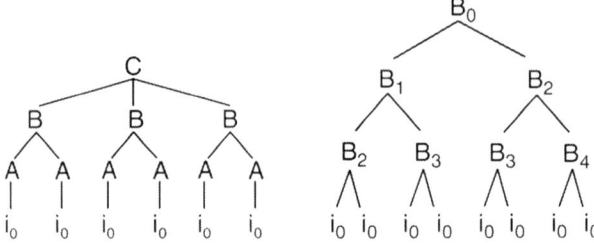

Fig. 1. Examples of sentences for the Royal Tree Problem (*left*) and the binary-functions problem (*right*)

4.1 Benchmark Problems

In the Royal Tree Problem, there is one function of arity n for each $n \in \{0, 1, 2, \ldots\}$. The function with arity 0 is i_0. The other functions are labeled A for arity 1, B for arity 2 and so on. The fitness function is defined recursively and in terms of "perfect" subtrees. A subtree is perfect if it is a full tree, i.e. all paths from the root to any leaf are equally long and the arity of any function is $n - 1$ where n is the arity its parent function. Figure 1 gives an example of a perfect tree of height 4. The fitness of a sentence is the fitness of the root. The fitness of a node is a weighted sum of the fitness values of its children. The weight of a child is

- *Full*
  if the child is a perfect subtree with a root function of arity $n - 1$ where n is the arity of the function in the current node.
- *Partial*
  if the child is not a perfect subtree but it has the correct root function.
- *Penalty*
  otherwise.

If the node is itself the root of a perfect tree, the final sum is multiplied by *Complete*. In our experiments, we follow the choices of Punch, Zongker and Goodman [22] and set *Full* $= 2$, *Partial* $= 1$, *Penalty* $= \frac{1}{3}$ and *Complete* $= 2$. The structure of this problem lies in perfect subtrees and child nodes having a function of arity that is one unit smaller than the function of their parent. Coincidentally, this results in a very strong dependence on the depth, since the optimal sentence subtree of height d has one function of arity $d - 1$ in the root, $d - 1$ functions of arity $d - 2$ on the second level and so on (see Fig. 1).

In the binary-functions problem there are various functions B_i, all of which have arity two. Similar to the Royal Tree Problem, there is only one input variable i_0. The fitness of a sentence is again the fitness of the root. The fitness of a node is 1 if the node is an input variable. Otherwise the fitness is a combination of the fitness values of its two children C_1 and C_2; it is computed according to:

$$\begin{cases} f(C_1) + f(C_2) & \text{if both children are inputs or no constraints violated} \\ \eta f(C_1) + f(C_2) & \text{if } C_1 \text{ violates first constraint, } C_2 \text{ doesn't violate first} \\ & \text{constraint and second constraint not violated} \\ f(C_1) + \eta f(C_2) & \text{if } C_2 \text{ violates first constraint, } C_1 \text{ doesn't violate first} \\ & \text{constraint and second constraint not violated} \\ \eta f(C_1) + \eta f(C_2) & \text{if both children violate first constraint or the second} \\ & \text{constraint is violated} \end{cases}$$

where the indicated constraints are:

1. The index of the parent function is smaller than the indices of its children.
2. The index of the left child is smaller than the index of the right child.

In our experiments we have used $\eta = 0.25$. The structure of this optimization problem is both local and overlapping. The local aspect lies in the fact that only a node and its direct descendants are dependent on each other. However, since these dependencies hold for each node, the structure is recursive throughout the tree and thus the dependencies are inherently overlapping.

4.2 Results

For both problems, we have computed convergence graphs for both algorithms and three different population sizes. The curves in the graphs are averaged over 25 runs. For the Royal Tree problem (see Fig. 2, left), GP achieves reasonable results for population sizes 500 and 1,000, but does not reliably find the optimum. The EDA identifies the optimal solution in every single run and for all population sizes. The Royal Tree problem is a benchmark problem for GP that features a depth-wise layered structure. Clearly, the use of depth

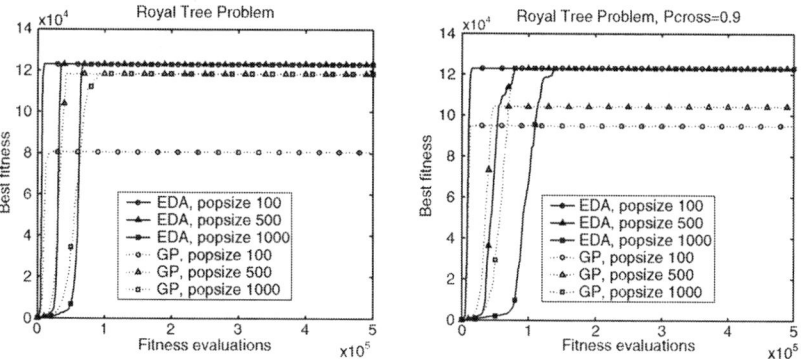

Fig. 2. Results for the Royal Tree problem

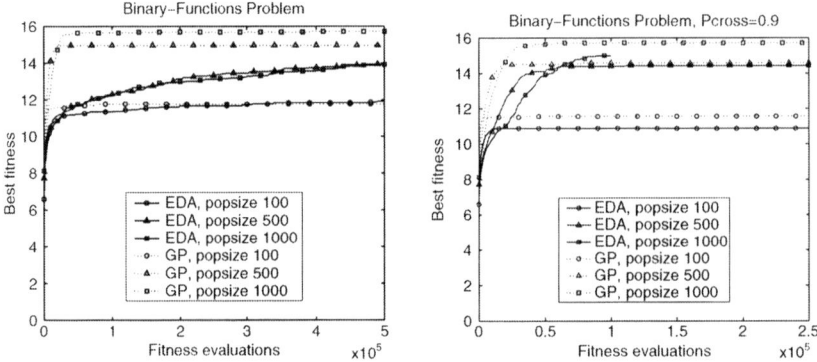

Fig. 3. Results for the binary-functions problem

information therefore renders the EDA particularly appropriate for problems of this kind. Still, this result demonstrates the feasibility of our EDA to GP. Figure 2 (right) shows the results when recombination is performed in only 90% of the cases, and copying a random parent is performed in the remaining 10%. Although GP is in this case also able to reliably find the optimum this is only the case for a population size of 1,000 whereas the EDA is still able to reliably find the optimum for all population sizes. Although lowering the probability of recombination normally speeds up convergence, in this case the EDA is only hampered by it because the distribution can perfectly describe the optimum and therefore using the distribution more frequently will improve performance. Moreover, copying introduces spurious dependencies that are estimated in the distribution and will additionally hamper optimization.

Figure 3 shows the results for the binary-functions problem (maximum fitness is 16). This problem has dependencies that are much harder for the distribution to adequately represent and reliably reproduce. Very large subfunctions are required to this end. Moreover, the multiple (sub)optima slow down convergence for the EDA as can be seen in Fig. 3 on the left. Crossover in GP is much more likely to reproduce large parts of parents. Hence, crossover automatically biases the search towards one of these solutions, allowing for faster convergence. The use of copying leads to faster convergence of the EDA. However, the deficiency of the distribution with respect to the dependencies in the problem still hamper the performance of the EDA enough to not be able to improve over standard GP. Additional enhancements may be required to make the distribution used more suited to cope with dependencies such as those encountered in the binary-functions problem, after which an improvement over standard GP may be expected similar to the improvement seen for the Royal Tree problem.

4.3 Examples of Learned Production Rules

In Fig. 4 the learned probability distributions during a run of the EDA on the Royal Tree problem are presented in a tabulated fashion. The figure shows the distributions at the beginning, in generation 50, and at the end of the run. A population of size 100 was used and the probability of recombination was set to 1.0. Clearly, the distribution can already be made specific enough by setting the probabilities because of the perfect dependency on the depth of the Royal Tree problem. Hence, no new production rules are introduced into the grammar at any point during the run and the probabilities converge towards a configuration that uniquely describes the optimal sentence.

Top left (beginning):

	D					
	0	1	2	3	4	5
i_0	0.167	0.167	0.167	0.167	0.167	1.000
$A(8)$	0.167	0.167	0.167	0.167	0.167	0.000
$B(8,8)$	0.167	0.167	0.167	0.167	0.167	0.000
$C(8,8,8)$	0.167	0.167	0.167	0.167	0.167	0.000
$D(8,8,8,8)$	0.167	0.167	0.167	0.167	0.167	0.000
$E(8,8,8,8,8)$	0.167	0.167	0.167	0.167	0.167	0.000

Center left (after 50 generations):

	D					
	0	1	2	3	4	5
i_0	0.000	0.000	0.029	0.140	0.158	1.000
$A(8)$	0.000	0.000	0.027	0.131	0.318	0.000
$B(8,8)$	0.000	0.000	0.079	0.362	0.146	0.000
$C(8,8,8)$	0.000	0.000	0.642	0.143	0.121	0.000
$D(8,8,8,8)$	0.000	0.992	0.011	0.106	0.133	0.000
$E(8,8,8,8,8)$	1.000	0.008	0.011	0.116	0.123	0.000

Bottom left (upon convergence):

	D					
	0	1	2	3	4	5
i_0	0.000	0.000	0.000	0.000	0.000	1.000
$A(8)$	0.000	0.000	0.000	0.000	1.000	0.000
$B(8,8)$	0.000	0.000	0.000	1.000	0.000	0.000
$C(8,8,8)$	0.000	0.000	1.000	0.000	0.000	0.000
$D(8,8,8,8)$	0.000	1.000	0.000	0.000	0.000	0.000
$E(8,8,8,8,8)$	1.000	0.000	0.000	0.000	0.000	0.000

Right (binary-functions, after 35 generations):

	D				
	0	1	2	3	4
i_0	0.000	0.000	0.185	0.346	1.000
$B_0(8,8)$	0.082	0.000	0.020	0.000	0.000
$B_1(8,8)$	0.002	0.000	0.009	0.000	0.000
$B_2(8,8)$	0.002	0.002	0.027	0.000	0.000
$B_3(8,8)$	0.006	0.011	0.059	0.000	0.000
$B_4(8,8)$	0.001	0.097	0.106	0.110	0.000
$B_5(8,8)$	0.000	0.158	0.105	0.185	0.000
$B_6(8,8)$	0.000	0.000	0.000	0.000	0.000
$B_0(B_1(8,8),8)$	0.582	0.000	0.032	0.000	0.000
$B_0(i_0,8)$	0.000	0.000	0.004	0.000	0.000
$B_6(i_0,8)$	0.000	0.412	0.114	0.000	0.000
$B_6(8,i_0)$	0.000	0.295	0.121	0.000	0.000
$B_6(i_0,i_0)$	0.000	0.000	0.000	0.249	0.000
$B_1(B_2(8,8),8)$	0.189	0.000	0.034	0.000	0.000
$B_1(i_0,8)$	0.000	0.000	0.001	0.018	0.000
$B_3(i_0,8)$	0.000	0.013	0.092	0.000	0.000
$B_3(i_0,i_0)$	0.000	0.000	0.000	0.058	0.000
$B_2(i_0,8)$	0.000	0.003	0.026	0.000	0.000
$B_2(B_3(8,8),8)$	0.018	0.010	0.046	0.000	0.000
$B_2(i_0,i_0)$	0.000	0.000	0.000	0.019	0.000
$B_0(B_2(8,8),8)$	0.096	0.000	0.011	0.000	0.000
$B_1(B_3(8,8),8)$	0.023	0.000	0.006	0.000	0.000
$B_0(B_6(8,8),8)$	0.000	0.000	0.000	0.000	0.000
$B_0(B_0(8,8),8)$	0.000	0.000	0.000	0.000	0.000
$B_0(i_0,i_0)$	0.000	0.000	0.000	0.015	0.000

Fig. 4. The probability distribution at the beginning (*top left*), after 50 generations (*center left*) and upon convergence after 139 generations (*bottom left*) of an example run of the EDA on the Royal Tree problem. Also, the probability distribution is shown after 35 generations in an example run of the EDA on the binary-functions problem (*right*). For brevity, only the expansion trees E_i of each production rule are shown and the production rule index lists are dropped for the symbols 8

In Fig. 4 the learned probability distribution in generation 35 during a run of the EDA on the binary-functions problem is also presented in a tabulated fashion. The population size used in this experiment is 10,000 and the probability of recombination is set to 0.9. In generation 35, the average fitness was 10.1379 whereas the best fitness was 13 (optimal fitness is 16). Clearly in this case the probability distribution based on the initial grammar cannot be made specific enough to describe the features of good sentences nor can it perfectly describe the optimal sentence. Therefore, additional production rules are added by the greedy search algorithm. As can be seen, the most important rules are added first and they have a high probability for at least one depth. The rules that are added afterwards mainly serve to make the distribution more specific by reducing the size of the default tables that have been created by the substitution transformations. The rules themselves have only a very low probability. One of the most important rules regarding the root of the tree has already been established (i.e. $B_0(B_1(S, S), S)$). Although other important rules have been discovered, they have not yet received high probabilities as they do not occur frequently in the population. As the search progresses towards the optimum, other important rules such as $B_2(B_3(S, S), S)$ will obtain more prominent probabilities as well to more specifically describe a smaller subset of high-quality sentences.

5 Discussion and Conclusions

In this chapter we have proposed a probability distribution over trees to be used in an EDA for GP. The distribution basically associates a probability with each production rule in a context-free grammar. More involved production rules or subfunctions can be introduced using transformations in which one production rule is expanded into another production rule. This allows the probability distribution to become more specific and to express a higher order of dependency. We have performed experiments on two benchmark problems from the literature. The results indicate that our EDA for GP is feasible. It should be noted however that learning advanced production rules using the greedy algorithm proposed in this chapter can take up a lot of time, especially if the number of production rules and the arity of the subfunctions increase. To speed up this greedy process, only a single rule can be randomly selected into which to expand each production rule from the base grammar instead of expanding each production rule from the base grammar into each currently available production rule. Although this significantly reduces the number of candidate distributions in the greedy algorithm, it also significantly improves the running time. Moreover, since the greedy algorithm is iterative and the probability distribution is estimated anew each generation, the most important subfunctions are still expected to emerge.

Because our approach to estimating probability distributions over trees does not fix or bound the tree structure beforehand, our approach can be seen as a more principled way of identifying arbitrarily-sized important sub-functions than by constructing subfunctions randomly (e.g. GLiB [2]) or by evolving them (e.g. ADFs [12] and ARL [24]). The use of the EDA principle may therefore offer an exiting new direction for GP that allows to detect and exploit substructures in a more principled manner for enhanced optimization performance, thus paving the way for novel future successes in EC.

References

1. T. Alderliesten, P. A. N. Bosman, and W. Niessen. Towards a real-time minimally–invasive vascular intervention simulation system. *IEEE Transactions on Medical Imaging*, 26(1):128–132, 2007.
2. P. J. Angeline and J. B. Pollack. The evolutionary induction of subroutines. In *Proceedings of the Fourteenth Annual Conference of the Cognitive Science Society*, pages 236–241, Lawrence Erlbaum, Hillsdale, NJ, 1992.
3. P. A. N. Bosman. *Design and Application of Iterated Density–Estimation Evolutionary Algorithms*. PhD thesis, Utrecht University, Utrecht, The Netherlands, 2003.
4. P. A. N. Bosman, J. Grahl, and F. Rothlauf. SDR: A better trigger for adaptive variance scaling in normal edas. In D. Thierens et al., (eds.), *Proceedings of the Genetic and Evolutionary Computation Conference – GECCO-2007*, pages 492–499, ACM, New York, 2007.
5. W. Buntine. Operations for learning with graphical models. *Journal of Artificial Intelligence Research*, 2:159–225, 1994.
6. W. Buntine. A guide to the literature on learning probabilistic networks from data. *IEEE Transactions on Knowledge and Data Engineering*, 8:195–210, 1996.
7. N. L. Cramer. A representation for the adaptive generation of simple sequential programs. In John J. Grefenstette, (ed.), *Proceedings of the First International Conference on Genetic Algorithms and their Applications*, Carnegie-Mellon University, pages 183–187, Lawrence Erlbaum, Hillsdale, NJ, 1985.
8. R. Etxeberria and P. Larrañaga. Global optimization using Bayesian networks. In A. A. O. Rodriguez et al., (eds.), *Proceedings of the Second Symposium on Artificial Intelligence CIMAF-1999*, pages 332–339. Institute of Cybernetics, Mathematics and Physics, 1999.
9. G. Harik, E. Cantú-Paz, D. E. Goldberg, and B. L. Miller. The gambler's ruin problem, genetic algorithms, and the sizing of populations. *Evolutionary Computation*, 7(3):231–253, 1999.
10. E. E. Korkmaz and G. Üçoluk. Design and usage of a new benchmark problem for genetic programming. In *Proceedings of the 18th International Symposium on Computer and Information Sciences ISCIS-2003*, pages 561–567, Springer, Berlin Heidelberg New York, 2003.
11. J. R. Koza. *Genetic Programming*. MIT, Cambridge, MA, 1992.
12. J. R. Koza. *Genetic Programming II: Automatic Discovery of Reusable Programs*. MIT, Cambridge, MA, 1994.
13. P. Larrañaga and J. A. Lozano. *Estimation of Distribution Algorithms. A New Tool for Evolutionary Computation*. Kluwer, London, 2001.
14. S. L. Lauritzen. *Graphical Models*. Clarendon, Oxford, 1996.

15. J. A. Lozano, P. Larrañaga, I. Inza, and E. Bengoetxea. *Towards a New Evolutionary Computation.* Springer, Berlin Heidelberg New York, 2006.
16. M. Pelikan. *Bayesian optimization algorithm: From single level to hierarchy.* PhD thesis, University of Illinois at Urbana–Champaign, Urbana, Illinois, 2002.
17. M. Pelikan and D. E. Goldberg. Hierarchical BOA solves ising spin glasses and maxsat. In E. Cantú-Paz et al., (eds.), *Proceedings of the Genetic and Evolutionary Computation Conference – GECCO-2003*, pages 1271–1282, Springer, Berlin Heidelberg New York, 2003.
18. M. Pelikan, D. E. Goldberg, and E. Cantú-Paz. BOA: The Bayesian optimization algorithm. In W. Banzhaf et al., (eds.), *Proceedings of the Genetic and Evolutionary Computation Conference – GECCO-1999*, pages 525–532, Morgan Kauffman, San Francisco, California, 1999.
19. M. Pelikan, D. E. Goldberg, and F. G. Lobo. A survey of optimization by building and using probabilistic models. *Computational Optimization and Applications*, 21(1):5–20, 2002.
20. M. Pelikan, D. E. Goldberg, and K. Sastry. Bayesian optimization algorithm, decision graphs and occam's razor. In L. Spector et al., (eds.), *Proceedings of the Genetic and Evolutionary Computation Conference – GECCO-2001*, pages 519–526, Morgan Kauffman, San Francisco, California, 2001.
21. M. Pelikan, K. Sastry, and E. Cantú-Paz. *Scalable Optimization via Probabilistic Modeling.* Springer, Berlin Heidelberg New York, 2006.
22. W. F. Punch, D. Zongker, and E. D. Goodman. The royal tree problem, a benchmark for single and multiple population genetic programming. In P. J. Angeline and K. E. Kinnear, Jr., (eds.), *Advances in Genetic Programming 2*, pages 299–316. MIT, Cambridge, MA, USA, 1996.
23. J. Rissanen. Hypothesis selection and testing by the MDL principle. *The Computer Journal*, 42(4):260–269, 1999.
24. J. P. Rosca and D. H. Ballard. Discovery of subroutines in genetic programming. In P.J. Angeline and K. E. Kinnear, Jr., (eds.), *Advances in Genetic Programming 2*, chapter 9, pages 177–202, MIT, Cambridge, MA, 1996.
25. R. P. Salustowicz and J. Schmidhuber. Probabilistic incremental program evolution. *Evolutionary Computation*, 5(2):123–141, 1997.
26. K. Sastry and D. E. Goldberg. Probabilistic model building and competent genetic programming. In R. L. Riolo and B. Worzel, (eds.), *Genetic Programming Theory and Practise*, chapter 13, pages 205–220, Kluwer, Dordrecht 2003.
27. Y. Shan, R. I. McKay, R. Baxter, H. Abbass, D. Essam, and H. X. Nguyen. Grammar model-based program evolution. In *Proceedings of the 2004 Congress on Evolutionary Computation – CEC2004*, IEEE, Piscataway, New Jersey, 2004.
28. D. Thierens. Scalability problems of simple genetic algorithms. *Evolutionary Computation*, 7(4):331–352, 1999.
29. D. Thierens and D. E. Goldberg. Mixing in genetic algorithms. In S. Forrest, (ed.), *Proceedings of the fifth conference on Genetic Algorithms*, pages 38–45, Morgan Kaufmann, USA, 1993.

The Automated Design of Artificial Neural Networks Using Evolutionary Computation

Jae-Yoon Jung and James A. Reggia

Department of Computer Science and UMIACS, University of Maryland, College Park, MD 20742, USA, jung@cs.umd.edu, reggia@cs.umd.edu

Summary. Neuroevolution refers to the design of artificial neural networks using evolutionary algorithms. It has been one of the promising application areas for evolutionary computation, as neural network design is still being done by human experts and automating the design process by evolutionary approaches will benefit developing intelligent systems and understanding "real" neural networks. The core issue in neuroevolution is to build an efficient, problem-independent encoding scheme to represent repetitive and recurrent modules in networks. In this chapter, we have presented our descriptive encoding language based on genetic programming and showed experimental results supporting our argument that high-level descriptive languages are a viable and efficient method for development of effective neural network applications.

1 Introduction

Most development of neural networks today is based upon manual design. A person knowledgeable about the specific application area specifies a network architecture and activation dynamics, and then selects a learning method to train the network via connection weight changes. While there do exist non-evolutionary methods for automatic incremental network construction [32], these generally presume a specific architecture, do not automatically discover network modules appropriate for specific training data, and have not enjoyed widespread use. This state of affairs is perhaps not surprising, given that the general space of possible neural networks is so large and complex that automatically searching it for an optimal network architecture may in general be computationally intractable or at least impractical for complex applications [6, 33].

Neuroevolution refers to the design of artificial neural networks using evolutionary algorithms, and it has attracted much research because (1) successful approaches will facilitate wide-spread use of intelligent systems based on artificial neural networks; and (2) it will shed lights on our understanding of how "real" neural networks have been evolved [21, 45]. Recent successes using

J.-Y. Jung and J.A. Reggia: *The Automated Design of Artificial Neural Networks Using Evolutionary Computation*, Studies in Computational Intelligence (SCI) **92**, 19–41 (2008)
www.springerlink.com

evolutionary computation methods as design/creativity tools in electronics, architecture, music, robotics, and other fields [3, 5, 30] suggest that creative evolutionary systems could have a major impact on the effectiveness and efficiency of designing neural networks. This hypothesis is supported by an increasing number of neuroevolutionary methods that search through a space of weights and/or architectures without substantial human intervention, trying to obtain an optimal network for a given task (reviewed in [2, 42, 43, 51]). Specifically, evolutionary algorithms have been successfully applied to many aspects of neural network design, including connection weights, learning rules, activation dynamics, and network architectures.

Evolution of connection weights in a fixed network architecture has been done successfully since the earliest stages of neuroevolution (e.g., [13, 14, 35, 46]). With either binary or real representations, all weights are encoded into a chromosome, then appropriate weights are searched for using genetic operations and selection. A problem in encoding all weights is that it may not be efficient in a large scaled network, with dense connectivity. A hybrid training approach that combines an evolutionary algorithm for global search with other local learning algorithms (e.g., backpropagation) for the fine tuning of weights has also been successful in many cases (e.g., [1,4,51]).

In many experiments, learning rules and activation dynamics have been predefined by the neural network designer and remain fixed throughout the evolution process. But if there is little knowledge about the problem domain, it is reasonable to include them as a part of evolution, since a small change in parameters or learning/activation rules can lead to quite different performance by the network. Encoding learning parameters into chromosomes can be considered as a first step towards the evolution of learning rules. In [22], the learning rate as well as the network architecture were encoded and evolved at the same time. Recently, Abraham encoded learning parameters and activation functions (e.g., sigmoid, tanh, logistic, and so on) and tried to evolve an optimal network for three time series [1]. Another interesting approach related to the learning rule is to use evolutionary algorithms to discover new learning rules. Chalmers assumed that a learning rule must be some linear combination of local information, and tried to find the needed coefficients using a genetic algorithm [10]. In [38], genetic programming was used instead of a standard genetic algorithm, since symbolic learning rules as well as their coefficients can be evolved within genetic programming.

Finally, there has been substantial interest in evolving neural network architectures (structures). There are roughly two different schemes that have been used to encode an architecture: direct and indirect encoding schemes. With the direct encoding scheme, connection information is explicitly specified in a connectivity matrix [33, 34]. It is simple and all possible network architectures within a fixed matrix size can be represented. However, for a large neural network, searching for the optimal architecture with a direct encoding scheme can be impractical since the size of the space increases exponentially with the network size. On the other hand, only some essential features of the network architecture are specified with an indirect encoding scheme [9,20,29,47]. For example, a network is represented by the size of each neuron block and their connections to the other blocks [22]. But this approach assumes that other features are predefined (e.g., all linked blocks are assumed to be fully connected in this example), so it searches a limited space.

Many previous studies involving evolution of neural networks start with randomly generated network architectures, undertake evolution without any specific bounds on the range of structures permitted, and/or are limited to an evolutionary process that occurs on a single level (e.g., only at the level of numbers of individual nodes and their connections, versus only at the higher level of whole network layers and their interconnecting pathways). In this chapter we describe our recent work developing a non-procedural language to specify the initial networks and the search space, and to permute virtually any aspects of neural networks to be evolved [25, 26]. This is done by using a high-level descriptive language that represents modules/layers and inter-layer connections (rather than individual neurons and their connections) in a hierarchical fashion that can be processed by genetic programming methods [3,30], and in this way our approach permits the designer to incorporate domain knowledge efficiently. This domain knowledge can greatly restrict the size of the space that the evolutionary process must search, thereby making the evolutionary process much more computationally tractable in specific applications without significantly restricting its applicability in general. Our system does not a priori restrict whether architecture, layer sizes, learning method, etc. form the focus of evolution as many previous approaches have done, but instead allows the user to specify explicitly which aspects of networks are to evolve, and permits the simultaneous evolution of both intra-modular and inter-modular connections.

2 Descriptive Encoding Methodology

We refer to our encoding scheme as a *descriptive encoding* since it enables users to describe the target space of neural networks that are to be considered in a natural, non-procedural and human-readable format. A user writes a text file like the ones shown later in this paper to specify sets of modules (layers) with appropriate properties, their range of legal evolvable property values, and allowable inter-module connectivity ("pathways"). This input description does *not* specify individual neurons, connections, nor their weights.[1] The specification of legal property values affects the range of valid genetic operations. In other words, a description file specifies the initial population and environment variables, and restricts the space to be searched by genetic operators during evolution.

---

[1] Individual neurons, connections and weights *can* be specified by creating layers/modules containing single neurons, but this does not take advantage of the language's ability to compactly describe large scale networks.

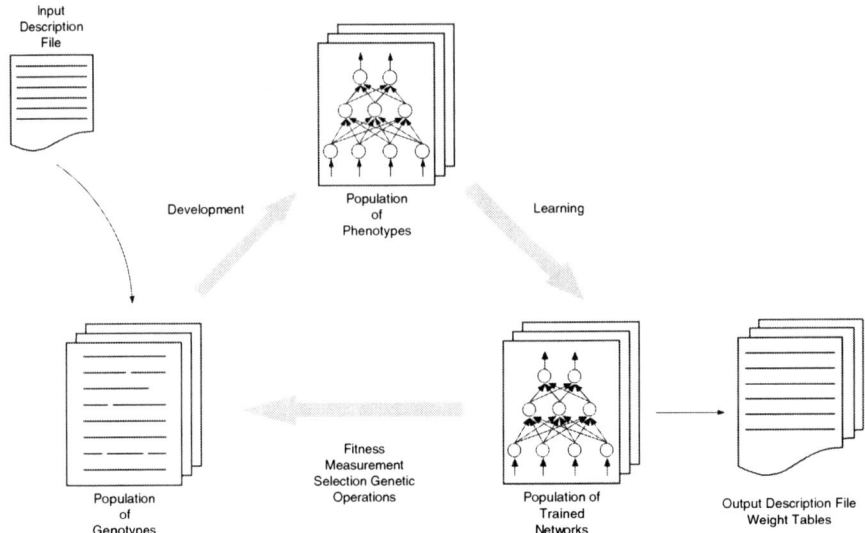

Fig. 1. The iterative three-step development, learning, and evolution procedure used in our system. The input description file (*upper left*) is a human-written specification of the class of neural networks to be evolved (the space to be searched) by the evolutionary process; the output description file (*lower right*) is a human-readable specification of the best specific networks obtained

The evolutionary process in our system involves an initialization step plus a repeated cycle of three stages, as shown in Fig. 1. First, the text description file prepared by the user is parsed and an initial random population of chromosomes (genotypes) is created within the search space represented by the description (leftmost part of Fig. 1). During the development stage, a new population of realized networks (phenotypes) is created or "grown" from the genotype population. Each phenotype network has actual and specific individual nodes, connection weights, and biases. The learning stage involves training each phenotype network, assuming that the user specifies one or more learning rules in the description file, making use of an input/output pattern file that contains training data. As evolutionary computation is often considered to be less effective for local, fine tuning tasks [17, 51], we adopt neural network training methods to adjust connection weights. After the training stage, each individual network is evaluated according to user-defined fitness criteria and genetic operators are applied to the genotypes. Fitness criteria may reflect both network performance (e.g., mean squared error) and a penalty for a large network (e.g., total number of nodes), or other measures. The end result of an evolutionary run consists of two things. First, an output description is produced (see Fig. 1, bottom right). This file uses the same syntax as the input description file to specify the most fit specific network architectures discovered by the evolutionary process. Second, another file gives a table of all the weights found by the final learning process using these specific network architectures.

2.1 Language Description

The basic unit of our encoding is a module that we call a *layer*, which is defined as an array of network nodes that share common properties. In other words, individual neurons are not the atomic unit of evolution, but sets of neurons are. The description of a layer/module starts with an identifier LAYER, which is followed by an optional layer name and a list of properties. Properties of a layer can be categorized into three groups: structure (e.g., BIAS, SIZE, NUM_LAYER), dynamics (e.g., ACT_RULE, ACT_INIT, ACT_MIN, ACT_MAX), and connectivity (e.g., CONNECT, CONNECT_RADIUS, CONNECT_INIT, LEARN_RULE). The order of declared properties in a layer description does not matter in general. Individual properties can be designated to be evolvable within some range, or to be fixed. Each property has its own default value for simplicity: if some properties are missing in the description file, they will be replaced with the default values during the initialization stage and considered as being constant throughout the evolutionary process (i.e., the chromosome is in fact more strict than the description; it requires all default, fixed, and evolvable properties to be present in some form). Layer properties used in this article are illustrated in Table 1. The meaning of each property is fairly straightforward, but the [CONNECT_RADIUS r] property with $0.0 \leq r \leq 1.0$ needs more explanation. It defines the range of the connectivity from each node in a source layer to the nodes in a target layer. For example, if $r = 0.0$, each node in the source layer is connected to just

Table 1. Example module/layer properties

Property	What it specifies about a module/layer
BIAS	Whether to use bias units and their initial value ranges if so
SIZE	Number of nodes in the current layer
NUM_LAYER	Number of layers of this type
ACT_RULE	Activation rule for nodes in the layer
ACT_INIT	Initial activation value for nodes
ACT_MIN	Minimum activation value
ACT_MAX	Maximum activation value
CONNECT	Direction of connections starting from this layer
CONNECT_RADIUS	Range of connectivity from 0.0 to 1.0 (see text)
CONNECT_INIT	Initial (range of) weights in the current connections
LEARN_RULE	Learning rule for the current connections

a single node in the matching position of the target layer. If r is a positive fraction less than one, each source node connects to the matching destination layer node and its neighbor nodes out to a fraction r of the radius of the target layer; thus, if $r = 1.0$, the source and target layers are fully connected. While these connectivity properties are basically intended to specify connections between two layers (i.e., an inter-modular connection), intra-modular connections such as self-connectivity can also be designated using the same properties. For example, one can specify that each node in a layer named *layer*1 connects to itself by using a combination of [CONNECT *layer*1] and [CONNECT_RADIUS 0.0]. The user can also change the default value of each property for a specific problem domain by declaring and modifying property values in the evolutionary part of a description file, which will be explained later.

Figure 2a illustrates part of an input description file written in our language for evolving a recurrent network. Each semantic block, enclosed in brackets [ ], starts with a type identifier followed by an optional name and a list of properties about which the user is concerned in the given problem. A network may contain other (sub)networks and/or layers recursively, and a network type identifier (SEQUENCE, PARALLEL, or COLLECTION) indicates the conceptual arrangement of the sub-networks contained in this network. If a network module starts with the SEQUENCE identifier, the sub-networks contained in this module are considered to be arranged in a sequential manner (e.g., like a typical feed-forward neural network). Using the PARALLEL identifier declares that the sub-networks are to be arranged in parallel, and the COLLECTION identifier indicates that an arbitrary mixture of sequential and parallel layers may be used and evolved. The COLLECTION identifier is especially useful when there is little knowledge about the appropriate relationships

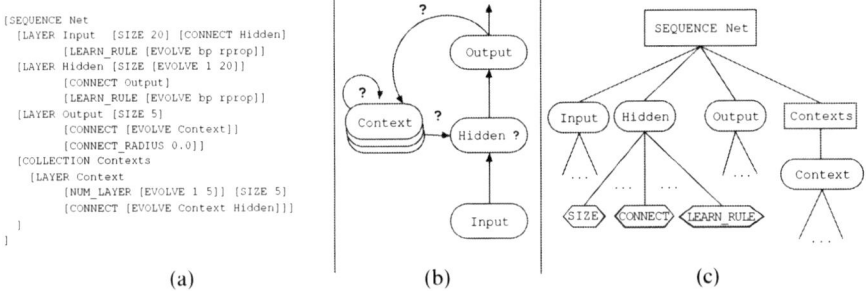

(a)   (b)   (c)

Fig. 2. (a) The first part of a description file specifies the network structure in a top-down hierarchical fashion (other information in the description file, such as details of the evolutionary process, is not shown). (b) A schematic illustration of the corresponding class of recurrent networks that are described in (a). *Question marks* indicate architecture aspects that are considered evolvable. (c) Part of the tree-like structure corresponding to the genotype depicted in (a). Each *rectangle*, *oval*, and *hexagon* designates a network, layer, and property, respectively

between the layers being evolved. As described earlier, we define a layer as a set (sometimes one or two dimensional, depending on the problem) of nodes that share similar properties, and it is the basic module of our network representation scheme. For example, the description in Fig. 2a indicates that a sequence of four types of layers are to be used: input, hidden, output, and context layers, as pictured in Fig. 2b. Properties fill in the details of the network architecture (e.g., layer size and connectivity) and in general specify other network features including learning rules and activation dynamics.

Most previous neuroevolution research has focused a priori on some limited number of network features (e.g., network weights, number of nodes in the hidden layer) assuming that the other features are fixed, and this situation has impeded past neuroevolutionary models from being used more widely in different environments. To overcome this limitation, we let users decide which fixed properties are necessary to solve their problems, and which other factors should be evolved, from a set of supported properties that span many aspects of neural networks. Unspecified properties are replaced with default values and are treated as being fixed after initialization. So, for example, in the description of Fig. 2a, the input layer has a fixed, user-assigned number of 20 nodes and is connected to the hidden layer, while the single hidden layer has an evolvable SIZE within the range 1–20 nodes. The EVOLVE attribute indicates that the hidden layer's size will be randomly selected initially and is to be modified within the specified range during the evolution process. Note that the learning rules to be used for connections originating from both input and hidden layers are also declared as an evolvable property (in this case, a choice between two variants of backpropagation). The description in Fig. 2a also indicates that one to five context layers are to be included in the network; this is the main architectural aspect that is to be evolved in this example. These context layers are to be ordered arbitrarily, all contain five neurons, and they evolve to have zero or more inter-layer output connections to either other context layers or to the hidden layer (Fig. 2b). Finally, the output layer evolves to propagate its output to one or more of the context layers, where the CONNECT_RADIUS property defines one-to-one connectivity in this case. Since the number of layers in the context network may vary from 1 to 5 (i.e., LAYER context has an evolvable NUM_LAYER property), this output connectivity can be linked to any of these layers that were selected in a random manner during the evolution process. Figure 2b depicts the corresponding search space schematically for the description file of Fig. 2a, and the details of each individual genotype (shown as question marks in the picture) will be assigned within this space at the initialization step and forced to remain within this space during the evolution process. Note that since the genotype structure is a tree, as shown in Fig. 2c, fairly standard tree-manipulation genetic operators as used in GP [3, 30] can be easily applied to them with our approach.

The remainder of the description file consists of information about training and evolution processes (not shown in Fig. 2a). A *training block* specifies the file name where training data is located and the maximum number of training epochs. Default property values may also be designated here, like the LEARN_RULE to be used. In other words, when a property value is specified in this block, the default value of that property is changed accordingly and affects all layers which have that property. An *evolution block* can also be present and specifies parameters affecting fitness criteria, selection method, type and rate of genetic operations, and other population information (illustrated later).

3 Evaluation Results

While the approach to evolutionary design of neural networks described above may seem plausible, it remains to actually establish that the combination of search space restriction via a high-level language, developmental encoding, modular network organization, and integrated use of evolution and network learning can be used effectively in practice. Although our system is intended ultimately to evolve any aspect of neural networks, here we focus on evolving network architectures, evaluating our system on two different problems in which the task is to discover an architecture that both solves the problem and also sheds light on the modular nature of the solution, as a first step in establishing such effectiveness. Our goal is simply to show that when one uses a high-level descriptive language (with all of its benefits as outlined in the Introduction), it is still possible to discover interesting neural network solutions to problems using evolutionary computation methods. While we do not claim that these results are specifically due to our language, we believe our language facilitates the specification of search spaces, supports analysis of initial conditions and final results, makes the tasks much more efficient for the human investigator, and encourages the emergence of modularity in network solutions (via the use of built-in cost functions).

3.1 Module Formation in a Feed-Forward Network

Many animal nervous systems have parallel, almost structurally independent sensory, reflex, and even cognitive systems, a fact that is sometimes cited as contributing to their information processing abilities. For example, biological auditory and visual systems run in parallel and are largely independent during their early stages [27], the same is true for segmental monosynaptic reflexes in different regions of the spinal cord [27], and the cerebral cortex is composed of regional modules with interconnecting pathways [36]. Presumably evolution has discovered that such *partitioning* of neural networks into parallel multi-modular pathways is both an effective and an efficient way to support parallel processing when interactions between modules are not necessary. However, the factors driving the evolution of modular brain architectures

Table 2. Training data for a 2-partition problem

Input	Output	Input	Output	Input	Output	Input	Output
0 0 0 0	0 0	0 1 0 0	1 0	1 0 0 0	1 0	1 1 0 0	0 0
0 0 0 1	0 1	0 1 0 1	1 1	1 0 0 1	1 1	1 1 0 1	0 1
0 0 1 0	0 1	0 1 1 0	1 1	1 0 1 0	1 1	1 1 1 0	0 1
0 0 1 1	0 0	0 1 1 1	1 0	1 0 1 1	1 0	1 1 1 1	0 0

having components interconnected by distinct pathways have long been uncertain and currently remain a very active area of discussion and investigation in the neurosciences [8, 15, 28, 48].

Inspired by such modular neurobiological organization, and as a first test problem for our descriptive encoding system, we examined whether an evolutionary process could discover the existence and details of n independent neural pathways between subsets of input and output units that are implied by training data. We call such problems *n-partition problems*. Table 2 gives an example when $n = 2$ of a 2-partition problem. The goal is to evolve a minimal neural network architecture that can learn the given mapping from four input units to two output units, all of which are assumed to be standard logistic neurons in this case. The data in Table 2 implicitly represent a 2-partition problem in that the correct value of the first output depends only on the values of the first two input units (leftmost columns in the table), while the correct value of the second output depends only on the values of the remaining two input units. Thus, two parallel independent pathways from inputs to outputs are implied, and in this specific example each output is arranged to be the exclusive-or function of its corresponding inputs. A human designer would recognize from this information both that hidden units are necessary to solve the problem (since exclusive-or is not linearly separable), and that two separate parallel hidden layers are a natural minimal architecture. Of course, given just the training data in Table 2 and not knowing a priori the input/output relationships, such a design would not be evident in advance to an evolutionary algorithm, nor most likely to a person, and the question being asked here is whether an evolutionary process would discover it when given a suitable descriptive encoding.

The description file that we used to evolve an architecture to solve the 2-partition problem of Table 2 is shown in Fig. 3a. It specifies the initial architecture for the 2-partition problem, in which only the number of input and output nodes are fixed while other aspects, such as inter-layer connectivity and hidden layers' structure as shown in Fig. 3b, are randomly created during initialization and evolve. In this example we separate the input neurons into groups that form the basis for the distinct pathways, but note that the learning algorithm makes no use of this fact. The descriptive encoding also specifies the boundaries of the search space, indicating that there can be 0–10 hidden layers organized in any possible interconnecting architecture and the number

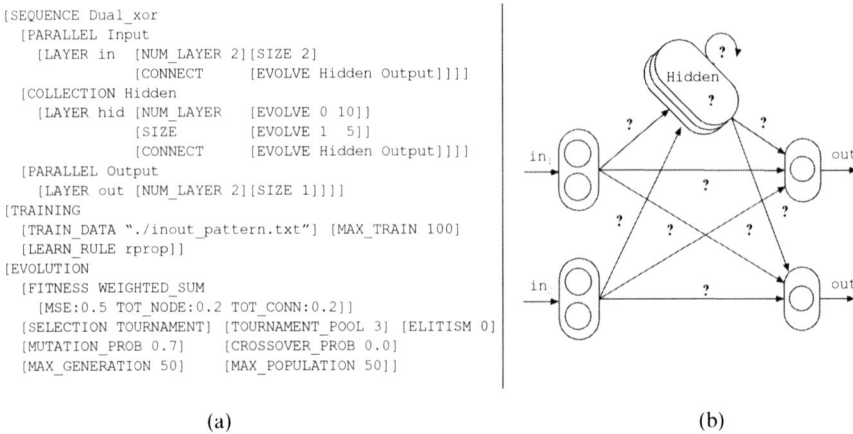

```
[SEQUENCE Dual_xor
  [PARALLEL Input
    [LAYER in  [NUM_LAYER 2][SIZE 2]
               [CONNECT      [EVOLVE Hidden Output]]]]
    [COLLECTION Hidden
      [LAYER hid [NUM_LAYER   [EVOLVE 0 10]]
                 [SIZE        [EVOLVE 1  5]]
                 [CONNECT     [EVOLVE Hidden Output]]]]
    [PARALLEL Output
      [LAYER out [NUM_LAYER 2][SIZE 1]]]]
  [TRAINING
    [TRAIN_DATA "./inout_pattern.txt"] [MAX_TRAIN 100]
    [LEARN_RULE rprop]]
  [EVOLUTION
    [FITNESS WEIGHTED_SUM
      [MSE:0.5 TOT_NODE:0.2 TOT_CONN:0.2]]
    [SELECTION TOURNAMENT]  [TOURNAMENT_POOL 3] [ELITISM 0]
    [MUTATION_PROB 0.7]     [CROSSOVER_PROB 0.0]
    [MAX_GENERATION 50]     [MAX_POPULATION 50]]
```

(a)   (b)

Fig. 3. (a) The initial network description and (b) a sketch of the space of networks to be searched for the 2-partition problem of Table 2

of nodes (SIZE property) in each hidden layer may vary from 1 to 5. The NUM_LAYER and SIZE properties in the hidden structure and all CON-NECT properties are declared as evolvable (indicated by question marks in Fig. 3b), so these properties and only these properties can be modified by genetic operators during evolution.

As explained earlier, a descriptive encoding generally provides additional information about the training and evolutionary processes to be used, and we illustrate that here. This user-defined information follows the network part of the description (Fig. 3a, top), setting various parameter values to control the training and evolutionary procedure. In this case, each phenotype network is to be trained for 100 epochs with the designated input/output pattern file that encodes the information from Table 2. The default learning rule is defined as a variant of backpropagation (RPROP [41]). Note that there is no issue of generalization in learning the boolean function here since all inputs and the correct outputs for them are given a priori. The EVOLUTION part of the description (Fig. 3a, bottom) indicates that a weighted sum method of three criteria to be used: mean squared error (MSE, e), total number of network nodes (n), and total number of layer-to-layer connections (c). MSE reflects the output performance of the network, and the other two criteria are adopted as penalties for larger networks. These three criteria are reciprocally normalized and then weighted with coefficients assigned in the description file. More specifically, the fitness value of the ith network, $Fitness_i$ that is described here is

$$Fitness_i = w_1 \left(\frac{e_{max} - e_i}{e_{max} - e_{min}} \right) + w_2 \left(\frac{n_{max} - n_i}{n_{max} - n_{min}} \right) + w_3 \left(\frac{c_{max} - c_i}{c_{max} - c_{min}} \right),$$

(1)

where $x_{min}(x_{max})$ denotes the minimum (maximum) value of criterion x among the population, and the coefficients w_1, w_2, and w_3 are empirically defined as 0.5, 0.2, and 0.2, respectively. In words, the fitness of an individual neural network is increased by lower error (a behavioral criterion), or by fewer nodes and/or connections (structural criteria). An implicit hypothesis represented in the fitness function is that minimizing the latter two structural costs may lead to fewer modules and independent pathways between them in evolved networks. Note that the EVOLUTION part of the description (Fig. 3a) specifies the coefficients in the fitness function above, and it also specifies tournament selection with a pool size of 3 as the selection method, a mutation rate of 0.7, and that no crossover and no elitism are to be used. Operators in this case can mutate layer size and direction of an existing inter-layer connection, and can add or delete a new layer or connection.

We ran a total of 50 simulations with the fixed population size (50) and the fixed number of generations of 50. Between simulations, the only changes are the initial architectures plus the initial value of connection weights that are assigned randomly in the range from -1.0 to 1.0, as used in [40]. For all runs, each final generation contained near-optimal networks that both solved the 2-partition problem (i.e., MSE ~ 0.0) and had a small number of nodes and connections. Converged networks can be categorized into two groups identified by their connectivity pattern as depicted in Fig. 4. The first group of networks, found during 44% of the runs, showed a dual independent pathway where each input pair has their own hidden layer and a direct connection to the corresponding output node (Fig. 4a and 4b). Ignoring the dotted line connections which have near zero weight values shown in Fig. 4a, this is an optimal network for the 2-partition problem in terms of the total number of network nodes and connections. In the second group of networks, found during 52% of the runs, input layers share a single hidden layer, without having direct connections to the corresponding output nodes. Such solutions require

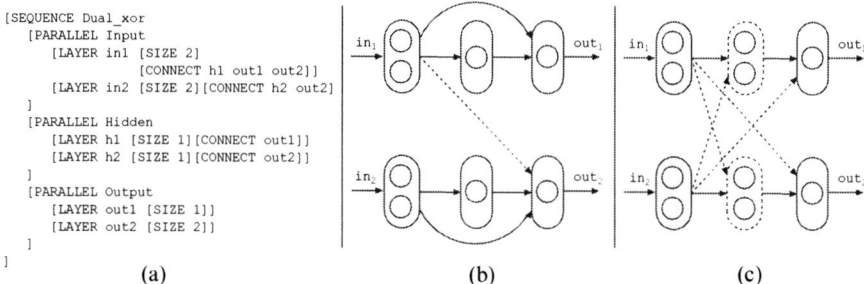

(a)   (b)   (c)

Fig. 4. Typical network architectures found during evolution for the 2-partition problem are depicted. *Dotted lines* show connectivity with near-zero weights. (**a**) Final output description file having two independent pathways. (**b**) Conceptual network architecture described by (**a**). (**c**) Dual pathway network without direct input-to-output connections. Implicit hidden sub-layers are indicated by *dotted ovals*

four hidden nodes, rather than two, to be an optimal network. Inspection of the connection weights shows that this model implicitly captures/discovers two distinct pathways embedded in the explicit hidden layer as illustrated in Fig. 4c (near zero weight connections that do not affect the performance are pruned). This type of network is also an acceptable near-optimal solution in terms of the number of connections needed for XOR problems and is sometimes used to illustrate layered solutions to single XOR problems in textbooks (e.g., [23]). The remaining 4% of the runs did not converge on just one type of network as described above, but both types are found in the final population. Thus, the evolutionary process generally discovered that "minimal cost" solutions to this problem involve independent pathways. While the networks considered here are very simple relative to real neurobiological systems, the frequent emergence of distinct and largely independent pathways rather than more amorphous connectivity during simulated evolution raises the issue of whether parsimony pressures may be an underrecognized factor in evolutionary morphogenesis, as outlined at the beginning of this section (see [45] for further discussion).

Without changing the evolutionary part of the description file, we tested n-partition problems for $n = 2, 3, 4,$ or 5 (the latter requires $2^{2n} = 1,024$ patterns for training). A typical input/output description file and network structure for $n = 5$ are illustrated in Fig. 5. Table 3 summarizes the experimental results. Minimum hidden nodes is the smallest number of hidden nodes found during an experiment, and numbers in the parentheses show the theoretically minimum number of nodes possible In an n-partition problem

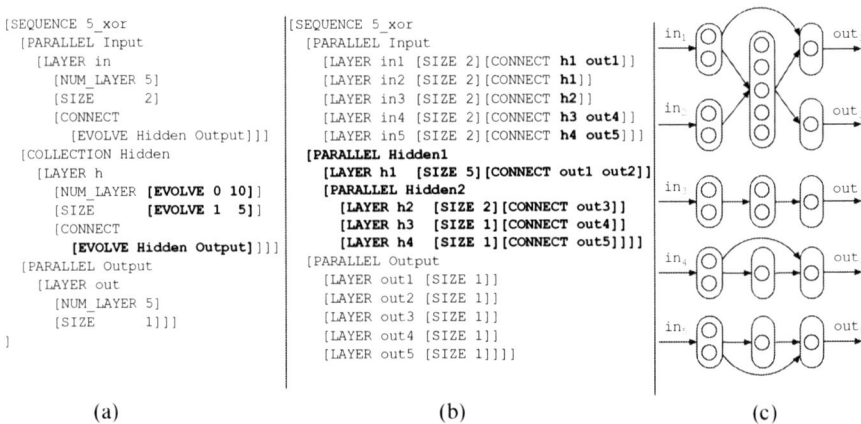

(a)   (b)   (c)

Fig. 5. A typical example of a final evolved network for the 5 XOR partition problem. (a) Initial network description. Properties to be evolved are in *bold font*. (b) Final description file produced as output by the system. All EVOLVE properties have been replaced by the specific choices in *bold font*. Only SIZE and CONNECT properties are shown. (c) Depicted network architecture. Connections that have near-zero weights are pruned

Table 3. Parallel n-partition problem results

N	# of patterns	Minimum hidden nodes	Minimum connections	Minimum MSE	Fully connected MSE	Average time
2	16	2 (2)	7 (6)	0.00000	0.20823	2,250
3	64	4 (3)	9 (9)	0.00021	0.20710	2,848
4	256	5 (4)	14 (12)	0.00073	0.22224	3,471
5	1,024	9 (5)	19 (15)	0.00236	0.20821	6,273

involving exclusive-OR relations, at least n hidden nodes are necessary even if direct connections from input to output are allowed. The Minimum connections column shows the minimum number of layer-to-layer connections found in the best individual. Again assuming direct connectivity from input to output and without increasing the number of hidden nodes, the best possible number of connections in an n-partition problem is $3n$ (e.g., input to output, input to the corresponding hidden layer, hidden to output). For each n partition problem, we also compared the best MSE results gathered from each evolutionary simulation with that of standard fully connected, single hidden layer backpropagation networks.

These latter networks have a single fixed hidden layer size of n, which is the theoretically minimal (optimal) number for each partition problem, initial weights randomly chosen from -1.0 to 1.0 (same as in the evolutionary simulations), and the MSE results averaged over 50 runs. With all other conditions set to be the same, post-training errors with the evolved networks are significantly less for each problem (p values on t-test were less than 10^{-5} for each of the four comparisons). More importantly, the fully connected networks sometimes produced totally wrong answers (i.e., absolute errors in output node values were more than 0.5), while this problem did not occur with the evolved networks. This shows the value of searching the architectural space even if it is believed that a fully connected network can theoretically approximate any function [12] ([49,50] for general discussion). The Average time column in Table 3 shows the mean time (seconds) needed for a single evolutionary run. This result shows that our system can identify the partial relationships between input and output patterns and represent them within an appropriate modular architecture.

3.2 Learning Word Pronunciations Using Recurrent Networks

Recurrent neural networks have long been of interest for many reasons. For example, they can learn temporal patterns and produce a sequence of outputs, and are widely found in biological nervous systems [27]. They have been applied in many different areas (e.g., word pronunciation [39], and learning formal grammars [18]) and several models of recurrent neural networks have been proposed (see [23]). Here we establish that our high-level descriptive

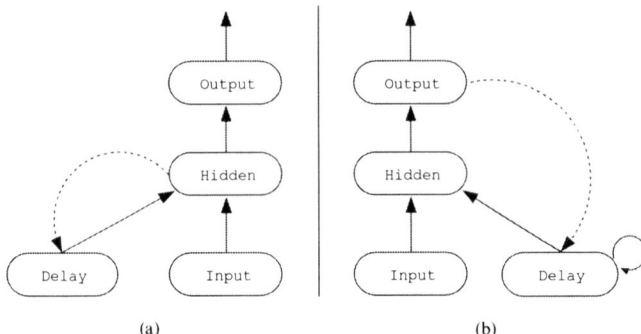

Fig. 6. The Elman (**a**) and Jordan (**b**) network architectures. *Dotted lines* show the backward/recurrent one-to-one connections that essentially represent a copying of the output at one time step to a delay layer that serves as input at the next time step

language is sufficiently powerful to support the evolution of recurrent networks in situations involving multi-objective optimization.

Two well-known, partially recurrent architectures that let one use basic error backpropagation from feed-forward nets essentially unchanged (because feedback connection weights are fixed and unlearnable) are often referred to as Elman networks and Jordan networks. Elman [16] suggested a recurrent network architecture in which a copy of the contents of the hidden layer (saved in the delay layer) acts as a part of the input data in the next time step, as shown in Fig. 6a. Jordan [24] proposed a similar architecture except that the content of the output layer is fed back to the delay layer where nodes possibly also have a "decaying" self-connection, as shown in Fig. 6b. These networks were originally proposed for different purposes: the Elman architecture for predicting the next element in a temporal sequence of inputs, and the Jordan architecture for generating a temporal sequence of outputs when given a single fixed input pattern. However, little is known about how to select the best recurrent network architecture for a given sequence processing task and, to our knowledge, no systematic experimental comparison between these different recurrent neural network architectures has ever been undertaken, except for some specific application comparisons (e.g., [37]).

In this context, our second problem is to find appropriate recurrent networks to produce a sequence of phoneme outputs, given a *fixed* input of the corresponding input word pattern. For example, for the word *apple*, a fixed pattern of the five letters A P P L E is the input, and the correct output temporal sequence of phonemes would be /ae/, /p/, and /l/, followed by an end of word signal. This challenging task was originally tackled in [39] using Jordan networks, and here we expand the size of input data (total of 230, two to six phoneme words selected randomly from the NetTalk corpus [44]), and focus on finding the optimal architecture of delay layers and their connectivity.

```
[SEQUENCE Psg_problem
  [LAYER Input    [SIZE 156] [CONNECT Hidden]]
  [LAYER Hidden   [SIZE 52]
    [[CONNECT DelayH:1] [CONNECT_RADIUS 0.0]
     [LEARN_RULE NONE]]
    [[CONNECT Output]]]
  [SEQUENCE DelayH
    [LAYER [NUM_LAYER [EVOLVE 0 4]] [SIZE 52]
      [[CONNECT FWD] [CONNECT_RADIUS 0.0]
       [LEARN_RULE NONE]]
      [[CONNECT [EVOLVE Hidden Output]]]]]
  [LAYER Output   [SIZE 52]
    [[CONNECT DelayO:1] [CONNECT_RADIUS 0.0]
     [LEARN_RULE NONE]]]
  [SEQUENCE DelayO
    [LAYER [NUM_LAYER [EVOLVE 0 4]] [SIZE 52]
      [[CONNECT FWD] [CONNECT_RADIUS 0.0]
       [LEARN_RULE NONE]]
      [[CONNECT [EVOLVE Hidden Output]]]]]
]
```

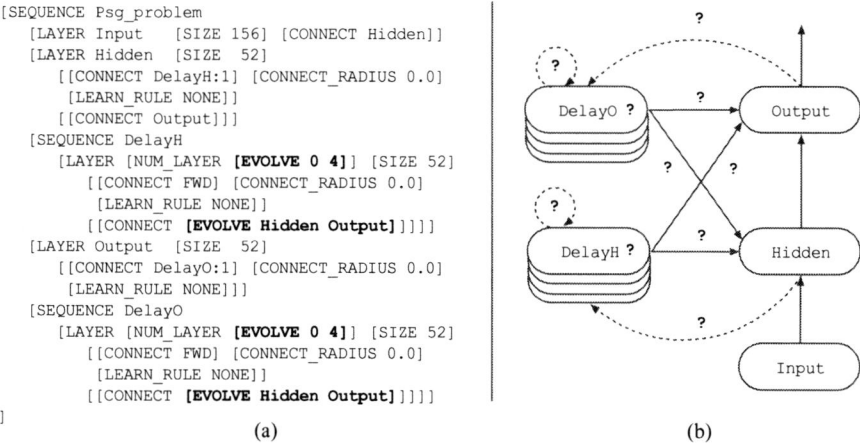

(a)   (b)

Fig. 7. (a) The network description file for the phoneme sequence generation (psg) task. FWD, delayH:1, and delayO:1 mean to make a connection to the next layer in the same network block, to the first layer in the delayH network block, and to the first layer in the delayO network block, respectively. If such a block does not exist, the corresponding connectivity properties are ignored. The evolvable properties are in *bold font*. (b) A schematic illustration of the space of neural network architectures corresponding to the description file in (a) that are to be searched for the phoneme sequence generation problem. *Dotted lines* designate non-trainable, one-to-one feedback connections; *solid lines* indicate weighted, fully connected pathways trained by error backpropagation. Note that the Elman and Jordan networks of Fig. 6 are included within this space as special cases

The question being asked is whether the high-level, modular developmental approach supported by our language can identify the "best" recurrent architecture to use, or at least clarify the tradeoffs.

Figure 7 gives the descriptive encoding and a corresponding schematic representation of the space of networks that we used for this problem. The fixed part of the structure is a feed-forward, three layer network consisting of input, hidden, and output layers (depicted on the right side of Fig. 7b). The size of the input layer is decided by the maximum length of a word in our training data and the encoding representation, and the output layer size is 52 since we use a set of 52 output phonemes, including the end of a word signal. The number of hidden nodes is arbitrarily set to be the same as the output layer size. Note that hidden and delay layers may have connections to different destination layers with different configurations (e.g., CONNECT_RADIUS and LEARN_RULE), such that each set of properties for connections has been separated from the other by using double brackets in the description of Fig. 7a. As shown here, the space of architectures to be searched by the evolutionary process consists of varying numbers of delay layers that receive recurrent one-to-one feedback connections (dotted arrows) from either the hidden layer (delayH) or the output layers (delayO). In either case, 0–4 delay layers may be evolved, but however many are evolved in each feedback pathway, they must be organized in a serial fashion, thus representing feedback delays of 0–4 time steps. Both feedback pathways from output and hidden layers may have zero layers, which means there are four possible architectures being considered during evolution: (1) feed-forward network only without delays; (2) hidden layer feedback only; (3) output layer feedback only; and (4) both types of feedback. In addition, for each class where a feedback pathway exists, it may have a varying amount of delay (1–4 time steps) and may provide feedback to the output layer, the hidden layer, or both. Each delay layer is sequentially connected to the adjacent delay layer by a one-to-one, fixed connection of weight 0.5, which acts as a decaying self-connection. Thus a total of 169 architectures are considered by the evolutionary process.[2]

We based fitness criteria on two cost measures or objectives: root mean squared error (RMSE) for performance, which was adjusted to be comparable with previous results [39], and the total sum of absolute weight values to penalize larger networks. The latter unbounded measure simply adds together the absolute values of all weights in the network after training. We used it rather than the number of nodes and connections as in n-partition problems, since the latter vary stepwise in this experiment while the summed weights are a continuous measure. This summed absolute weights measure is especially useful here as it can potentially discriminate between two different architectures having the same numbers of node and connections (e.g., Jordan vs Elman networks). Similar weight minimization fitness criteria have been used previously evolving neural networks and can be viewed as a "regularization term" that acts indirectly on an evolutionary time scale rather than directly during the learning process (see [45] for discussion). We adopted a multi-objective evolutionary algorithm (SPEA [52]) based on these two fitness criteria, which enables one to get a sense of the tradeoffs in performance and parsimony among the different good architectures found during evolution. This also illustrates that our system consists of components that can be expanded or plugged in depending on the specific problem. The population size was decreased to 25 because of the large computational expense of learning and evolution, and the archive in which non-dominated individuals are stored externally in SPEA was set to be the same size as the population. The maximum number of generations was fixed at 50, and all networks trained for 200 epochs with RPROP, a variant of error backpropagation which has been shown to be very effective in previous research [39]. For genetic operations, only mutating the number of layers and their connectivity are allowed, specified by default and applied within the range of property values designated in the network description file.

---

[2] No delay: 1, delayH only: 4 delays × 3 directions, delayO only: 4 × 3, both delays: (4 × 3) × (4 × 3) = 169.

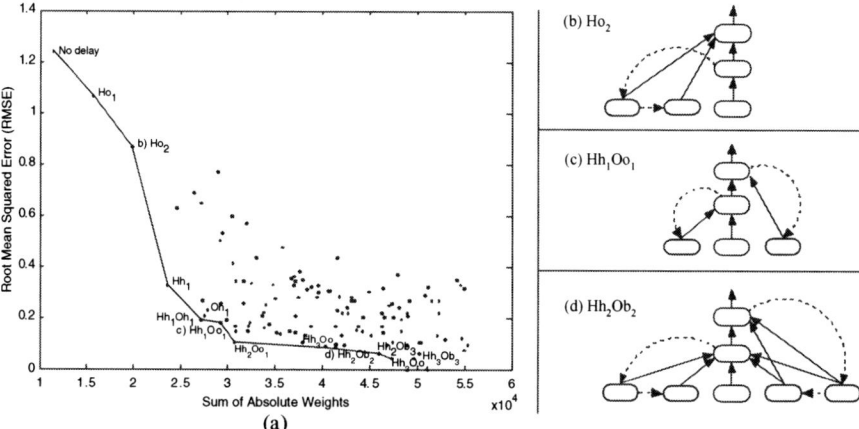

Fig. 8. **(a)** The performance/weights result of networks from all final generations are depicted. Each point represents one network architecture's values averaged over all evolutionary runs (most points are not labeled). The points on the *solid line* represent the Pareto optimal set, and the labels on some of these latter points designate the type of network that they represent. For example, label Hh_3Oh_1 means that the network represented by that node has both hidden (H) and output (O) delay layers, while there are three hidden and one output delay layers, in both cases connected to the hidden (h) layer (see text). **(b)**–**(d)** Example of evolved network architectures. Evolved layers are shown in *bold ovals*

We ran a total of 100 runs, randomly changing the initial network architectures and their weights in each run. The results are shown in Fig. 8, averaged over the same architectures (i.e., each point in Fig. 8 represents a network having a specific number of hidden and output delays, and a specific layer-to-layer connectivity, with the RMSE and weight sum averaged over all runs). In a label "$Hc_\#Oc_\#$" in Fig. 8, $\#$ indicates the number of hidden (H) or output (O) delays, and c designates the destination of delay outputs, either to the hidden (h), output (o), or both (b) layers. For example, "Hh_1Ob_2" means that there is one hidden delay layer (connected back to the hidden layer) and two output delay layers connected back to both hidden and output layer. Figure 8b–d shows some other examples of evolved network architectures. An example of the final network descriptions for Elman and Jordan networks is illustrated in Fig. 9.

Several observations can be made from Fig. 8. First, and surprising to us, feed-forward only networks without delays still remain in the final Pareto optimal set (upper left). The Pareto optimal set in this context consists of "non-dominated" neural networks for which no other neural network has been found during evolution that is better on all of the objective fitness criteria. Thus, feed-forward networks are included in the Pareto optimal set because of their quite small weight values, even though their performance is poor relative

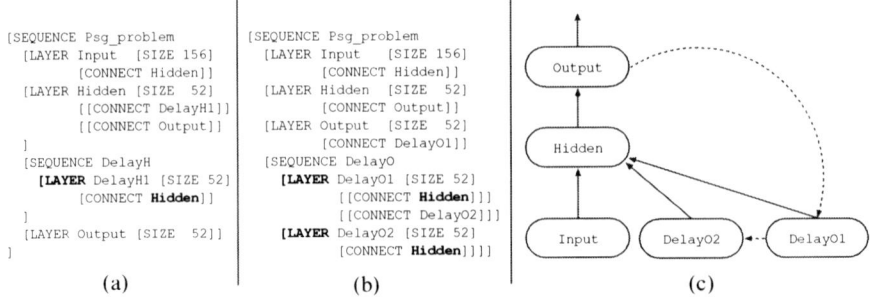

```
[SEQUENCE Psg_problem          [SEQUENCE Psg_problem
  [LAYER Input   [SIZE 156]      [LAYER Input    [SIZE 156]
     [CONNECT Hidden]]              [CONNECT Hidden]]
  [LAYER Hidden [SIZE 52]        [LAYER Hidden  [SIZE 52]
     [[CONNECT DelayH1]]            [CONNECT Output]]
     [[CONNECT Output]]          [LAYER Output  [SIZE 52]
  ]                                 [CONNECT DelayO1]]
  [SEQUENCE DelayH               [SEQUENCE DelayO
    [LAYER DelayH1 [SIZE 52]       [LAYER DelayO1 [SIZE 52]
       [CONNECT Hidden]]              [[CONNECT Hidden]]]
  ]                                   [[CONNECT DelayO2]]]
  [LAYER Output [SIZE 52]]       [LAYER DelayO2 [SIZE 52]
]                                     [CONNECT Hidden]]]]
```

| (a) | (b) | (c) |

Fig. 9. The final network description of (**a**) an Elman network with single delay (labeled "Hh_1" in the text) and (**b**) a Jordan-like network with double delays (labeled "Oh_2"). Only SIZE and CONNECT properties are shown. The evolved properties (including the number of layers) are in *bold font*. (**c**) An illustration of the Jordan network specified in (**b**). *Dotted lines* designate fixed, one-to-one connections

to the other types. Following the Pareto optimal front downward, we see that networks with one or two hidden delay layers connected to the output layer (labeled "Ho_1" and "Ho_2") are the next Pareto-front points (upper left of Fig. 8). This type of network in which delays are connected to the output layer does not provide good performance in general. A big increase in performance occurs however with networks having only hidden delay layers connected to the hidden layer (bottom left): an Elman network (labeled "Hh_1" in Fig. 8 and depicted in Fig. 9a) performs much better and is on the Pareto front. A Jordan network (labeled "Oh_1", lower right of "Hh_1" in Fig. 8) performs even better at the cost of increased weights. Finally, networks with increasing numbers of delay layers that combine hidden and output delays generally performed progressively better, although at the cost of increasing numbers of weights and connections (bottom right in Fig. 8). Surprisingly, "Hh_1Oh_1" on the Pareto front of Fig. 8 performs better than the original Elman (Hh_1) and Jordan (Oh_1) networks with smaller total weight values than the latter, and would be a very good choice for an architecture for this problem (and one that was not evident prior to the evolutionary process). Summarizing, the Pareto-optimal front in Fig. 8 and, more generally, the correlations between architectures and performance given in Table 4, explicitly lay out the tradeoffs for the human designer selecting an architecture. From a practical point of view, which Pareto optimal architecture one would adopt depends on the relative importance one assigns to error minimization vs network size in a specific application. A very reasonable choice would be networks such as Hh_2Oo_1 or Hh_2Ob_2 (Fig. 8d) that produce low error by combining features from both Jordan and Elman networks while still being constrained in size. These results show that our evolutionary approach using a high-level descriptive language can be applied effectively in generating and evaluating alternative neural networks even for complex temporal tasks requiring recurrent networks.

Table 4. Representative results for the phoneme sequence generation problem

Architecture	RMSE	Absolute weight sum	PCT[a]
No delay	1.239	11,410.4	23.3
Ho_1	1.066	15,711.7	43.2
Ho_2	0.869	19,892.6	62.3
Oo_1	0.768	28,909.3	70.5
Ob_1	0.599	30,502.8	82.1
Hb_2	0.531	29,421.4	85.9
Ho_1Ob_1	0.501	29,243.5	87.5
Hh_1	0.328	23,604.5	94.6
Hh_2	0.270	27,232.6	96.4
Hh_3	0.255	29,250.4	96.7
Oh_1	0.232	27,893.6	97.3
Hb_3Oo_4	0.219	48,989.6	97.6
Hh_2Oh_1	0.210	27,540.5	97.8
Hh_1Oh_1	0.191	27,090.4	98.2
Hh_1Oo_1	0.180	29,290.6	98.4
Hh_3Oo_1	0.152	32,109.6	98.9
Hh_2Oo_1	0.107	30,718.3	99.4
Hh_3Oo_2	0.107	37,946.3	99.4
Hh_2Ob_2	0.088	40,383.5	99.6
Hh_2Ob_3	0.062	45,920.2	99.8
Hh_3Ob_3	0.044	49,589.8	99.9

[a] PCT = Percentage of phonemes generated completely correctly [39]

4 Discussion

Recent advances in neuroevolutionary methods have repeatedly been successful in creating innovative neural network designs [2, 7, 11, 19, 20, 31, 42, 43, 51],
but these successes have had little practical influence on the field of neural
computation. We believe this is partially because of a dilemma: the general space of neural network architectures and methods is so large that it
is impractical to search efficiently, yet attempting to avoid this problem by
hand-crafting the evolution of neural networks on a case-by-case basis is very
labor intensive and thus also impractical.

In this context, we explored the hypothesis that a high-level descriptive language can be used effectively to support the evolutionary design
of task-specific neural networks. This approach addresses the impracticality
of searching the enormous general space of neural networks by allowing a
designer to easily restrict the search space to architectures and methods that
appear a priori to be relevant to a specific application, greatly reducing the size
of the space that an evolutionary process must search. It also greatly reduces
the time needed to create a network's design by allowing one to describe the
class of neural networks of interest at a very high level in terms of possible modules and inter-module pathways, rather than in terms of individual

neurons and their connections. Filling in the "low level" details of individual networks in an evolving population is left to an automated developmental process (the neural networks are "grown" from their genetic encoding) and to well-established neural learning methods that create connection weights prior to fitness assessment.

In this chapter, we have presented experimental results suggesting that evolutionary algorithms can be successfully applied to the automated design of neural networks, addressing the dilemma stated above using a novel encoding scheme. We showed that human-readable description files could guide an evolutionary process to produce near-optimal modular solutions to n-partition problems. We also showed that this approach could not only create effective recurrent architectures for temporal sequence generation, but that it could simultaneously indicate the tradeoffs in the costs of architectural features versus network performance (via multi-objective evolution). All that was needed to guide the evolutionary processes in both of these applications was short description files like those illustrated in Figs. 3a and 7a.

References

1. Abraham A (2002) Optimization of evolutionary neural networks using hybrid learning algorithms. In: Proceedings of the 2002 International Joint Conference on Neural Networks (IJCNN '02), IEEE, vol. 3, pp. 2797–2802
2. Balakrishnan K, Honavar V (2001) Evolving neuro-controllers and sensors for artificial agents. In: Advances in the Evolutionary Synthesis of Intelligent Agents, MIT, pp. 109–152
3. Banzhf W, Nordin P, Keller RE, Francone FD (1997) Genetic Programming: An Introduction. Morgan Kaufmann, San Francisco, CA
4. Belew RK, McInerney J, Schraudolph NN (1991) Evolving networks: Using the genetic algorithm with connectionist learning. CSE Technical Report #CS90-174, Computer Science and Engineering Department, UCSD
5. Bentely PJ, Corne DW (2001) Creative Evolutionary Systems. Morgan Kaufmann, San Francisco, CA
6. Blum AL, Rivest RL (1992) Training a 3-node neural network is np-complete. Neural Networks 5(1):117–127
7. Bonissone PP, Subbu R, Eklund N, Kiehhl TR (2006) Evolutionary algorithms + domain knowledge = real-world evolutionary computation. IEEE Transactions on Evolutionary Computation 10(3):256–280
8. Brown M, Keynes R, Lumsden A (2001) The Developing Brain. Oxford University Press, Oxford
9. Cangelosi A, Parisi D, Nolfi S (1994) Cell division and migration in a genotype for neural networks. Network: Computation in Neural Systems 5(4):497–515
10. Chalmers DJ (1990) The evolution of learning: An experiment in genetic connectionism. In: Touretsky DS, Elman JL, Sejnowski TJ, Hinton GE (eds.) Proceedings of the 1990 Connectionist Summer School, Morgan Kaufmann, pp. 81–90, URL citeseer.ist.psu.edu/chalmers90evolution.html

11. Chong SY, Tan MK, White JD (2005) Observing the evolution of neural networks learning to play the game of othello. IEEE Transactions on Evolutionary Computation 9(3):240–251
12. Cybenko G (1989) Approximation by superpositions of a sigmoidal function. Mathematics of Control, Signals, and Systems 2(4):303–314
13. De Garis H (1991) GenNETS: Genetically programmed neural nets using the genetic algorithm to train neural nets whose inputs and/or output vary in time. In: Proceedings of the International Joint Conference on Neural Networks (5th IJCNN '91), IEEE, Singapore, vol. 2, pp. 1391–1396
14. Dill FA, Deer BC (1991) An exploration of genetic algorithms for the selection of connection weights in dynamical neural networks. In: Proceedings of the IEEE 1991 National Aerospace and Electronics Conference (NAECON '91), IEEE, New York, NY, Dayton, OH, vol. 3, pp. 1111–1115
15. Dimond S, Blizard D (1977) Evolution and Lateralization of the Brain. New York Academy of Sciences, New York
16. Elman JE (1990) Finding structure in time. Cognitive Science 14(2):179–211
17. Ferdinando AD, Calabretta R, Parisi D (2001) Evolving modular architectures for neural networks. In: French R, Sougné J (eds.) Proceedings Sixth Neural Computation and Psychology Workshop Evolution, Learning, and Development
18. Giles CL, Miller CB, Chen D, Sun GZ, Chen HH, Lee YC (1992) Extracting and learning an *unknown*, grammar with recurrent neural networks. In: Moody JE, Hanson SJ, Lippmann RP (eds.) Advances in Neural Information Processing Systems 4, Morgan Kaufmann, Denver, CO, pp. 317–324
19. Gruau F (1995) Automatic definition of modular neural networks. Adaptive Behavior 3:151–183
20. Gruau F, Whitley D, Pyeatt L (1996) A comparison between cellular encoding and direct encoding for genetic neural networks. In: Proceedings of the Sixth International Conference on Genetic Programming, Stanford University Press
21. Grushin A, Reggia JA (2005) Evolving processing speed asymmetries and hemispheric interactions in a neural network model. Neurocomputing 65:47–53
22. Harp S, Samad T, Guha A (1989) Towards the genetic synthesis of neural networks. In: Proceedings of the Third International Conference on Genetic Algorithms, Morgan Kaufmann, pp. 360–369
23. Haykin S (1999) Neural Networks: A Comprehensive Foundation. Prentice Hall, Upper Saddle River, NJ
24. Jordan MI (1986) Attractor dynamics and parallelism in a connectionist sequential machine. In: Proceedings of the Eighth Conference of the Cognitive Science Society, Erlbaum, pp. 531–546
25. Jung JY, Reggia JA (2004) A descriptive encoding language for evolving modular neural networks. In: Genetic and Evolutionary Computation – GECCO-2004, Part II, Springer, Lecture Notes in Computer Science, vol. 3103, pp. 519–530
26. Jung JY, Reggia JA (2006) Evolutionary design of neural network architectures using a descriptive encoding language. IEEE Transactions on Evolutionary Computation 10:676–688
27. Kandel E, Schwartz J, Jessel T (1991) Principles of Neural Science. Appleton and Lange, Norwalk, CT
28. Killackey H (1996) Evolution of the human brain: A neuroanatomical perspective. In: Gazzaniga M (ed.) The Cognitive Neurosciences, MIT, pp. 1243–1253

29. Kitano H (1994) Neurogenetic learning: An integrated method of designing and training neural networks using genetic algorithms. Physica D 75:225–238
30. Koza J, Bennett F, Andre D, Keane M (1999) Genetic Programming III: Darwinian Invention and Problem Solving. Morgan Kaufmann, San Francisco, CA
31. Lehmann KA, Kaufmann M (2005) Evolutionary algorithms for the self-organized evolution of networks. In: GECCO '05: Proceedings of the 2005 conference on Genetic and evolutionary computation, ACM, New York, NY, USA, pp. 563–570, DOI http://doi.acm.org/10.1145/1068009.1068105
32. Mehrotra K, Mohan CK, Ranka S (1997) Elements of Artificial Neural Networks. MIT, Cambridge, MA
33. Miller GF, Todd PM, Hegde SU (1989) Designing neural networks using genetic algorithms. In: Proceedings of third International Conference on Genetic algorithms (ICGA89), pp. 379–384
34. Mitchell M (1996) An Introduction to Genetic Algorithms. MIT, Cambridge, MA
35. Montana D, Davis L (1990) Training feedforward neural networks using genetic algorithms. In: Proceedings of eleventh International Joint Conference on Artificial Intelligence, Morgan Kaufmann, pp. 370–374
36. Mountcastle V (1998) The Cerebral Cortex. Harvard University Press, Cambridge, MA
37. Pérez-Ortiz J, Calera-Rubio J, Forcada M (2001) A comparison between recurrent neural architectures for real-time nonlinear prediction of speech signals. In: Miller D, Adali T, Larsen J, Hulle MV, Douglas S (eds.) Neural Networks for Signal Processing XI, Proceedings of the 2001 IEEE Neural Networks for Signal Processing Workshop (NNSP '01), IEEE Signal Processing Society, pp. 73–81
38. Radi A, Poli R (1998) Genetic programming can discover fast and general learning rules for neural networks. In: Koza JR, Banzhaf W, Chellapilla K, Deb K, Dorigo M, Fogel DB, Garzon MH, Goldberg DE, Iba H, Riolo R (eds.) Genetic Programming 1998: Proceedings of the Third Annual Conference, Morgan Kaufmann, University of Wisconsin, Madison, Wisconsin, USA, pp. 314–323
39. Radio MJ, Reggia JA, Berndt RS (2001) Learning word pronunciations using a recurrent neural network. In: Proceedings of International Joint Conference on Neural Networks (IJCNN '01), vol. 1, pp. 11–15
40. Riedmiller M (1994) Advanced supervised learning in multi-layer perceptrons: from backpropagation to adaptative learning algorithms. Computer Standards & Interfaces 16(3):265–278
41. Riedmiller M, Braun H (1993) A direct adaptive method for faster backpropagation learning: The rprop algorithm. In: Proceedings of 1993 IEEE International Conference on Neural Networks, vol. 1, pp. 586–591
42. Ruppin E (2002) Evolutionary autonomous agents: A neuroscience perspective. Nature Reviews Neuroscience 3(2):132–141
43. Saravanan N, Fogel D (1995) Evolving neural control systems. IEEE Expert 10:23–27
44. Sejnowski T, Rosenberg C (1987) Parallel networks that learn to pronounce english text. Complex Systems 1:145–168
45. Shkuro Y, Reggia JA (2003) Cost minimization during simulated evolution of paired neural networks leads to asymmetries and specialization. Cognitive Systems Research 4(4):365–383

46. Srinivas M, Patnaik LM (1991) Learning neural network weights using genetic algorithms- improving performance by search-space reduction. In: 1991 IEEE International Joint Conference on Neural Networks, IEEE, Singapore, vol. 3, pp. 2331–2336
47. Stanley KO, Miikkulainen R (2002) Evolving neural network through augmenting topologies. Evolutionary Computation 10(2):99–127
48. Tooby J, Cosmides L (2000) Toward mapping the evolved functional organization of mind and brain. In: Gazzinga M (ed.) The New Cognitive Neurosciences, MIT, pp. 1167–1178
49. Wolpert DH, Macready WG (1997) No free lunch theorems for optimization. IEEE Transactions on Evolutionary Computation 1(1):67–82
50. Wolpert DH, Macready WG (2005) Coevolutionary free lunches. IEEE Transactions on Evolutionary Computation 9(6):721–735
51. Yao X (1999) Evolving artificial neural networks. Proceedings of the IEEE 87(9):1423–1447
52. Zitzler E, Thiele L (1999) Multiobjective evolutionary algorithms: A comparative case study and the strength pareto approach. IEEE Transactions on Evolutionary Computation 3(4):257–271

A Versatile Surrogate-Assisted Memetic Algorithm for Optimization of Computationally Expensive Functions and its Engineering Applications

Yoel Tenne[1] and S.W. Armfield[2]

[1] The School of Aerospace, Mechanical and Mechatronic Engineering,
The University of Sydney, Australia, joel.tenne@aeromech.usyd.edu.au
[2] The School of Aerospace, Mechanical and Mechatronic Engineering,
The University of Sydney, Australia, armfield@aeromech.usyd.edu.au

Summary. A core element of modern engineering is optimization with computationally expensive 'black-box' functions. We identify three open critical issues in optimization of such functions: (a) how to generate accurate surrogate-models (b) how to handle points where the simulation fails to converge and (c) how to balance exploration vs. exploitation.

In this work we propose a novel surrogate-assisted memetic algorithm which addresses these issues and analyze its performance. The proposed algorithm contains four novelties: (a) it autonomously generates an accurate surrogate-models using multiple cross-validation tests (b) it uses interpolation to generate a global penalty function which biases the global search away from points where the simulation did not converge (c) it uses a method based on the Metropolis criterion from statistical physics to manage exploration vs. exploitation and (d) it combines an EA with the trust-region derivative-free algorithm. The proposed algorithm generates global surrogate-models of the expensive objective function and searches for an optimum using an EA; it then uses a trust-region local-search which combines local quadratic surrogate-models and sequential quadratic programming with the trust-region framework, and converges to an accurate optimum of the true objective function. While the global–local search approach has been studied previously, the improved performance in our algorithm is due to the four novelties just mentioned.

We present a detailed performance analysis, in which the proposed algorithm significantly outperformed four variants of a real-coded EA on three difficult optimization problems which include mathematical test functions with a complicated landscape, discontinuities and a tight limit of only 100 function evaluations. The proposed algorithm was also benchmarked against a recent memetic algorithm and a recent EA and significantly outperformed both, showing it is competitive with some of the best available algorithms.

Lastly, the proposed algorithm was also applied to a difficult real-world engineering problem of airfoil shape optimization; also in this experiment it consistently performed well, and in spite of a many points where the simulation did not converge,

Y. Tenne and S.W. Armfield: *A Versatile Surrogate-Assisted Memetic Algorithm for Optimization of Computationally Expensive Functions and its Engineering Applications*, Studies in Computational Intelligence (SCI) **92**, 43–72 (2008)
www.springerlink.com

and the tight limit on function evaluations it obtained an airfoil which satisfies the requirements very well.

Overall, the proposed algorithm offers a solution to important open issues in optimization of computationally-expensive black-box functions: generating accurate surrogate-models, handling points where the simulation does not converge and managing exploration–exploitation to improve the optimization search when only a small number of function evaluations are allowed, and it efficiently obtains a good approximation to an optimum of such functions in difficult real-world optimization problems.

1 Introduction

Optimization is a core element of modern engineering. Typically a baseline design exists and an optimization algorithm is applied so as to yield new and hopefully better designs. Mathematically, one seeks the global optimum $x_{\mathbf{g}}$ of an objective function f which measures some merit value of a candidate design, i.e.

$$x_{\mathbf{g}} : \min\{f(\boldsymbol{x}), \quad \boldsymbol{x} \in \mathcal{F} \subseteq \mathcal{R}^n\}, \tag{1}$$

where $\boldsymbol{x}$ represents a candidate design and $\mathcal{F}$ is the prescribed search domain. Often $\mathcal{F}$ is defined by vectors of lower and upper bounds, $\boldsymbol{x}_{\mathrm{L}}$ and $\boldsymbol{x}_{\mathrm{U}}$, respectively.

Manufacturing a prototype of a candidate design and then evaluating it by laboratory experiments is costly and time consuming. To avoid these drawbacks engineers increasingly substitute laboratory experiments with *computer experiments*, i.e. accurate computer simulations which model real-world physics and yield the required merit value [3,41]. For example, in the aerospace industry an accurate aerodynamics computer simulation can replace expensive wind tunnel experiments.

While it is beneficial to use computer experimentss in optimization, they introduce four major difficulties:

- No expression is available for f or its derivatives, i.e. f is a *black-box function*.
- A candidate solution is evaluated by running the computer simulation, which is typically computationally intensive, i.e. f is *expensive*. Thus, only a small number of function evaluations (simulation calls) are allowed.
- The underlying real-world physics and/or numerical solution yield a multimodal f.
- The computer simulation may not converge for some candidate designs, so $f(\boldsymbol{x})$ is *undefined* (or discontinuous) at some points. We denote a point $\boldsymbol{x} \in \mathcal{F}$ as *defined* if $f(\boldsymbol{x})$ is defined but otherwise the point is *undefined*, and $\bar{\mathcal{E}} \subseteq \mathcal{R}^n$ is the set of undefined points.

These four difficulties pose a formidable challenge in the development of algorithms for optimization of computationally expensive black-box functions.

Although activity in this research field is continuously increasing, we identify three fundamental and critical issues which are unresolved (a) how to generate accurate surrogate models (b) how to handle the undefined points and (c) how to allocate the limited number of calls to the expensive objective function between the global search and the local search, i.e. managing exploration–exploitation. In this study propose methods which address these issues and which fill a gap with respect to previous studies. Specifically, we describe a surrogate-assisted memetic algorithm as a combined global–local search approach where these three issues are addressed as follows: (a) to generate accurate global surrogate models we propose a method which chooses an optimal kernel and hyper-parameter for a radial basis functions surrogate model (b) to handle undefined points we propose a penalty function which is generated by interpolation and which biases the global search away from undefined points and (c) to manage the allocation of function evaluations between the global search and the local search we propose a method inspired by the Metropolis criterion from statistical physics. For the local search we implement an efficient trust-region derivative-free algorithm which converges to an accurate minimizer and which has been modified to handle undefined points. The three proposed methods and the hybridization of the EA with the trust-region local search algorithm are *the main novelties in our study with respect to previous studies*. Performance analysis with both mathematical test functions and a real-world engineering problem of airfoil optimization shows the proposed algorithm obtains a good approximation to an optimum of computationally expensive black-box functions and it offers a solution to the open issues mentioned.

This chapter is organized as follows: Sect. 2 reviews the current state-of-the-art in optimization of computationally expensive black-box functions and briefly outlines the proposed memetic algorithm. Next, the detailed description of the proposed algorithm is presented in Sect. 3. This is followed by Sect. 4 which provides a detailed performance analysis and finally Sect. 5 summarizes this chapter.

2 Optimization of Computationally Expensive Black-Box Functions

2.1 A Review of Approaches Based on Evolutionary Algorithms

Extensive reviews on optimization of computationally expensive black-box functions are [2, 17, 21, 48]; this section focuses on approaches in evolutionary optimization.

EAs perform well in difficult optimization problems, and this has motivated their application to optimization of computationally expensive black-box functions. However, EAs can require many function evaluations to converge and so several approaches have been studied so as to reduce the number of evaluations of the computationally expensive objective function they require. One such approach is *fitness inheritance*, where only a portion of the offspring are evaluated with the computationally expensive objective function and the rest inherit their fitness from their parents [23, 46]. A second approach is that of *hierarchical* or *variable-fidelity* optimization which uses several computer simulations of varying computational intensity (fidelity); promising candidate solutions migrate from low- to high-fidelity simulations and vice versa [14, 44]. A third and recent approach is that of *surrogate-assisted* optimization. A *surrogate model* (also known as a *meta model* or a *response surface*) is an approximation of the objective function which is computationally cheaper to evaluate. The surrogate model is generated by interpolating points where f has been evaluated. During the optimization the surrogate model substitutes for the expensive objective function, i.e. the value of a candidate solution is obtained by evaluating the surrogate model instead of the expensive objective function, thus yielding large computational savings. Recent studies have demonstrated the effectiveness of this approach when combining EAs with various surrogate models, e.g. quadratic least-squares (Response Surface Methodology) [25], Kriging [5, 38] and radial basis functions [31, 55]. Typically, the surrogate model is inaccurate since it approximates the objective function based only a small number of sample points; Fig. 1 shows an example. If it is significantly inaccurate, i.e. its landscape is completely different to the true objective function, the EA may converge to a false optimum, i.e. which is not an optimum of the true objective function [22, 32]. Thus, *the accuracy of the surrogate model must be ensured*.

To accelerate the convergence of EAs and to obtain an accurate approximation to an optimum in a small number of function evaluations *memetic algorithms* (or *hybrids*) combine an accurate local search with an EA. The local search is used either as an-EA operator or as a separate stage [33, 39]. Such combinations require *efficient allocation of the limited function evaluations between the EA (exploration) and the local search (exploitation)*.

Lastly, and as mentioned in Sect. 1, the objective function may be undefined at some points, i.e. for which the computer simulation fails to converge.

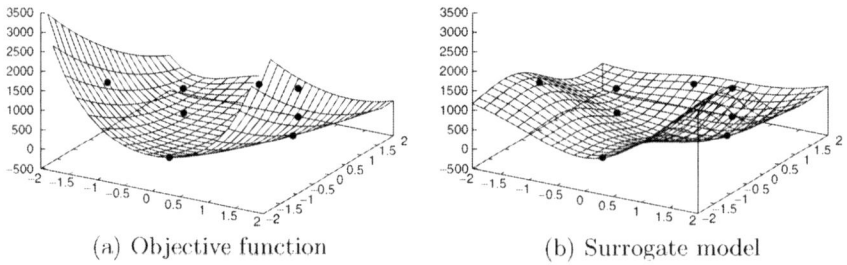

(a) Objective function   (b) Surrogate model

Fig. 1. A radial basis functions surrogate model for the Rosenbrock function based on eight interpolation points. The surrogate model has a false optimum

There seems to be little treatment of this real-world problem [15]; in [37] a clustering approach is used to classify candidate solutions as likely to be defined or not and in [8] the search region of the constrained optimization problem is reduced in case an undefined point is encountered. Thus, *an algorithm for optimization with computer simulations must handle undefined points.*

Accordingly, we identify three critical and open issues in optimization of computationally black-box functions: (a) how to generate accurate surrogate models (b) how to handle undefined points and (c) how to balance exploration vs exploitation.

2.2 Outline of the Proposed Algorithm

To address these important open issues we propose a novel surrogate-assisted memetic algorithm. An optimization iteration of the proposed algorithm operates in six steps:

S.1: A space-filling sample of points is generated using Latin Hypercube sampling (LHS) and the objective function is evaluated at these points. These points are added to a cache of previously evaluated points.

S.2: A proposed method generates an accurate global surrogate model for the objective function and a penalty function; the latter increases the predicted function value in regions containing undefined points. The surrogate model and the penalty function are combined to a resultant model.

S.3: An evolutionary algorithm (EA) searches for the global optimum of the resultant surrogate model; due to the penalty function the global search is biased to regions with defined points. It is stopped when the population is sufficiently clustered or when the number of generations reaches an upper bound.

S.4: A proposed method uses the Metropolis criterion to decide whether to initiate the local search and how many function evaluations to allocate to it.

S.5: The local search, if initiated, is used to the extent decided by the Metropolis criterion. To ensure convergence to a true optimum of the expensive objective function we use a trust-region derivative-free algorithm which combines sequential quadratic programming with the trust-region approach.

S.6: If termination criteria are met the algorithm stops; otherwise, new points obtained in the current optimization iteration are added to the cache and the algorithm returns to Step 1 to begin another iteration.

The proposed algorithm offers four novelties:

- It generates accurate surrogate models without the user being required to prescribe the basis function and hyper-parameter.

- It offers a systematic framework for handling undefined points both in the global and local search.
- It manages the allocation of function evaluation between the global search and local search using a Metropolis criterion.
- It combines a surrogate-assisted EA using global surrogate models with the trust-region derivative-free algorithm using local surrogate models. This ensures it converges to a true optimum of the expensive objective function.

Figure 2 shows the work flow of the proposed memetic algorithm.

3 A Detailed Description of the Proposed Memetic Algorithm

3.1 Generating the Global Surrogate Models

Since it is expensive to evaluate the objective function, the proposed memetic algorithm caches all points evaluated with the latter during the optimization search. To ensure the cache contains a space-filling sample of points, thus improving the accuracy of the global surrogate models (described below), an optimization iteration begins by sampling a small number of points globally using the Latin Hypercube method [26]. n_{LHSfir} points are sampled in the first iteration and n_{LHSsuc} in successive ones, where $n_{\mathrm{LHSsuc}} \leqslant n_{\mathrm{LHSfir}}$ and both are prescribed parameters. The objective function is evaluated at these new points and they are added to the cache; the cache now contains k_{d} defined points and k_{ud} undefined points.

Based on all the cached points the algorithm generates (a) a surrogate model for the objective function (b) a penalty function which increases the predicted function value around undefined points encountered and (c) a resultant surrogate model.

Generating Accurate Surrogate Models

To ensure the accuracy of the surrogate model (Sect. 2.1) we propose a method which generates surrogate models having a low interpolation error and which are numerically stable. The method uses radial basis functions (RBFs), as they have outperformed many other interpolants in term of computational efficiency and accuracy [16, 18, 20].

An RBF surrogate model is a linear sum of basis functions (or kernels); common basis functions are given in Table 1. Given a set of interpolation points $\boldsymbol{x}_i, i = 1 \ldots k$ where the values $g(\boldsymbol{x}_i)$ of some objective function $g(\boldsymbol{x})$ are known (the notation $f(\boldsymbol{x})$ is reserved for the computationally expensive

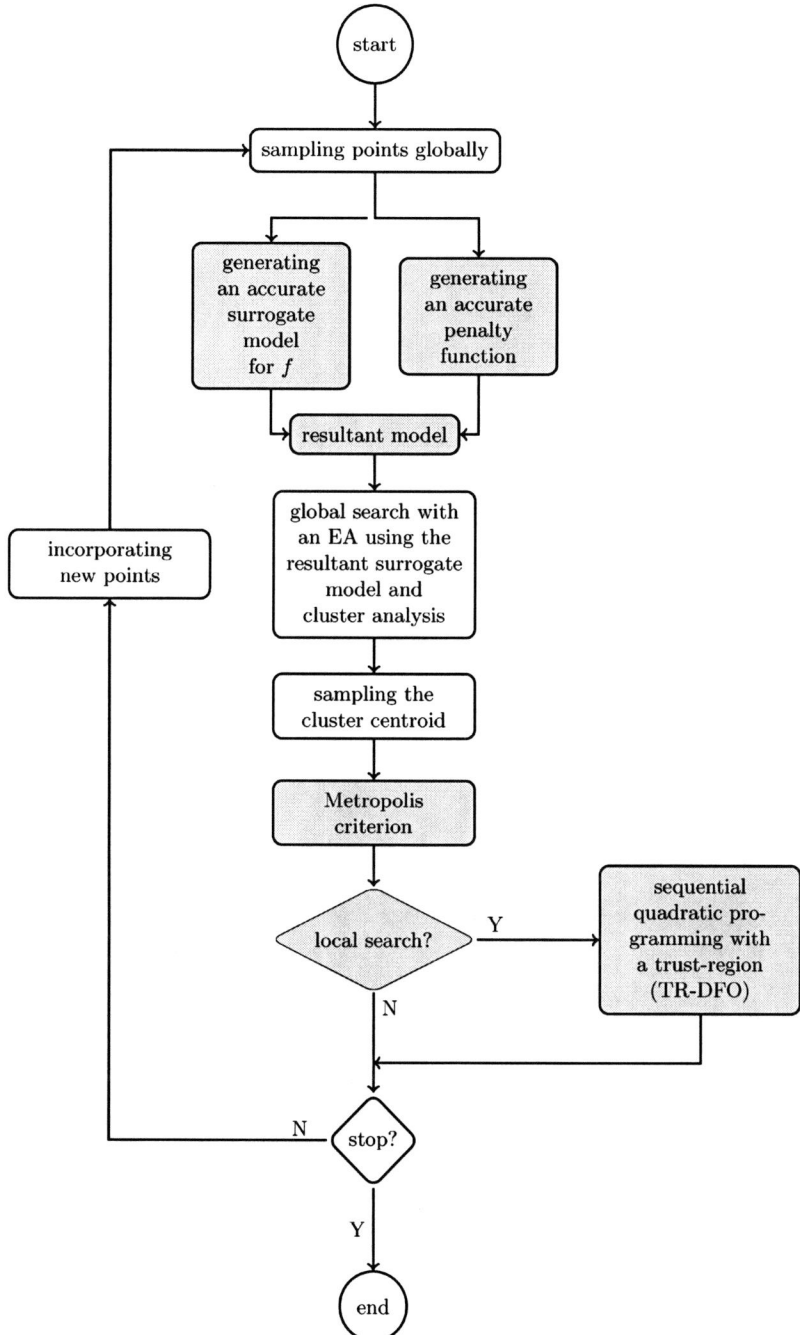

Fig. 2. The work flow of the proposed memetic algorithm. *Shaded blocks* show where the proposed methods are applied

Table 1. Common radial basis functions

Name	Expression $\phi_i(\boldsymbol{x})$
Gaussian	$\exp(-\alpha\|\boldsymbol{x} - \boldsymbol{x}_i\|_2^2)$
Inverse multiquadric	$1/\sqrt{\|\boldsymbol{x} - \boldsymbol{x}_i\|_2^2 + \alpha^2}$
Linear	$\|\boldsymbol{x} - \boldsymbol{x}_i\|_2$
Multiquadric	$\sqrt{\|\boldsymbol{x} - \boldsymbol{x}_i\|_2^2 + \alpha^2}$

$\boldsymbol{x}_i$ is an interpolation point (where the objective function value is known)
α is a prescribed hyper-parameter

objective function) and a basis-function type, then the RBF surrogate model for $g(\boldsymbol{x})$ is

$$S(\boldsymbol{x}) = \left\{ \sum_{i=1}^{k} \lambda_i \phi_i(\boldsymbol{x}) \right\} + \nu, \tag{2}$$

where the coefficients λ_i, ν are obtained from the linear system

$$\begin{pmatrix} \boldsymbol{\Phi} & 1 \\ 1^{\mathrm{T}} & 0 \end{pmatrix} \begin{pmatrix} \boldsymbol{\lambda} \\ \nu \end{pmatrix} = \begin{pmatrix} g(\boldsymbol{x}_i) \\ 1 \end{pmatrix}, \tag{3}$$

where

$$\boldsymbol{\Phi}^{k \times k} : \Phi_{i,j} = \phi_j(\boldsymbol{x}_i), \tag{4}$$

is the interpolation matrix for the given type of basis-functions and interpolation points.

This accuracy of the RBF surrogate model is affected by (a) the basis-functions type (b) the hyper-parameter α (c) the condition number of the interpolation matrix $\boldsymbol{\Phi}$ (numerical stability). Typically, the user prescribes a priori the basis-functions type and hyper-parameter value and does explore other options to obtain a more accurate surrogate model.

Accordingly, we propose the following method which autonomously chooses the basis-function type *and* adjusts the hyper-parameter to obtain an accurate RBF surrogate model. Our method builds on the leave-one-out cross-validation strategy [28, 45]; however, our method uses cross-validation with *multiple* excluded points *and* increases the cross-validation residual if the surrogate model is numerically unstable, i.e. its interpolation matrix is ill-conditioned. The method is provided with the following parameters: (a) a set of candidate basis-functions types (b) three hyper-parameters $\alpha_{\text{low}} < \alpha_{\text{int}} < \alpha_{\text{hig}}$ per basis-function type; the values can be different for each basis-function type (c) a set of cross-validation ratios (as explained below).

$\gamma_i \in (0,1), i = 1\ldots n_{cv}$ and (d) κ_{max} the maximal condition number of the interpolation matrix for which no penalty is applied.

Next, starting from $i = 1$ then for each combination of a candidate basis-functions type and hyper-parameter the following steps are performed:

S.1: A surrogate model $\widehat{S}_i$ is generated based on the subset of points $\{x_i\}, i = 1\ldots\gamma_i k$.

S.2: An *adjusted cross-validation residual* r_i is calculated, which we define as

$$r_i = \sum_{j=\gamma_i k+1}^{k} |g(x_j) - \widehat{S}_i(x_j)| \max\left\{1, \frac{\kappa_i}{\kappa_{max}}\right\}, \tag{5}$$

where κ_j is the condition number of the interpolation matrix $\boldsymbol{\Phi}$ for the current combination of basis function, hyper-parameter and $\gamma_i k$ points. Thus r_i measures how well the surrogate model based on the latter three predicts the objective function values at the cross-validation points; it is increased for numerically unstable surrogate models.

S.3: The above steps are repeated for $i = 2\ldots n_{cv}$.

S.4: For the current combination of basis-function and hyper-parameter the total cross-validation residual $\widehat{r}$ is calculated, which we define as

$$\widehat{r} = \sum_{i=1}^{n_{cv}} r_i. \tag{6}$$

After repeating these steps for each combination of basis-function and hyper-parameter then the basis-function having the smallest total cross-validation residual is chosen for further processing.

Next, the method refines the hyper-parameter to further improve the accuracy of the surrogate model. Given the total cross-validation residuals corresponding to α_{low}, α_{int}, α_{hig} with the chosen kernel, then a quadratic is fitted to predict the dependency of $\widehat{r}$ on α (for the chosen kernel). If the quadratic is positive-definite than its minimizer $\alpha^\star$ is obtained and $\widehat{r}(\alpha^\star)$ is calculated. Given α_{low}, α_{int}, α_{hig} and $\alpha^\star$ then the hyper-parameter which has yielded the smallest cross-validation residual is taken as the optimal value.

The values of α_{low}, α_{int}, α_{hig} are set to avoid ill-conditioning of the interpolation matrix. For this, it follows from Table 1, they must depend on the typical squared distance between points in $\mathcal{F}$. Accordingly, their values are calculated anew at each iteration based on d_m the mean distance between the cached points

$$d_m = \frac{1}{k^2} \sum_{i=1}^{k} \sum_{j=1}^{k} \|x_i - x_j\|_2. \tag{7}$$

To avoid ill-conditioning of $\boldsymbol{\Phi}$, the proposed method sets $\alpha_{low} = \sqrt{0.05 d_m}$, $\alpha_{int} = \sqrt{0.1 d_m}$ and $\alpha_{hig} = \sqrt{0.5 d_m}$ for the quadric and multiquadric kernels, and $\alpha_{low} = 0.5/d_m^2$, $\alpha_{int} = 1/d_m^2$ and $\alpha_{hig} = 2/d_m^2$ for the Gaussian kernel.

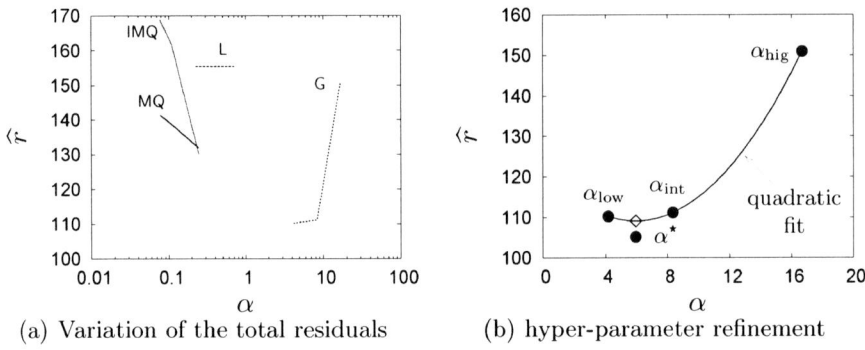

(a) Variation of the total residuals   (b) hyper-parameter refinement

Fig. 3. An example generating an accurate surrogate model. The objective function is the Rosenbrock function with ten points in $\mathcal{F} = [-1, 1]$. The candidate basis-functions were Gaussian (G), inverse multiquadric (IMQ), linear (L) and multiquadric (MQ). (**a**) shows the variation of the total cross-validation residual $\hat{r}$ with the hyper-parameter α; it follows the Gaussian kernel yields the most accurate surrogate models. (**b**) shows how hyper-parameter was refined by fitting a quadratic for the given α_{low}, α_{int} and α_{hig}, obtaining its minimizer $\alpha^\star$ (indicated by *diamond*), which yielded a lower total cross-validation (shown as *filled circle* below the *diamond*). The combination of the Gaussian kernel and $\alpha^\star$ yielded the lowest total cross-validation residual, i.e. the most accurate surrogate model

At each iteration of the memetic algorithm the proposed method is used twice: once for generating the surrogate model of the computationally expensive objective function with the defined points and then for generating the penalty function with both defined and undefined points, as explained in the following subsection. The proposed method has several merits:

- It yields an accurate surrogate model.
- It allows great flexibility such that different kernels can be used in different stages of the optimization. Also, different kernels can be used for generating the surrogate model of the objective function and the penalty function.
- It autonomously selects the most suitable kernel and hyper-parameter.

Figure 3 shows an experiment with the proposed method.

Handling the Undefined Points in the Global Search

We propose to handle undefined points by biasing the global search away from them, i.e. by penalizing the predicted function value in their vicinity. The method proposed here extends an earlier study described in [50,51].

Recalling the cache contains k_{d} defined points and k_{ud} undefined ones, then a penalty function $\mathcal{S}_{\mathrm{p}}(\boldsymbol{x})$ is generated such that

$$\mathcal{S}_{\mathrm{p}}(\boldsymbol{x}) = \begin{cases} 0 & \text{if } \boldsymbol{x} = \boldsymbol{x}_{\mathrm{d},\,i}, i = 1 \ldots k_{\mathrm{d}} \\ \kappa_{\mathrm{p}} & \text{if } \boldsymbol{x} = \boldsymbol{x}_{\mathrm{ud},\,i-k_{\mathrm{d}}}, i = k_{\mathrm{d}} + 1 \ldots k_{\mathrm{d}} + k_{\mathrm{ud}}, \end{cases} \tag{8}$$

where κ_p is a positive parameter. The method from the previous subsection is used to generate $\mathcal{S}_p(\boldsymbol{x})$ such that

$$\mathcal{S}_p(\boldsymbol{x}) = \sum_{i=1}^{k_d+k_{ud}} \lambda_{p,i}\phi_i(\boldsymbol{x}) + \nu_p, \tag{9}$$

where the kernel type and coefficients and determined as described earlier. Due to the interpolation conditions (8) the value of $\mathcal{S}_p(\boldsymbol{x})$ at a point $\boldsymbol{x}$ corresponds to the probability of $\boldsymbol{x}$ to be undefined, i.e. the set of points where $\mathcal{S}_p(\boldsymbol{x}) > 0$ is an approximation of $\bar{\mathcal{E}}$.

At each iteration the proposed memetic algorithm generates $\mathcal{S}_p(\boldsymbol{x})$ and $\mathcal{S}_o(\boldsymbol{x})$ a surrogate model for the objective function using only the cached points which are defined, i.e. with the interpolation conditions

$$\mathcal{S}_o(\boldsymbol{x}_{d,i}) = f(\boldsymbol{x}_{d,i}), \quad i = 1 \ldots k_d. \tag{10}$$

$\mathcal{S}_p(\boldsymbol{x})$ and $\mathcal{S}_o(\boldsymbol{x})$ are combined to a resultant surrogate model $\mathcal{S}_r(\boldsymbol{x})$ which will be used during the EA global search, where

$$\mathcal{S}_r(\boldsymbol{x}) = \mathcal{S}_o(\boldsymbol{x})\big(1 + \max\{0, \mathcal{S}_p(\boldsymbol{x})\}\big), \tag{11}$$

so the max$\{\}$ operator clips negative penalties; these can occur due to oscillations in $\mathcal{S}_p(\boldsymbol{x})$ induced by interpolation with adjacent points. Figure 4 shows an example of the proposed method.

3.2 Global Search

After generating the $\mathcal{S}_r(\boldsymbol{x})$ surrogate model an EA is used to search for the latter's global optimum. We use the EA from [6] with real-coded variables as this is suitable for the engineering problems we consider. It is set to use linear ranking, stochastic universal sampling (SUS), intermediate recombination, elitism with a generation gap g_{gap} and the breeder-genetic-algorithm mutation operator with probability p_m [29]. During the global search the computationally surrogate model is used so the EA uses a large population size s_{pop} and a large number of generations g_{max}, which both assist in locating the global optimum. The initial EA population is generated by Latin hyper-cube sampling. The EA is stopped when (a) the number of generations is g_{max} or (b) $\boldsymbol{x}_{EA}$ the best point in the population has a predicted function value $\mathcal{S}_r(\boldsymbol{x}_{EA})$ which is better than the best objective function value found so far (i.e. better than the best cached point) *and* the population is clustered around $\boldsymbol{x}_{EA}$, indicating search convergence. To check the latter condition the population is clustered at the end of each generation; if the cluster whose centroid is $\boldsymbol{x}_{EA}$ contains a minimum of $\gamma_{pop}s_{pop}$ points, where $\gamma_{pop} \in (0,1]$, then the condition is met. For clustering we use Törn's density algorithm as it is efficient and does not require user-set parameters [52].

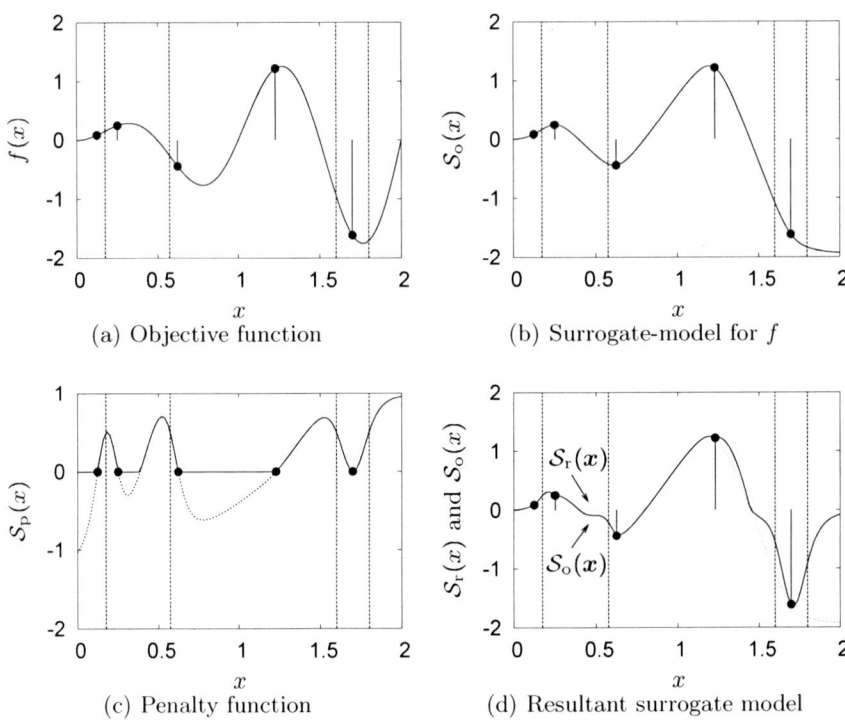

Fig. 4. An example of the proposed method for handling undefined points. (**a**) shows the objective function $f(x) = x \sin(2\pi x)$. We artificially set it to be undefined at the four points indicated by the *dashed lines*; it is evaluated at five defined points and the four undefined ones. (**b**) shows the surrogate model $\mathcal{S}_o(x)$ for $f(x)$; it is generated using only the defined points. (**c**) shows the penalty function $\mathcal{S}_p(x)$; negative penalties (*dotted line*) are clipped at $\mathcal{S}_p(x) = 0$. (**d**) compares the resultant surrogate model $\mathcal{S}_r(x)$ to $\mathcal{S}_o(x)$; $\mathcal{S}_r(x)$ approximates $f(x)$ well and is increased around undefined points

3.3 Managing Exploration vs Exploitation

x_{EA}, the best point found by the EA, is evaluated with the computationally expensive objective function yielding $f(x_{EA})$. To obtain an even better candidate solution we use an accurate local search. A naive approach is to initiate the local search from x_{EA} unconditionally at each iteration. However, since only a small number of calls to the expensive objective function are allowed this approach quickly exhausts the optimization resources without ensuring a good optimum is found.

Accordingly, we propose the following method which efficiently allocates function evaluations between the global search and the local search. The method uses four metrics: (a) f_b the best objective function value found so far (b) $f(x_{EA})$ (c) m_{tot} the total number of function evaluations allowed and

(d) m_{cur} the current number of function evaluations, i.e. already used. The method is based on the following heuristics:

- If $f(\boldsymbol{x}_{\mathrm{EA}}) < f_{\mathrm{b}}$ then the surrogate model is accurate since its predicted optimum is indeed better than the best objective function value found so far. Thus, we trust $\boldsymbol{x}_{\mathrm{EA}}$ to be a good approximation to the global optimum and allow the local search to use all the remaining calls to the computationally expensive objective function.
- If $f(\boldsymbol{x}_{\mathrm{EA}}) \geqslant f_{\mathrm{b}}$ than the surrogate model is inaccurate since its predicted optimum is not better than the best objective function value found so far. Accordingly, we reduce our confidence in $\boldsymbol{x}_{\mathrm{EA}}$ to be a good approximation to the global optimum and we reserve calls to the objective function for sampling points globally so as to improve the global surrogate models. However, $\boldsymbol{x}_{\mathrm{EA}}$ may be in the neighbourhood of a good optimum so we may still initiate the local search but with the following conditions:
  - As more points are cached and the global surrogate models improve in accuracy the probability of initiating the local search (given that $f(\boldsymbol{x}_{\mathrm{EA}}) \geqslant f_{\mathrm{b}}$) decreases during the optimization search.
  - Since $f(\boldsymbol{x}_{\mathrm{EA}}) \geqslant f_{\mathrm{b}}$ and if the optimization search is in its early stages then a small portion of the remaining calls to the expensive objective function are allocated to the local search to ensure more exploration. Similarly, if the optimization search is in its late stages then a larger portion of the remaining calls will be allocated to the local search to emphasize exploitation.

Since each iteration of the proposed memetic algorithm begins by sampling a prescribed number of points globally (exploration), varying the number of calls allocated to the local search controls the exploration vs exploitation of the optimization search.

To implement these heuristics we propose a method based on the Metropolis criterion from statistical physics [27]. The criterion states that the probability of a physical system to change its state depends on the difference between its current energy level and the energy level of a candidate state. Due to the laws of thermodynamics a system tends to accept lower energy states, but it can occasionally accept higher energy states in a semi-random process. Given T the temperature of the system and ΔE the difference between its current and the candidate energy level then the probability $p(\Delta E, T)$ of the system to accept the candidate state is

$$p(\Delta E, T) = \begin{cases} 1 & \text{if } \Delta E < 0, \\ \exp{(-\Delta E/T)} & \text{if } \Delta E \geqslant 0. \end{cases} \tag{12}$$

It follows: (a) reducing T make the system more 'greedy', i.e. it tends to accept only better states and (b) the probability p is strictly positive.

The Metropolis criterion has been used in the simulated annealing optimization method [4, 24]; however, the novelty in the proposed method is the use of the criterion as a mechanism for managing exploration–exploitation. The proposed method achieves this by determining in each iteration m_{loc} the number of calls to the computationally expensive objective function allocated to the local search, where

$$m_{\text{loc}} = l(m_{\text{tot}} - m_{\text{cur}}), \quad l \in (0,1), \tag{13}$$

and l is the fraction of calls allocated. To determine l we define a measure of merit for $f(\boldsymbol{x}_{\text{EA}})$

$$\theta_f = (f(\boldsymbol{x}_{\text{EA}}) - f_{\text{b}})/(|f_{\text{b}}| + \varepsilon), \tag{14}$$

where $\varepsilon \ll 1$ is added to avoid singularity in θ_f when $f_{\text{b}} = 0$, and define the fraction of calls to the objective function already used

$$\theta_{\text{m}} = m_{\text{cur}}/m_{\text{tot}}. \tag{15}$$

Both θ_f and θ_{m} measure relative quantities and hence are problem independent. l is determined using the proposed Metropolis criterion such that

$$l = \begin{cases} 1 & \text{if } f(\boldsymbol{x}_{\text{EA}}) < f_{\text{b}} \quad (\theta_f < 0), \\ \exp(-5\,\theta_f\,\theta_{\text{m}}) & \text{with probability } p(\theta_f, \theta_{\text{m}}) \text{ otherwise,} \end{cases} \tag{16}$$

where we define $p(\theta_f, \theta_{\text{m}})$ as

$$p(\theta_f, \theta_{\text{m}}) = \exp(-\theta_f/(10\,\theta_{\text{m}})). \tag{17}$$

Accordingly, if $f(\boldsymbol{x}_{\text{EA}})$ is better than the current best objective function found so far (best cached point) then all remaining calls to the objective function are allocated to the local search. Otherwise, the proposed Metropolis criterion determines whether to initiate the local search and if so what fraction of the remaining calls to allocate to it. With respect to (12), the merit measure θ_f is equivalent to ΔE and the fraction of remaining calls θ_{m} is equivalent to T, i.e. it controls the 'greediness' of the method; the weights 5 and 10 were set to account for magnitude differences between θ_f and θ_{m}. Figure 5 shows the dependency of p and l on θ_f and θ_{m}.

3.4 The Local Search

If the method described in the previous subsection yields $l > 0$ then the local search is initiated.

As explained below, the local search is initiated at most once per optimization iteration and it generates a sequential series of surrogate models.

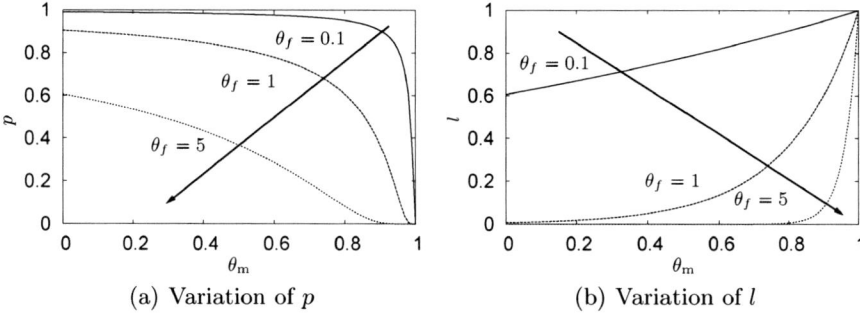

(a) Variation of p (b) Variation of l

Fig. 5. The variation of p the probability of initiating the local search and l fraction of calls allocated to the latter, when $f(\boldsymbol{x}_{EA}) > f_b$. (**a**) shows p decreases as $f(\boldsymbol{x}_{EA})$ becomes worse compared to f_b (θ_f increases) and as the optimization progresses (θ_m increases), i.e. the algorithm becomes 'greedier'. (**b**) shows l decreases as $f(\boldsymbol{x}_{EA})$ becomes worse compared to f_b but increases as the optimization progresses so as to promote exploration in early stages and exploitation in later ones

To improve the accuracy of the initial surrogate model the local search is supplied with $\boldsymbol{x}_{EA}$ and cached points adjacent to it, i.e.

$$\|\boldsymbol{x}_i - \boldsymbol{x}_{EA}\|_2 \leqslant n_1 \Delta_0, \quad n_1 \geqslant 1, \tag{18}$$

where n_1 is a parameter and Δ_0 is the initial trust-region radius (defined below).

For the local search we implemented a trust-region derivative-free optimization algorithm (TR-DFO) [7, 36, 49]. It combines sequential quadratic programming (SQP) with a trust-region framework, i.e. it extends the powerful trust-region approach to the case of black-box functions. This guarantees that the TR-DFO algorithm converges to a first order optimal point satisfying the Karush–Kuhn–Tucker conditions from *any* initial starting point (under mild conditions on the boundedness of the model gradient and Hessian) [1, 7]. The TR-DFO algorithm operates in five step:

S.1: Given a set of points with known function values it generates $\hat{f}(\boldsymbol{x})$ a quadratic surrogate model of the objective function by multivariate Lagrangian interpolation, i.e.

$$\hat{f}(\boldsymbol{x}) = \hat{c} + \boldsymbol{x}^{\mathrm{T}}\hat{g}(\boldsymbol{x}) + \frac{1}{2}\boldsymbol{x}^{\mathrm{T}}\hat{H}\boldsymbol{x}, \tag{19}$$

where $\hat{c}$, $\hat{g}$ and $\hat{H}$ are the constant, gradient and Hessian of the quadratic surrogate model, respectively. $\hat{f}$ is generated using the interpolation algorithm of Sauer–Xu [7, 42].

S.2: Given a current iterate $\boldsymbol{x}_c$ and a spherical *trust-region* where the approximation (19) is considered valid

$$\mathcal{T} = \{\boldsymbol{x} : \|\boldsymbol{x} - \boldsymbol{x}_c\|_2 \leqslant \Delta\}, \tag{20}$$

then a quadratic programming problem is solved to find $\boldsymbol{x}_m$ the constrained local optimum of $\hat{f}$ in $\mathcal{T}$. $\boldsymbol{x}_m$ is obtained by the projected conjugate-gradient algorithm of Dennis–Vicente [9].

S.3: The agreement between the predicted and actual reduction is determined by the measure of merit

$$\rho = \frac{f(\boldsymbol{x}_c) - f(\boldsymbol{x}_m)}{\hat{f}(\boldsymbol{x}_c) - \hat{f}(\boldsymbol{x}_m)}. \tag{21}$$

If $\rho = 1$ then $\hat{f}$ models f accurately, while smaller values represent poorer agreement.

S.4: Based on ρ the trust-region and the set of interpolation points are updated. Given the parameter $\eta_+ > 0$, then if $\rho < \eta_+$ a new point $\boldsymbol{x}_n$ is generated in $\mathcal{T}$ and replaces an existing interpolation point so as to improve the accuracy of $\hat{f}$. Also, given the parameters $\delta_+ > 1$, $\delta_- < 1$, $\Delta_{max} > 0$, the following updates are performed:

$$\boldsymbol{x}_c^{(t+1)} = \begin{cases} \boldsymbol{x}_m^{(t)} & \text{if } \rho \geqslant \eta_+, \\ \boldsymbol{x}_c^{(t)} & \text{if } \rho < \eta_+, \end{cases} \tag{22}$$

$$\Delta^{(t+1)} = \begin{cases} \min\left\{\delta_+ \Delta^{(t)}, \Delta_{max}\right\} & \text{if } \rho \geqslant \eta_+, \\ \delta_- \Delta^{(t)} & \text{if } \rho < \eta_+, \end{cases} \tag{23}$$

i.e. if the agreement is good then the trust-region is centred at the new constrained local optimum $\boldsymbol{x}_m$ and the trust-region radius is increased but otherwise it is decreased.

S.5: If termination criteria (as described below) are met the algorithm stops, but otherwise it returns to Step 1.

Two merits of this algorithm are (a) it does not use finite-differences to approximate derivatives and avoids the problems associated the them [12] (b) the quadratic models are both locally accurate and their associated quadratic programming problems are efficiently solved [35].

If the TR-DFO encounters an undefined $\boldsymbol{x}_m$ then the trust-region radius is temporarily reduced and a new constrained local optimum is sought [8]. This is repeated until a defined point is found or the temporary trust-region radius is deemed too small, where in the latter case the TR-DFO algorithm skips to Step 4. The trust-region radius is reduced since $\boldsymbol{x}_c$ is defined hence an $\boldsymbol{x}_m$ sufficiently adjacent to it is also likely to be defined. If the algorithm is unable to obtain $\boldsymbol{x}_m$ and $\boldsymbol{x}_n$ which are both defined then the TR-DFO iteration is unsuccessful since no progress has been made.

The TR-DFO algorithm is stopped when either (a) $\Delta < \Delta_{min}$ where $\Delta_{min} > 0$ is a parameter (b) $\|P(\nabla \hat{f}(\boldsymbol{x}_c))\|_2 \leqslant \epsilon_y$, where $P(\nabla \hat{f}(\boldsymbol{x}_c))$ is the projected gradient of the quadratic surrogate model at $\boldsymbol{x}_c$ onto the feasible domain $\mathcal{F}$ and $\epsilon_g > 0$ is a parameter (c) m_{loc} calls were made to the computationally expensive objective function in the current local search (d) both $\boldsymbol{x}_m$

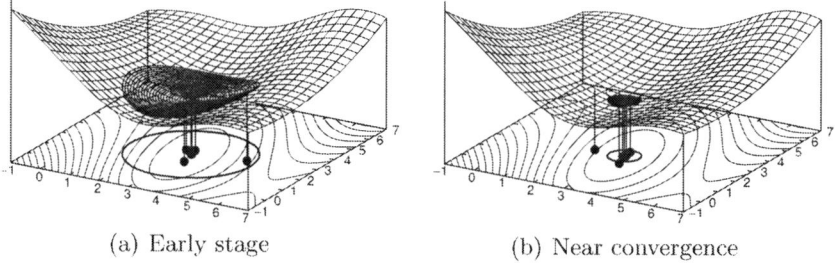

(a) Early stage   (b) Near convergence

Fig. 6. An example of the trust-region derivative-free algorithm (TR-DFO) minimization of the Branin function. The quadratic surrogate model is superimposed on the objective function surface; the local optimum, interpolation points and the trust-region are shown on the x–y plane. **(a)** shows an early stage of the optimization search when the trust-region is large. **(b)** shows a near convergence stage where the trust-region has been reduced and encloses the local optimum

and x_n are undefined. Thus, the TR-DFO algorithm may stop before using m_{loc} calls. Figure 6 shows an example of the algorithm operation.

3.5 Remarks on the Complete Algorithm

After the local search is stopped the current optimization iteration of the proposed memetic algorithm is completed. The algorithm now stops if (a) the best function value found f_b is sufficiently good, i.e. $f_b \leqslant f_{min}$, where $f_{min} \geqslant -\infty$ is a parameter (b) the number of function evaluations used is m_{tot}. Otherwise, another optimization iteration begins, as shown in Fig. 2. In this case, new points obtained during the current optimization iteration are cached and incorporated into the global surrogate models.

4 Performance Analysis

This section presents performance analysis of the proposed memetic algorithm using mathematical test functions and a real-world application of airfoil shape optimization.

4.1 Mathematical Test Functions

Performance of the proposed algorithm is analyzed using three well-known test functions: Rosenbrock [40] ($n = 2$), Rastrigin [53, p. 185–192] ($n = 5$) and Schwefel 7 [43, p. 292–293] ($n = 10$). These functions were chosen since they represent difficult test cases due to their complicated and multimodal landscape [11]; Table 2 gives the functions details.

For a problem of dimension n the feasible domain $\mathcal{F}$ was the hyper-cube defined by the lower and upper bounds $x_L = -5 \in \mathcal{R}^n$, $x_U = 5 \in \mathcal{R}^n$ for the

Table 2. Test functions details

Function	Definition	x_g	$f(x_g)$
Rosenbrock	$f(x) = \{10(x_2 - x_1{}^2)\}^2 + (1 - x_1)^2$	1	0
Rastrigin	$f(x) = \sum_{i=1}^{n} \{x_i^2 - A \cdot \cos(\omega x_i) + B\}$	0	0
Schwefel 7	$f(x) = \sum_{i=1}^{n} x_i \sin(\sqrt{|x_1|})$	$x_i \simeq 420.967$	$\simeq -418.983n$

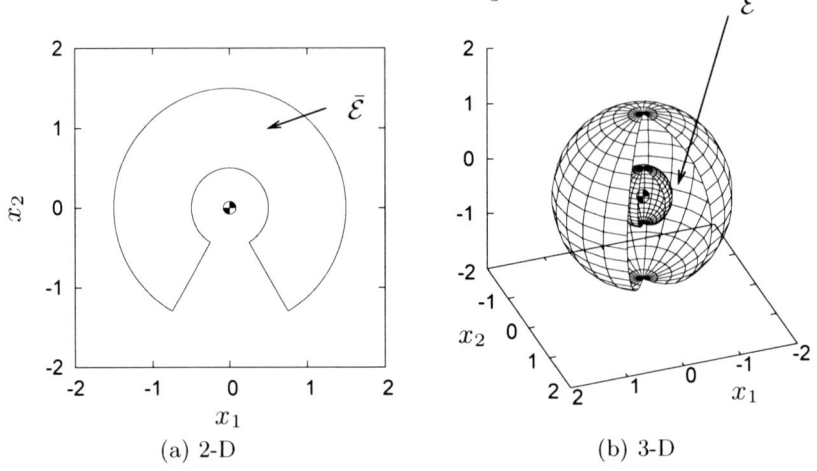

(a) 2-D   (b) 3-D

Fig. 7. The region of undefined points $\bar{\mathcal{E}}$ in two and three dimensions. The global optimum ☉ is at the origin

Rosenbrock and Rastrigin functions, and $x_L = -500 \in \mathcal{R}^n$, $x_U = 500 \in \mathcal{R}^n$ for the Schwefel 7 function. The bounds were chosen to provide a large search space and to increase the difficulty of locating the global optimum.

To present a difficult optimization scenario with respect to undefined points we artificially set $\bar{\mathcal{E}}$ the region of undefined points as the volume between two concentric hyper-spheres centred at the global optimum and each having a cutaway gap. These are readily defined in hyper-spherical coordinates using a radius r and $n - 1$ angular coordinates ϕ_i, $i = 1 \ldots n - 1$. We use an inner hyper-sphere with a radius of 0.5 and an outer one with a radius of 1.5; the cutaway gap is the region where $\phi_1 \in [60°, 120°]$. Thus, $\bar{\mathcal{E}}$ encloses the global optimum and leaves a small gap from which the global optimum can be approached. Figure 7 shows $\bar{\mathcal{E}}$ in two and three dimensions.

We restricted the number of objective function evaluations m_{tot} to 100, as a realistic setting in optimization of computationally expensive objective function. Each test function was minimized by the proposed algorithm and by four variants of the real-coded EA from [6]; each variant used a different combination of population size (s_{pop}) and mutation probability (p_m): EA 1: s_{pop} = 10, p_m = 0.01; EA 2: s_{pop} = 10, p_m = 0.1; EA 3: s_{pop} = 25, p_m = 0.01; EA 4: s_{pop} = 25, p_m = 0.1. For the four variants, at each generation all undefined points encountered were discarded and the recombination and mutation operators were repeated until all points in the population were defined. To obtain statistically significant results 20 tests were repeated for each function with each algorithm. Parameter settings are given in Table 3; they were set to values typical to an optimization of a real-world objective function and were not fine-tuned during the tests.

Table 4 provides the best, worst, mean and median values of the objective function obtained in the benchmark tests and Fig. 8 shows the convergence histograms, where peaks at smaller function values indicate better performance. The histogram bar-width was $(f_w - f_b)/25$, where f_w and f_b are the worst and best function values, respectively, obtained from all five algorithms. The bar-width was calculated separately for each test function.

Results show the proposed algorithm consistently obtained a good approximation to the global optimum and significantly outperformed the EAs with all three test functions. The good approximation of the global optimum was obtained in spite of the high multimodality of the test functions and the complicated shape of the region $\bar{\mathcal{E}}$.

During each optimization search the method described in Sect. 3.1 continuously updated the RBF type and the hyper-parameter used in the surrogate model for the objective function and of the penalty function. Figure 9 shows an example of these updates from one experiment with the Rosenbrock function. Also from the same experiment, Fig. 10 shows how the method described in Sect. 3.3 managed the exploration and exploitation of the proposed algorithm.

We also benchmark the proposed memetic algorithm against two recent algorithms: (a) the *Evolutionary Algorithm with n-dimensional Approximation* (EANA): a memetic algorithm which generates quadratic surrogate models of the objective function so as to simplify the objective function landscape; for the local search it uses the Local Evolutionary Search with Random Memorizing method [25] and (b) the *Improved Fast Evolutionary Programming* (IFEP): an EA where each parent generates two offspring, one with a Gaussian mutation and one with a Cauchy mutation [54]. As before, 20 tests were repeated for each test function. For the three algorithms, Table 5 shows the mean of both the optimum and the number of function evaluation; results for the EANA and IFEP algorithms are taken from [25]. For consistency with the reference, no undefined region was used in these tests for the proposed algorithm. The number of total function evaluations m_{tot} was increased for the Hartman 6 from 100 to 250 to obtain an accuracy comparable to the reference algorithms. It follows the proposed algorithm obtained an approximation to the global optimum comparable to that of the two reference algorithms, but in a fraction of the function evaluations. This comparison shows the merits of our proposed algorithm, i.e. with respect to recent algorithms it is comparable in accuracy but is significantly more efficient.

The efficiency of the proposed memetic algorithm in terms of computation time was also evaluated; the algorithm was run on an 800 MHz Intel Pentium 3 PC with 512 MB of memory, and was coded in OCTAVE version 2.1.69 [13].

Table 3. Parameter settings for the proposed memetic algorithm

	Overall algorithm operation	
m_{tot}	Total calls to the computationally expensive objective function[a]	100
f_{min}	Minimum required function value	$-\infty$
δ_a	Acceptance distance for adding new points to the cache	0.01

	Global surrogate models	
n_{LHSfir}	LHS points, 1st iteration	10
n_{LHSsuc}	LHS points, successive iterations	2
$\phi_i(\boldsymbol{x})$	Candidate RBF types	MQ, IMQ, linear, Gaussian
γ_i	Cross-validation ratio	0.5, 0.75, 0.9
	Hyper-parameter bounds, MQ, IMQ RBFs	$\sqrt{0.05d_m}, \sqrt{0.1d_m}, \sqrt{0.5d_m}$
	Hyper-parameter bounds, Gaussian RBF	$0.5/d_m^2, 1/d_m^2, 5/d_m^2$
κ_{max}	Maximal interpolation matrix condition number for which no penalty is applied	10^3
κ_p	Penalty parameter	0.5

	EA	
s_{pop}	Population size	100
g_{gap}	Generation gap	0.9
g_{max}	Maximum generations per iterations	100
p_m	Mutation probability	0.05

	Cluster analysis	
γ_{pop}	Population portion required in one cluster	0.5

	TR-DFO	
n_1	Distance measure for using cached points	2
η_+	Ratio for 'good' improvement	0.75
Δ_0	Initial trust-region radius	0.1
Δ_{min}	Minimum trust-region radius	10^{-3}
Δ_{max}	Maximum trust-region radius	2
δ_+	Trust-region radius increase factor	2
δ_-	Trust-region radius decrease factor	0.5
ϵ_g	Minimum projected gradient norm	10^{-3}

[a] In the benchmark tests mathematical test functions simulate expensive ones

Table 6 shows the timing statistics; it follows that time required by the proposed algorithm is negligible compared to a single evaluation of a computationally expensive objective function and hence it is computationally efficient.

Table 4. Statistical metrics for the benchmark tests

Function	Metric	Proposed	EA1	EA2	EA3	EA4
Rosenbrock	Best	**1.417e–03**	8.122e–01	1.079e+00	4.063e–01	1.241e+00
	Worst	**3.196e+00**	9.335e+02	4.124e+02	2.025e+03	4.555e+02
	Median	**8.354e–01**	1.496e+02	7.451e+01	1.974e+02	1.477e+02
	Mean	**1.151e+00**	2.393e+02	1.188e+02	2.748e+02	2.006e+02
Rastrigin	Best	**4.331e+00**	3.910e+01	3.364e+01	2.553e+01	2.446e+01
	Worst	**4.310e+01**	7.525e+01	7.965e+01	7.538e+01	7.538e+01
	Median	**1.918e+01**	5.828e+01	5.703e+01	5.153e+01	6.309e+01
	Mean	**2.118e+01**	5.563e+01	5.746e+01	5.144e+01	5.944e+01
Schwefel 7	Best	**−2.911e+03**	−1.477e+03	−1.427e+03	−1.729e+03	−1.651e+03
	Worst	**−1.066e+03**	−1.478e+02	−4.656e+02	−4.962e+02	−4.848e+02
	Median	**−2.021e+03**	−8.966e+02	−9.015e+02	−9.575e+02	−8.634e+02
	Mean	**−1.933e+03**	−8.911e+02	−9.288e+02	−1.028e+03	−9.590e+02

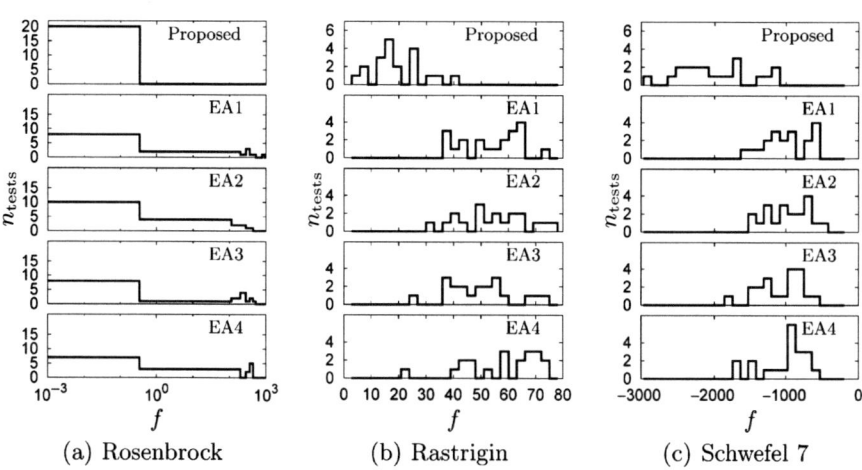

(a) Rosenbrock   (b) Rastrigin   (c) Schwefel 7

Fig. 8. Convergence histograms for the benchmark tests. n_{tests} (y-axis) is the number of tests and f (x-axis) is the function value

4.2 Aerodynamic Airfoil Shape Optimization

The proposed memetic algorithm was also applied to a real-world engineering problem of airfoil shape optimization. The problem is important since an aircraft using an efficient airfoil (which generates a low aerodynamic friction) requires less engine power and hence has a lower fuel consumption; this reduces pollution and operational expenses. In the experiment described the proposed algorithm optimized the airfoil shape of a small business jet (SBJ). From an optimization aspect, this is a suitable test for the proposed algorithm since evaluating a candidate airfoil with an aerodynamics simulation is

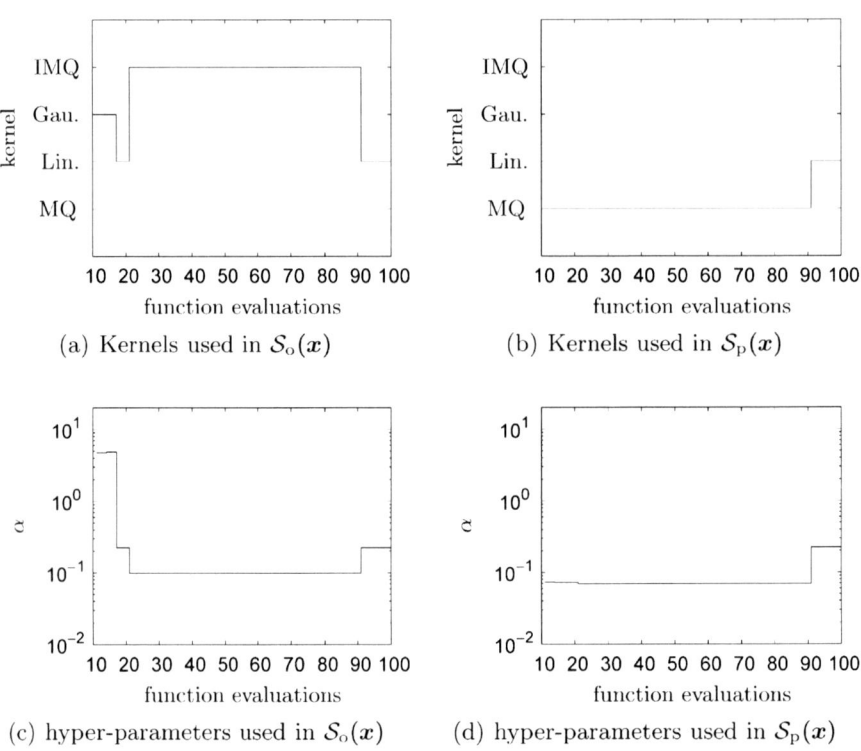

Fig. 9. An experiment with the Rosenbrock function showing the updating of the kernels and hyper-parameters during the optimization search

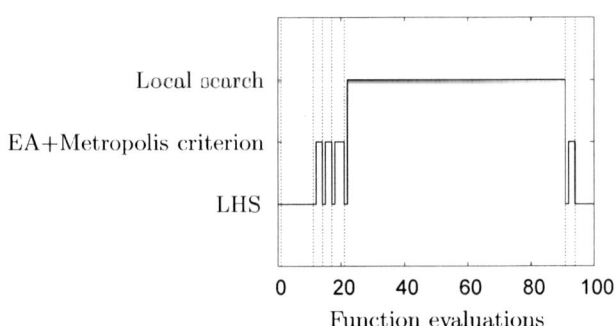

Fig. 10. An experiment with the Rosenbrock function showing allocation of function evaluations to different algorithm stages, i.e. how exploration and exploitation were managed. The *dashed lines* indicate a new optimization iteration. In the first four iterations the optimum obtained by the EA was unsatisfactory hence the local search was not initiated and calls were allocated to more exploration. A satisfactory optimum was obtained in the fifth iteration and the local search was initiated; it stopped before exhausting the remaining calls to the objective function. The remaining calls were then used for additional exploration

Table 5. Benchmarks against reference algorithms

Function	Proposed		IFEP		EANA	
	f_b	m_{tot}	f_b	m_{tot}	f_b	m_{tot}
Branin	3.979e–01	100	**3.979e–01**	4,500	3.979e–01	945
Hartman 3	−3.824e+00	100	**−3.860e+00**	6,000	−3.860e+00	1,297
Hartman 6	−3.180e+00	250	**−3.260e+00**	6,000	−3.313e+00	8,357

Table 6. Timing statistics of the proposed memetic algorithm

Function	Mean (s)	SD (s)[a]
Rosenbrock	4.944e+02	1.997e+02
Rastrigin	8.486e+02	8.362e+02
Schwefel 7	6.645e+02	4.298e+02

[a] Standard deviation

computationally expensive *and* the latter fails to converge for many candidate designs.

The goal is to obtain an airfoil which generates the required lift c_L while minimizing the aerodynamic friction (or drag) c_D; these depend on the aircraft flight condition (speed, altitude, angle of attack). Furthermore, the airfoil must be sufficiently thick to sustain the forces exerted on it during flight; in this optimization problem an airfoil is required to have a minimum of 2% thickness between 0.2 to 0.8 of its chord line. Accordingly, we defined the objective function

$$f = (|(c_{L,tar} - c_L)| + (10 \cdot c_D)) \times p_{air}, \qquad (24)$$

where $c_{L,tar}$ is the target lift coefficient, c_L and c_D are the candidate airfoil's lift and drag coefficients respectively, and p_{air} is the penalty term

$$p_{air} = \max \left\{ 1, \frac{t_{min,all}}{t_{min}} \right\}, \qquad (25)$$

where $t_{min,all}$, t_{min} are the prescribed minimum allowed thickness and the candidate airfoil thickness, respectively. The drag coefficient c_D is multiplied by ten to scale it with c_L. In this optimization problem the target lift coefficient was $c_{L,tar} = 1$. A candidate airfoil was represented as a vector of 11 geometrical design variables using the PARSEC method [30, 47]. Figure 11 shows the problem description and the PARSEC variables definition.

Calculation of the lift and drag coefficients for a candidate airfoil were performed by the aerodynamics simulation XFoil, a subsonic airfoil development system [10]. The business jet flight conditions were supplied to XFoil as fixed parameters, i.e. an atmospheric density at a flight altitude of 30 kft, a flight speed of 440 knots and an angle of attack of 2°. For the airfoil parametrization, bounds on the PARSEC variables were set to extremal values from previous

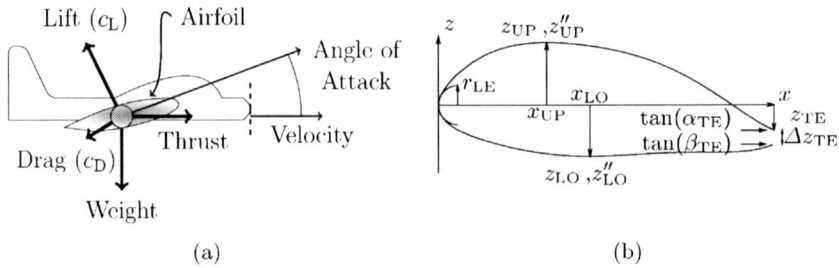

(a)   (b)

Fig. 11. (a) The forces acting on an aircraft during flight. (b) The PARSEC variables definition

Table 7. PARSEC variables and their bounds

Variable	Meaning	x_{Li}	x_{Ui}
r_{LE}	Leading-edge radius	0.002	0.030
x_{up}	Max. upper thickness location	0.2	0.7
z_{up}	Max. upper thickness	0.03	0.18
z_{up}''	Max. upper curvature	-0.9	0.0
x_{lo}	Max. lower thickness location	0.2	0.7
z_{lo}	Max. upper thickness	-0.13	0.02
z_{lo}''	Max. lower curvature	0.00	0.09
z_{TE}	Trailing edge height	-0.05	0.05
Δz_{TE}	Trailing edge thickness	0	0
β_{TE}	Trailing edge, lower angle°	-20	0
α_{TE}	Trailing edge, upper angle°	0	20

studies which have used the PARSEC parametrization [19, 34]; this was done to capture as many undefined points as may exists thus making the optimization more challenging. We set $\Delta z_{TE} = 0$ to ensure a closed airfoil shape. Table 7 gives the variables bounds. As in the previous subsection, the proposed memetic algorithm was benchmarked against the four EAs with five tests per algorithm. With the system configuration described in the previous subsection, an evaluation of a candidate airfoil requires approximately 20–60 s.

Table 8 provides the best, worst, mean and median values obtained for the objective function defined in (24). It follows the proposed algorithm obtained an airfoil shape which closely satisfies the optimization requirements, i.e. $c_L = 0.996$ with a low drag coefficient of $c_D = 0.008$, while satisfying the thickness requirement. As before, the proposed algorithm significantly outperformed the four EAs.

Figure 12 shows the convergence history in one experiment and compares an initial candidate airfoil with the optimal one obtained in the experiment. The optimal airfoil significantly differs from an initial candidate airfoil, which shows the proposed memetic algorithm efficiently handled this difficult optimization problem. During this experiment 13 undefined points were encountered, out of the 100 evaluated. Figure 13 shows the distribution of the

Table 8. Statistical metrics for objective function in the airfoil optimization

Metric	Proposed	EA1	EA2	EA3	EA4
Best	**8.280e−02**	1.259e−01	5.070e−01	3.179e−01	2.148e−01
Worst	**2.264e−01**	7.887e−01	7.309e 01	6.340e−01	7.897e−01
Median	**1.078e−01**	4.193e−01	5.575e−01	5.756e−01	3.360e−01
Mean	**1.398e−01**	4.273e−01	5.924e−01	5.058e−01	4.169e−01

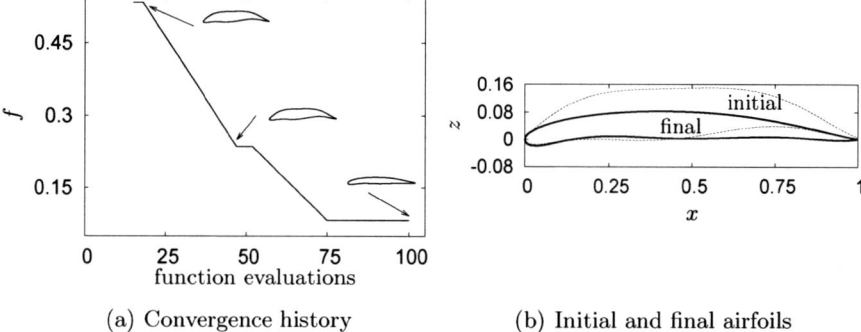

(a) Convergence history   (b) Initial and final airfoils

Fig. 12. Results from one airfoil optimization experiment

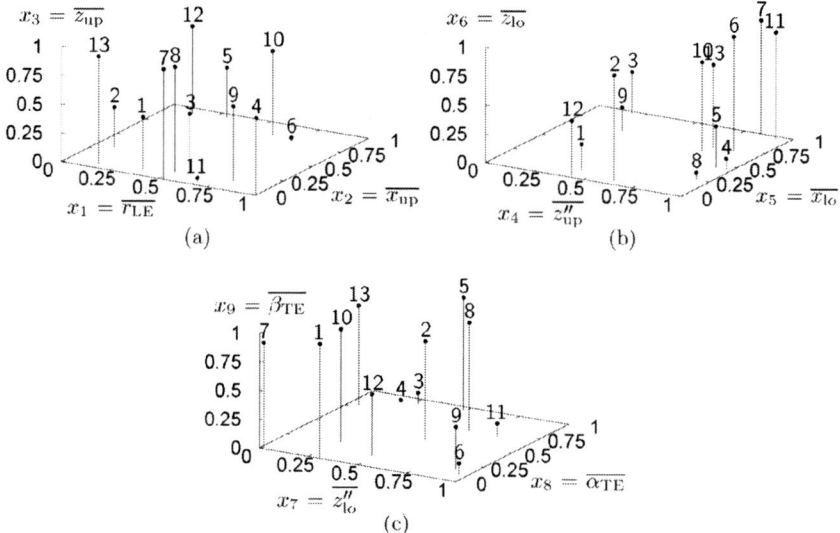

Fig. 13. The distribution of undefined points in one airfoil optimization experiment

undefined points for 9 out of the 10 variables (the 11th variable Δz_{TE} was fixed); variables are normalized to the range $[0, 1]$ and hence are denoted by an overline (e.g. $\overline{r_{\mathrm{LE}}}$). The undefined points are distributed in all dimensions which increases the optimization difficulty. In spite of the undefined points and small number of function evaluations the proposed algorithm obtained an airfoil which satisfies the requirements very well.

5 Summary

A core element of modern engineering is optimization with computationally expensive black-box functions. We have identified three open critical issues in optimization of such functions: (a) how to generate accurate surrogate models (b) how to handle undefined points and (c) how to balance exploration vs exploitation.

In this work we have proposed a novel memetic algorithm which addresses these issues and have analyzed its performance. The proposed algorithm contains four novelties: (a) it autonomously generates an accurate surrogate model using multiple cross-validation tests (b) it uses interpolation to generate a global penalty function which biases the global search away from undefined points and (c) it uses a method based on the Metropolis criterion from statistical physics to manage exploration vs exploitation and (d) it combines an EA with the trust-region derivative-free algorithm. Specifically, we describe a surrogate-assisted memetic algorithm as a combined global-local search approach; it generates global surrogate models of the computationally expensive objective function and searches for an optimum using an evolutionary algorithm; it then uses a trust-region derivative-free algorithm which combines local quadratic surrogate models and sequential quadratic programming with the trust-region framework, and converges to an accurate optimum of the computationally expensive objective function. While the global–local search approach has been studied previously, the improved performance in our algorithm is due to the four novelties just mentioned.

We conducted a detailed performance analysis, in which the proposed algorithm significantly outperformed four variants of a real-coded evolutionary algorithm on three difficult optimization problems which include mathematical test functions with a complicated landscape, discontinuities and a tight limit of only 100 function evaluations. The proposed algorithms was also benchmarked against a recent memetic algorithm and a recent EA and significantly outperformed both, showing it is competitive with some of the best available algorithms.

Lastly, the proposed algorithm was also applied to a difficult real-world engineering problem of airfoil shape optimization; also in this experiment it consistently performed well, and in spite of a many undefined points and the tight limit on function evaluations it obtained an airfoil which satisfies the requirements very well.

Overall, the proposed algorithm offers a solution to important open issues in optimization of computationally expensive black-box functions: generating accurate surrogate models, handling undefined points and managing exploration–exploitation to improve the optimization search when only a small number of function evaluations are allowed, and it efficiently obtains a good approximation to an optimum of such functions in difficult real-world optimization problems.

References

1. N. Alexandrov. Robustness properties of a trust region framework for managing approximations in engineering optimization. In *Proceedings of the Sixth AIAA/NASA/ISSMO Symposium on Multidisciplinary Analysis and Design*, number AIAA-96-4102-CP, pages 1056–1059. AIAA, 1996.
2. J. F. M. Barthelemy and R. T. Haftka. Approximation concepts for optimum structural design — a review. *Structural optimization*, 5:129–144, 1993.
3. A. J. Booker. Case studies in design and analysis of computer experiments. In *Proceedings of the Section on Physical and Engineering Sciences*, pages 244–248. American Statistical Association, 1996.
4. V. Černy. Thermodynamical approach to the traveling salesman problem: An efficient simulation algorithm. *Journal of Optimization Theory and Applications*, 45(1):41–51, 1985.
5. K. Chiba, S. Jeong, O. Shigeru, and H. Morino. Data mining for multidisciplinary design space of regional-jet wing. In *Proceedings the 2005 IEEE Congress on Evolutionary Computation–CEC 2005*, pages 2333–2340. IEEE, 2005.
6. A. Chipperfield, P. Fleming, H. Pohlheim, and C. Fonseca. *Genetic Algorithm TOOLBOX for use with MATLAB, Version 1.2*. Department of Automatic Control and Systems Engineering, University of Sheffield.
7. A. R. Conn, K. Scheinberg, and P. L. Toint. Recent progress in unconstrained nonlinear optimization without derivatives. *Mathematical Programming*, 79:397–414, 1997.
8. A. R. Conn, K. Scheinberg, and P. L. Toint. A derivative free optimization algorithm in practice. In *Proceedings of the Seventh AIAA/USAF/NASA/ISSMO Symposium on Multidisciplinary Analysis and Optimization*. AIAA, 1998.
9. J. E. Dennis and L. N. Vicente. Trust-region interior-point algorithms for minimization methods with simple bounds. In H. Fischer, B. Riedmuller, and S. Schaffler, editors, *Applied Mathematics and Parallel Computing, Festschrift for Klaus Ritter*, pages 97–107. Physica, Heidelberg, 1996.
10. M. Derla and H. Youngren. *XFOIL 6.9 User Primer*. Department of Aeronautics and Astronautics, Massachusetts Institute of Technology, Cambridge, MA, 2001.
11. J. G. Digalakis and K. G. Margaritis. An experimental study of benchmarking functions for genetic algorithms. *International Journal of Computer Mathematics*, 79(4):403–416, 2002.
12. L. C. W. Dixon. The choice of step length, a crucial factor in the performance of variable metric algorithms. In F. A. Lootsma, editor, *Numerical Methods for Non-linear Optimization*, chapter 10, pages 149–170. Academic, London New York, 1972.
13. J. W. Eaton. *Octave: An Interactive Language for Numerical Computations*, University of Wisconsin–Madison, Madison, 1997.
14. D. Eby, R. C. Averill, W. F. I. Punch, and E. D. Goodman. Evaluation of injection island GA performance on flywheel design optimization. In *Proceedings*

of the Third Conference on Adaptive Computing in Design and Manufacturing, Plymouth, England, 1998.

15. P. D. Frank. Global modeling for optimization. *SIAM SIAG/OPT Views-and-News*, 7:9–12, 1995.

16. R. Franke. Scattered data interpolation: Tests of some method. *Mathematics of Computation*, 38(157):181–200, 1982.

17. K. C. Giannakogolu. Design of optimal aerodynamic shapes using stochastic optimization methods and computational intelligence. *International Review Journal Progress in Aerospace Sciences*, 38(1):43–76, 2002.

18. R. L. Hardy. Theory and applications of the multiquadric-biharmonic method. *Computers and Mathematics with Applications*, 19(8/9):163–208, 1990.

19. T. L. Holst and T. H. Pulliam. Aerodynamic shape optimization using a real-number-encoded genetic algorithm. Technical Report 2001-2473, AIAA, 2001.

20. R. Jin, W. Chen, and T. W. Simpson. Comparative studies of metamodeling techniques under multiple modeling criteria. *Journal of Structural Optimization*, 23(1):1–13, 2001.

21. Y. Jin. A comprehensive survey of fitness approximation in evolutionary computation. *Soft Computing*, 9(1):3–12, 2005.

22. Y. Jin, M. Olhofer, and B. Sendhoff. A framework for evolutionary optimization with approximate fitness functions. *IEEE Transactions on evolutionary computation*, 6(5):481–494, 2002.

23. H.-S. Kim and S.-B. Cho. An efficient genetic algorithm with less fitness evaluation by clustering. In *Proceedings of 2001 IEEE Conference on Evolutionary Computation*, pages 887–894. IEEE, 2001.

24. S. Kirkpatrick, C. D. Gelatt, and M. P. Vecchi. Optimization by simulated annealing. *Science*, 220(4598):471–480, 1983.

25. K.-H. Liang, X. Yao, and C. Newton. Evolutionary search of approximated N-dimensional landscapes. *International Journal of Knowledge-Based Intelligent Engineering Systems*, 4(3):172–183, 2000.

26. M. D. McKay, R. J. Beckman, and W. J. Conover. A comparison of three methods for selecting values of input variables in the analysis of output from a computer code. *Technometrics*, 21(2):239–245, 1979.

27. N. Metropolis, A. W. Rosenbluth, M. N. Rosenbluth, A. H. Teller, and T. Edward. Equation of state calculations by fast computing machines. *Journal of Chemical Physics*, 21(6):1087–1092, 1953.

28. T. Mitchell and M. Morris. Bayesian design and analysis of computer experiments: Two examples. *Statistica Sinica*, 2:359–379, 1992.

29. H. Mühlenbein and D. Schlierkamp-Voosen. Predictive models for the breeder genetic algorithm I: Continuous parameter optimization. *Evolutionary Computations*, 1(1):25–49, 1993.

30. S. Obayashi. Airfoil shape optimization for evolutionary computation. In *Genetic Algorithms for Optimization in Aeronautics and Turbomachinery*, VKI Lecture Series 2000-07. Rhode Saint Genese, Belgium and Von Karman Institute for Fluid Dynamics, 2000.

31. Y.-S. Ong, P. B. Nair, and A. J. Keane. Evolutionary optimization of computationally expensive problems via surrogate modeling. *American Institute of Aeronautics and Astronautics Journal*, 41(4):687–696, 2003.

32. Y.-S. Ong, P. B. Nair, and K. Y. Lum. Max-min surrogate-assisted evolutionary algorithm for robust aerodynamic design. *IEEE Transactions on Evolutionary Computation*, 10(4):392–404, 2006.

33. Y.-S. Ong, Z. Zong, and D. Lim. Curse and blessing of uncertainty in evolutionary algorithm using approximation. In *Proceedings of the IEEE Congress on Evolutionary Computation–WCCI 2006*, pages 2928–2935. IEEE, 2006.
34. A. Oyama, S. Obayashi, and T. Nakamura. Real-coded adaptive range genetic algorithm applied to transonic wing optimization. In M. Schoenauer, K. Deb, R. Günter, X. Yao, E. Lutton, J. J. M. Guervós, and H.-P. Schwefel, editors, *The 6th Parallel Problem Solving from Nature International Conference–PPSN VI*, volume 1917 of *Lecture Notes in Computer Science*, pages 712–721. Springer, Berlin Heidelberg New York, 2000.
35. M. J. D. Powell. Direct search algorithms for optimization calculations. *Acta Numerica*, 7:287–336, 1998.
36. M. J. D. Powell. UOBYQA: Unconstrained optimization by quadratic approximation. *Mathematical Programming, Series B*, 92:555–582, 2002.
37. K. Rasheed, H. Hirsh, and A. Gelsey. Genetic algorithm for continuous design space search. *Artificial Intelligence in Engineering*, 11(3):295–305, 1996.
38. A. Ratle. Accelerating the convergence of evolutionary algorithms by fitness landscape approximations. In A. E. Eiben, B. T., M. Schoenauer, and H. P. Schwefel, editors, *Proceedings of the 5th International Conference on Parallel Problem Solving from Nature–PPSN V*, volume 1498 of *Lecture Notes in Computer Science*, pages 87–96. Springer, Berlin Heidelberg New York, 1998.
39. J.-M. Renderes and S. P. Flasse. Hybrid methods using genetic algorithms for global optimization. *IEEE Transcation on Systems, Man and Cybernetics–Part B*, 26(2):243–258, 1996.
40. H. H. Rosenbrock. An automated method for finding the greatest of least value of a function. *The Computer Journal*, 3:175–184, 1960.
41. J. Sacks, W. J. Welch, T. J. Mitchell, and H. P. Wynn. Design and analysis of computer experiments. *Statistical Science*, 4(4):409–435, 1989.
42. T. Sauer and Y. Xu. On multivariate Lagrange interpolation. *Mathematics of Computation*, 64(211):1147–1170, 1995.
43. H.-P. Schewefel. *Numerical Optimization of Computer Models*, volume 26 of *Interdisciplinary Systems Research*. Wiley, Chichester, 1981.
44. M. Sefrioui and J. Périaux. A hierarchical genetic algorithm using multiple models for optimization. In M. Schoenauer, K. Deb, G. Rudolph, X. Yao, E. Lutton, J. J. M. Guervós, and H.-P. Schwefel, editors, *Proceedings of the 6th International Conference on Parallel Problem Solving from Nature–PPSN VI*, volume 1917 of *Lecture Notes in Computer Science*, pages 879–888. Springer, Berlin Heidelberg New York, 2000.
45. T. W. Simpson, D. K. J. Lin, and W. Chen. Sampling strategies for computer experiments. *International Journal of Reliability and Applications*, 2(3):209–240, 2001.
46. R. E. Smith, B. Dike, and S. Stegmann. Fitness inheritance in genetic algorithms. In K. M. George, editor, *Proceedings of the 1995 ACM Symposium on Applied Computing–SAC 95*, pages 345–350. ACM, 1995.
47. H. Sobieckzy. Parametric airfoils and wings. In K. Fujii and G. S. Dulikravich, editors, *Recent Development of Aerodynamic Design Methodologies-Inverse Design and Optimization*, volume 68 of *Notes on Numerical Fluid Mechanics*, pages 71–88. Vieweg, Germany, 1999.
48. J. Sobieszczansk-Sobieski and R. Haftka. Multidisciplinary aerospace design optimization: Survey of recent developments. *Structural Optimization*, 14(1): 1–23, 1997.

49. Y. Tenne and S. W. Armfield. A memetic algorithm using a trust-region derivative-free optimization with quadratic modelling for optimization of expensive and noisy black-box functions. In S. Yang, Y.-S. Ong, and Y. Jin, editors, *Evolutionary Computation in Dynamic and Uncertain Environments*, volume 51 of *Studies in Computational Intelligence*, pages 389–415. Springer, Berlin Heidelberg New York, 2007.
50. Y. Tenne, S. Obayashi, and S. Armfield. Airfoil shape optimization using an algorithm for minimization of expensive and discontinuous black-box functions. In *Proceedings of AIAA InfoTec 2007*, number AIAA-2007-2874. AIAA, 2007.
51. Y. Tenne, S. Obayashi, S. Armfield, Y.-S. Ong, and R. Tapabrata. A surrogate-assisted memetic algorithm for handling computationally expensive functions with ill-defined constraints. In preparation.
52. A. Törn. Cluster analysis using seed points and density-determined hyperspheres as an aid to global optimization. *IEEE Transcation on Systems, Man and Cybernetics*, 7(8):610–616, 1977.
53. A. A. Törn and A. Žilinskas. *Global Optimization*, volume 350 of *Lecture Notes In Computer Science*. Springer, Berlin Heidelberg New York, 1989.
54. X. Yao, G. Lin, and Y. Liu. An analysis of evolutionary algorithms based on neighbourhood and step sizes. In P. Angeline, R. Reynolds, J. McDonnel, and R. Eberhart, editors, *Evolutionary Programming VI:Proceedings of the Sixth Annual Conference on Evolutionary Programming*, volume 1213 of *Lecture Notes on Computer Science*, pages 297–307. Springer, Berlin Heidelberg New York, 1997.
55. Z. Z. Zhou, Y.-S. Ong, P. B. Nair, A. J. Keane, and K. Y. Lum. Combining global and local surrogate models to accelerate evolutionary optimization. *IEEE Transactions On Systems, Man and Cybernetics-Part C*, 37(1):66–76, 2007.

Data Mining and Intelligent Multi-Agent Technologies in Medical Informatics

Fariba Shadabi and Dharmendra Sharma

School of Information Sciences and Engineering, The University of Canberra, Australia

Summary. In recent years there has been a rapid growth in the successful use of advanced data mining and knowledge discovery systems in many diverse areas such as science, medicine and commerce. The main contribution factor for the development of such systems in medicine has been the extra demands for more powerful yet flexible and transparent techniques to cope with the extensive amount of data and knowledge stored in clinical databases. Many of these techniques are currently under active development for use in the practice of clinical medicine. This chapter will consider a long standing problem of clinical knowledge discovery and decision making process, which is concerned with intelligently processing large amounts of clinical and medical data for knowledge discovery and decision support. The primary aim of this chapter will be to outline some of the potential and common challenges and expectations of intelligent knowledge discovery and data mining systems in medical sectors.

1 Introduction

Over the last two decades Information Technology (IT) has transformed many industries and similarly has great potential to facilitate efficiency and improvements in the healthcare sectors. Medical centres, hospitals and other medical facilities are using Artificial Intelligence (AI) systems to provide electronic education for medical students and patients, deliver health care to remote location, provides reasoning for knowledge management and scientific research, share information about medical diagnosis, and improve medical services. The purpose of this chapter is to look at some of the different directions AI is taking today in the healthcare sectors and discuss the potential and many challenges and expectations of AI-based data mining for knowledge discovery systems in the medical domains.

F. Shadabi and D. Sharma: *Data Mining and Intelligent Multi-Agent Technologies in Medical Informatics*, Studies in Computational Intelligence (SCI) **92**, 73–92 (2008)
`www.springerlink.com` © Springer-Verlag Berlin Heidelberg 2008

2 Artificial Intelligence in Medicine

From the early 1950s, scientists and healthcare professionals have been aware of the potential of computer systems in medicine. In 1959, Ledley and Lusted [44] published a paper entitled, "*The reasoning foundations of medical diagnosis*" [44]. In this paper they point out the potential for computer programs to assist physicians with diagnosis tasks. They applied their idea by using mathematical methods of Boolean algebra and symbolic logic to solve medical problems. Over the years AI systems, along with performing tasks that require reasoning with medical knowledge have played a significant role in the process of scientific research. Most significantly, AI systems have the capacity to learn, which has lead to the discovery of new phenomena and the creation of medical knowledge. Recently, AI systems have been involved in clinical decision support systems, mining medical databases, effective diagnostic assistance and treatment methodologies and intelligent systems for hospital, home and community health care. Diagnostic assistance is when AI systems help in developing a possible diagnosis for a patient based on their patient data. Image recognition and interpretation of medical films such as X-rays, CT and MRI scans, are areas were AI systems can indicate potential abnormal images. They can also be used in pharmacological sectors. For example, a learning system can be given examples of one or more drugs that weakly exhibit a particular activity, and based upon a description of the chemical structure of those compounds, the learning system can then suggest which of the chemical attributes are necessary for that pharmacological activity. One of the earliest computer-based knowledge discovery systems was DENDRAL [11, 45]. DENDRAL (DENDRitic Algorithm) was developed between 1965 and 1983 by a group of chemists, geneticists, and computer scientists. In particular, the DENDRAL project was started by Edward Feigenbaum, Joshua Lederberg (a Nobel Prize-winning geneticist and biochemist) and Bruce Buchanan as an effort to aid organic chemists in predicting the molecular structures of unknown organic compounds. DENDRAL uses AI methodology to manipulate the rules and decide which candidates can be considered as a plausible candidate structures for new or unknown chemical compounds.

Overall, AI can be considered as an attempt to build intelligent computer programs that behave like humans, furthermore it can also be described as a broad discipline with many applications, some of which are shown in Fig. 1.

3 Expert System Technology

The DENDRAL project described in Sect. 2 pioneered the use of rule-based reasoning strategy, which then grew into knowledge engineering tools, today known as expert systems. The terms expert system, knowledge-based system, or knowledge-based expert system are often used synonymously [84]. Expert systems are computer programs that are derived from AI.

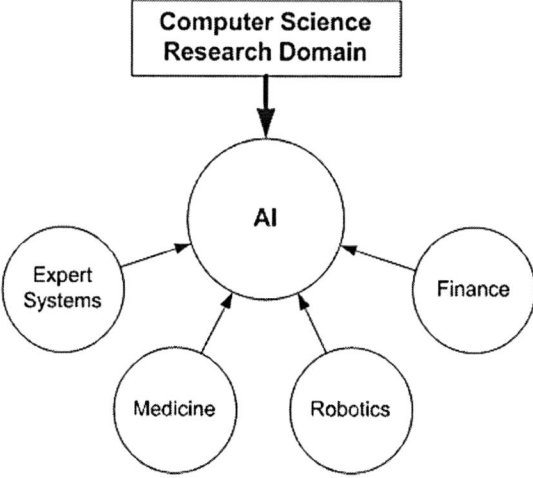

Fig. 1. Some of the application areas of artificial intelligence

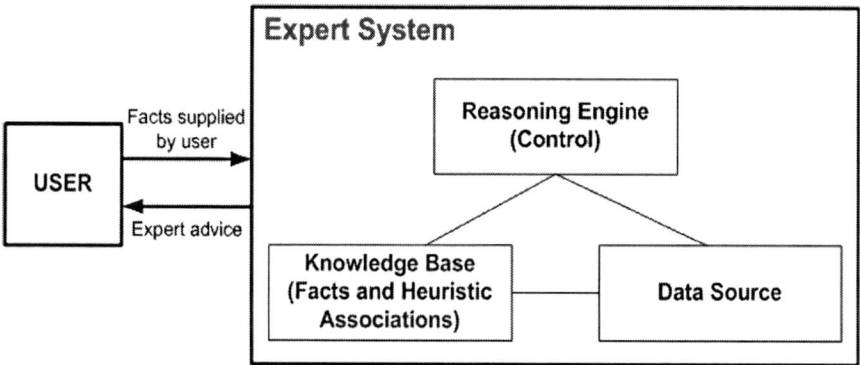

Fig. 2. Basic components of an expert system [29]

Expert systems are designed to act as an intelligent assistant to a user or human expert in one problem domain [29]. The general structure of an expert system is shown in Fig. 2. An expert system usually has three components:

1. External data sources.
2. A knowledge base (supplied by users), which stores the facts, related to the problem domain, their relations to one another, and perhaps some heuristic knowledge (less rigorous and more experiential). The knowledge within an expert system is generally represented in the form of a set of rules.
3. A reasoning engine, which operates on the knowledge base and external data sources (performs the knowledge based reasoning) to draw conclusions, answers questions, and gives advice.

3.1 Expert Systems in Medicine

The DENDRAL project led to the development of other knowledge-based systems such as META-DENDRAL [11] and MYCIN [71,72].

META-DENDRAL [11] was developed in the early 1970s. This inductive program uses a scientific theory and methodology similar to DENDRAL itself to discover molecular fragmentation rules. META-DENDRAL uses a heuristic search (plan, generate and test) to automatically generate rules for DEN-DRAL to use in justifying information about unknown chemical compounds. The learning strategy employed in META-DENDRAL was designed to be able to deal simultaneously with multiple concepts and to cope with noisy and incomplete chemical data. Although META-DENDRAL is no longer a productive tool, the learning and discovery ideas that META-DENDRAL introduced into rule-based expert systems have been applied to many new domains.

Another important program, carried out in the early 1970s, was MYCIN [71]. MYCIN is one of the most famous examples of early expert systems. MYCIN is an AI based interactive diagnostic program that was developed between 1972 and 1980 at Stanford University to assist practitioners in the selection of an appropriate therapy for patients with certain infectious diseases [58]. The system has the ability to prescribe drug treatment and clearly explain its reasoning as a set of IF-THEN rules. In 1979, nine researchers compared the performance of MYCIN system to that of physicians in a small range of infection cases. The study considered ten types of infection such as viral, tubercular, fungal, and bacterial. For each of these cases, appropriate therapies were obtained from MYCIN, a member of the Stanford Infectious Diseases faculty, a resident medical officer, and a medical student. The evaluation results revealed that MYCIN performed remarkably well, and in some cases, it even outperformed the human doctors [86]. Sadly, although MYCIN performed as well as some domain experts and produced high quality results, due to ethical and legal issues associated with the use of computers in medicine, it was never actually used as a production tool.

Although the MYCIN system was never actually used in diagnosis by clinicians, many other clinical decision support systems were created from this system. NEOMYCIN is an example of one of these support systems [14]. NEOMYCIN is a tree-based problem solving strategy that was developed upon a MYCIN type rule-based system [73]. NEOMYCIN trains doctors by taking them through a broad range of examples of diseases. The system employs a tree-based problem solving strategy that looks at known facts about different kinds of diseases. The diseases near the top of the tree usually belong to general classes of diseases. NEOMYCIN is then able to differentiate between two disease subclasses by moving down a tree from very general classes of diseases to more precise classes.

EMYCIN (Empty or Essential MYCIN) shell is another subsidiary of MYCIN [80,81]. EMYCIN is a domain-independent tool that can be used to build rule-based expert systems in which the user must supply only all the necessary knowledge about the task domain. The system was made by removing the medical knowledge base of the MYCIN expert system. The inference engine that applies the knowledge to the task domain is usually built into a shell. Overall, EMYCIN was designed to provide:

1. An easy way for knowledge base development and refinement.
2. Certainty factor associated with each predication (rather than reporting only true or false for all predications).
3. Better explanations for its behavior.

In the late 1970s a program called PUFF was developed from EMYCIN [1]. The program is able to diagnose the presence and seriousness of lung diseases. The PUFF system has the ability to automatically interpret pulmonary function test results and produce a clear report to be kept for future use. The PUFF system is one of the most successful examples of the areas in which expert systems are used and is still in routine use in the clinical laboratories around the world.

Quick medical reference (QMR) is another diagnostic decision support system that was developed in the 1970s at the University of Pittsburgh to help physicians diagnose adult disease [54]. However, it was only in the early 1980s that the major progress in the development and use of QMR in hospitals and health office practices occurred. The diagnosis process of QMR is based on historical diseases and findings. The QMR knowledge base contains information on more than 700 diseases and more than 4,500 clinical findings. Like many other earlier expert systems, this program is currently used by medical students and professionals as an interactive textbook.

In the 1980s, researchers made major progress in applying the AI-based systems to medical practice [15, 54, 74]. DXplain was developed at the Massachusetts General Hospital [5] and is a successful example of 1980s intelligence medical education and decision support systems. Given a list of symptoms, signs and laboratory information, DXplain can generate a hierarchical list of all possible diagnoses with clear explanation and suggestions. DXplain is still used by medical students and professionals as a medical education and reference system.

Many expert systems including MYCIN and QMR program were found to provide reliable, automatic methods to deal with the increasing amounts of clinical data. These systems have the ability to generate case-specific advice for clinical decision making [57, 83]. Although there are many variations, expert systems generally combine If–Then production rules with inference engines to create the knowledge base. This rule-based approach is easier to build, and it also has the ability to provide transparent suggestions to the physicians.

However, in the late eighties and early nineties researchers in knowledge-based systems realised that generating and maintaining a large rule-based knowledge acquisition tool can be very labor intensive [67, 68]. This shortcoming was one of the main reasons that limited the broader use of the QMR system. They also discovered that the knowledge acquired from experts alone was unsuitable for solving difficult problems. Hence, there was a need to alter the way in which knowledge was created and stored in expert system, especially for complex fields such as medicine that were known to be knowledge intensive. The newer systems generally required adaptations to changes, better flexibility and possibility of plug-in-able updates and knowledge reuse [12,82].

4 Data Mining Techniques for Supporting Medical Tasks

In recent years, knowledge-based systems have begun utilising a set of more powerful AI-based data mining techniques. AI-based techniques such as artificial neural networks (ANN) are able to accept numerous input variables and adapt their criteria to better match the data they analyse. Figure 3 outlines the data mining process in general and describe what data mining can offer.

In recent years there has been a rapid growth in the successful use of hybrid intelligent systems in many diverse areas such as science, medicine and commerce. The main contributing factor influencing the development of hybrid systems in medicine has been the demand for a more powerful yet flexible and robust technique. This is in order to cope with the extensive amount of data and knowledge stored in clinical databases, and more importantly, the complexity of interpretation.

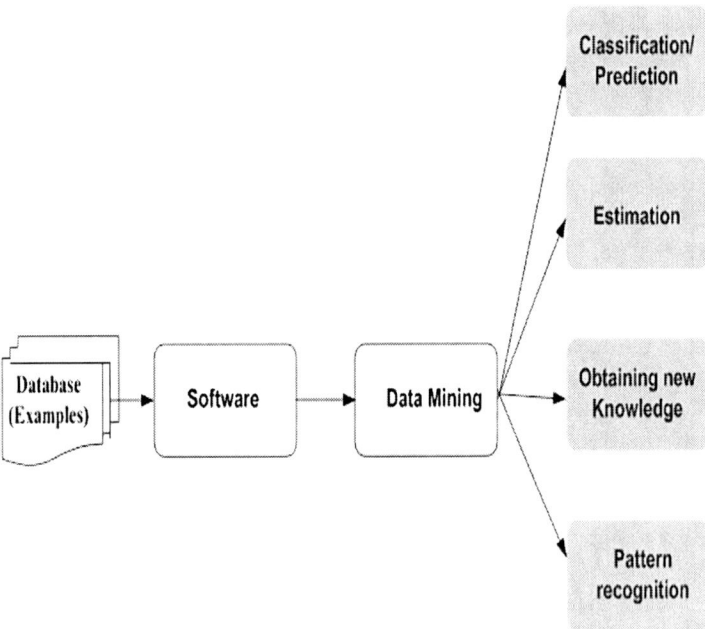

Fig. 3. What data mining can offer

An expert system in the healthcare sector ideally needs to be able to deal with ever increasing amounts of data, especially in the complex fields of medicine and biology. In most expert systems, information and knowledge are represented as a set of rules. Theses rules are usually easily understood by users; however generating and maintaining a large rule-based, knowledge-acquisition tool can be labor intensive. In fact, as mentioned previously, issues relating to the so-called, knowledge-acquisition, task have been raised over the years by many researchers in knowledge-based systems [67, 68]. Furthermore, the brittleness that can happen when a system is presented with noisy data, unusual values or incomplete data with missing values [34]. This can usually occur when a system tries to perform the data mining task by using only symbolic data mining algorithms such as decision tree models [9, 62, 63].

The idea of using more powerful methods for improving the learning performance of knowledge acquisition tools has been around for many years [27, 53]. Over the past few years, information researchers in data mining and knowledge-based systems have turned to a multi-strategy learning approach [40, 42]. Theses approaches facilitate data mining by combining the power of connectionist learning systems such as ANN with rule-based learning methodologies.

The following section provides an overview of a variety of hybrid intelligent techniques that might be utilised for extracting rules from ANN, particularly for clinical knowledge discovery and decision making processes.

4.1 Extracting Rules from ANNs: Examples of Hybrid Intelligent Techniques

An important reason for the extraction of explanations and rules from neural networks in expert systems is to directly add the extracted knowledge to the knowledge base system [28, 67–69]. Gallants Connectionist Expert Systems [26] and Matrix Controlled Inference Engine [28] are examples of early models where expert system rules are extracted from a neural network. The basic idea is to combine the explanation facility of rule-based systems with the ANN in order to generate the knowledge [21]. Coupling ANN and rule extraction algorithms can significantly improve the learning performance of knowledge acquisition tools, enhance the overall utility of ANN and reduce the brittleness of rule-based systems. Rule extraction algorithms can be categorized into four categories, namely the *decompositional, pedagogical, eclectic* and *compositional* algorithms [4, 78]. These classification schemes are based on the approach used to study and analyse the underlying ANN architecture or/and the classification given by the network for the processed input vectors.

Decompositional (local) methods start by extracting rules from each +unit (hidden and output) in a trained neural network. The rules extracted at the individual unit level are then combined to form a global relationship and the final rule base for the ANN architecture as a whole. Some examples of this style of algorithm are KT [24], Subset [79], COMBO [43] and RX [70], RULEX [2,3].

KT [24] is the earliest implementation of this style of algorithm. This approach focuses on mapping individual (hidden and output) units into conventional Boolean rules. The following example of rules generated from KT is taken from FU [24]:

If $(0 \leq Output \leq Threshold_1) \Rightarrow no,$ *and*

If $(Threshold_2 \leq Output \leq 1) \Rightarrow yes,$ *where* $(Threshold_1 < Threshold_2)$

Similar to the KT algorithm is the Subset algorithm by Towel and Shavlik (1993). The Subset algorithm searches for subsets of incoming weights that exceed the bias of a particular unit. In many situations, this style is capable of delivering a set of rules that yield exact representation of the underling ANN architecture, however as reported by Towel and Shavlik (1993), the computational time for algorithms such as Subset and KT may grow exponentially with the number of inputs. Consequently, this can impose great restriction on network architecture and training procedures.

Towel and Shavlik (1993) tried to reduce the complexity of rule searches by introducing a second rule extraction approach, namely M of N, which is an alternative to the if–then form of rules. The stages of the M of N approach are outlined below:

1. Generate an ANN and train it using back propagation. With each hidden and output unit, form groups of similarly weighted links.
2. Set the link weights of all group members to the average of the group.
3. Eliminate any groups that do not significantly affect whether the unit will be active or inactive.
4. Holding all link weights constant, optimise the bias of all hidden and output units using the back propagation algorithm.
5. Form a single rule for each hidden and output unit. The rule consists of a threshold given by the biases and weighted antecedents specified by the remaining links.
6. Where possible, simplify rules to eliminate superfluous weights and thresholds.

The Boolean expression or rules that are generated by this algorithm states:

If $(M$ *of* N *antecedents/conditions,* $a_1, a_2, \ldots, a_n,$ *are true)* *Then*
(the conclusion b is true).

This approach can be useful for problems that can be expressed in the form of the M of N rules. The most impressive feature of the M of N algorithm is its ability to analyse clusters of similarly weighted links. Most of the previously reported algorithms conduct searches by exploring and testing a space of conjunctive rules against the network in order to find out whether they are valid rules or not. However, the M of N algorithm starts by analysing each cluster of similarly weighted links as whole and eliminating groups that have little significant effect on the result. In the final stage it creates rules from the remaining clusters through an exhaustive search. Approaching the rule extraction problem in this way can reduce the search space.

Similarly in the late 1990s, Taha and Ghosh [75] proposed three new techniques with different power of rule extraction, namely BIO-RE (Binarized Input–Output Rule Extraction), Partial-RE and Full-RE for extracting rules from trained feed-forward ANN. BIO-RE is a binary rule extraction technique that generates a truth table from the binary inputs and simplifying the resultant Boolean function. However in practice, creating a truth table that represents all possible valid input–output mapping of the trained network based on this approach can only work for small size problems. Another limitation to this approach is the requirement for Binarized Input–Output variables. If the original inputs are not binary, they have to be converted to binary values with great amount of care. Partial-RE and Full-RE use a similar link rule extraction approach to the Subset algorithm with some improvements and simplifications. As the name suggest, theses two rule extraction techniques can be used to extract partial or full rule sets from trained feed-forward. The Partial-RE algorithm is well suited for larger networks where time complexity can be a real issue. These algorithms provide users with some degree of control over the extracted rule set and information they obtain from the learner.

The second category of rule extraction algorithms is pedagogical. The core idea in the pedagogical (global) approach is to treat the trained neural network as a black box. It aims to extract rules that map inputs directly into outputs [78]. The pedagogical algorithm uses a trained neural network to only generate test data for the rule generation algorithm. In this strategy the target concept is computed by the network and the input vectors are the actual network's input vectors [17].

At the end of 1980s, Saito and Nakano [67] proposed a medical diagnosis expert system based on a multi-layer neural network. They implemented a pedagogical algorithm and used it to observe the changes in the levels of input and output units. In order to explain the inference process in the system, they presented a rule extraction routine that characterises the output classes directly from the inputs of simple networks. They were particularly interested in implementing a routine for extracting compact and propositional rules from a simple network. Interestingly, their research revealed that even for a relatively simple problem domain, the number of extracted rules can be large. VIA [76], TREPAN [16, 18], STARE [87] are other examples of this style of algorithm.

TREPAN [16, 18] is one of the most significant developments in the pedagogical approach. The TREPAN algorithm is a relatively new data mining approach that operates on neural networks and uses the M of N rule extraction strategy. Figure 4 shows the TREPAN algorithm described in [20] for converting a Neural Network to an M of N type Decision Tree.

TREPAN

Input: trained network, training set used for network

Initialize the tree as a leaf node

While stopping criteria not met

Pick the most promising leaf node to expand

Draw a sample of instances

Select a splitting test for the node

For each possible outcome of the test make a new leaf node

Return: extracted decision tree

Fig. 4. The TREPAN algorithm

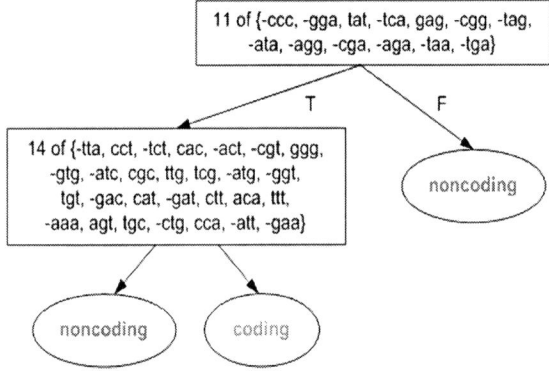

Fig. 5. A representative tree for the DNA coding domain [16]

The TREPAN algorithm is similar to standard decision tree algorithms such as See5 [64], and C4.5 [62], which learn directly from a training set. Figure 5 that has been taken from [16], shows the tree extracted by TREPAN from the DNA coding network. This tree represents the function that the network has learnt. In this scenario, the system generated a decision tree by recursively partitioning the input space in order to recognise protein-coding regions in *E. coli* DNA sequences. The features in the coding domain represent the 64 codons (three-letter words) that can be formed from the DNA bases: fa, c, g, tg. Each Boolean feature indicates the presence or absence of a given codon in-frame in the sequence being classified. Negated features are preceded by a minus sign. The class labels indicate whether a given, fixed-length sequence is predicted to encode a protein or not.

TREPAN does not require any specialised neural network architecture or training algorithm since, like other pedagogical algorithms, it operates on datasets and not the actual underlying architecture. In fact, Craven and Shavlik [20] believe that TREPAN is so general and flexible it (theoretically) can be applied to wide variety of networks, including ensembles [16].

Johansson and Niklasson [38] performed comparative study between the accuracy of ANN approaches and traditional linear approaches. Not surprisingly, their results clearly revealed the superiority of ANN. Furthermore, in a separate study, Johansson and Niklasson [39] exploited the potential of neural networks in decision-making problem domains where comprehensibility was also considered to be an important factor for decision makers. In this study they successfully used the TREPAN algorithm [16, 19] to create decision trees from trained ANN models. Their results revealed that the decision tree extracted by the TREPAN algorithm had a higher accuracy rate on unseen data than the trees created by the standard tool See5 [64]. However, they noted that the trees created by TREPAN were still as complicated as the trees generated by See5 [64]. This shortcoming was also noted by Boz [8]. This shortcoming was the motivation for another study by Johansson et al. [37] where they proposed a novel method called G-REX based on Genetic Programming (GP) for rule extraction. The genetic programming rule extraction algorithms attempt to incorporate ideas of natural evolution for classifications as well as optimisation tasks (for detailed discussion on genetic algorithms see [31] and [55]. In their primary study they used a relatively small advertising dataset. The results showed that the rules extracted by G-REX generally outperformed the TREPAN and See5 in terms of both accuracy and comprehensibility. However, the G-REX algorithms can be computationally expensive, even for relatively small datasets.

Further supports for improving the comprehensibility of rules from overly complex models, come from Quinlan [7, 22, 41, 61]. Blackmore and Bossomaier [7] used genetic algorithms to evolve rules directly from trained neural networks. Figure 6 taken from [7], shows a rule set example generated by a GP algorithm. Their experimental results revealed that the rules extracted by genetic algorithm approach outperform various decision tree based algorithm such as WEKA *J48.PART* [85] in terms of both accuracy and comprehensibility. However, due to the nature of GP algorithms, the computational time grows considerably for the large neural networks. A similar shortcoming for the TREPAN algorithm was reported by Golea [32].

The third rule extraction category is eclectic algorithms. The methods from this category, like the decompositional category, carefully examine the ANN at the level of individual units; they also extract rules at the global relationship within trained neural networks. DEDEC [77] is an example of this style of algorithm.

The third rule extraction category is eclectic algorithms. The methods from this category, like the decompositional category, carefully examine the ANN at the level of individual units; they also extract rules at the global relationship within trained neural networks. DEDEC [77] is an example of this style of algorithm.

1. IF behavioural = Condition3, on, AND region = Condition 1, off, AND Age = Condition3, on,
      THEN =outcome1
2. IF mn_problems – Condition3, off, AND reporter – Condition1, on, AND Marital status – Condition2, on,
      THEN =outcome1
3. IF longterm_stressors = Condition2, on,
      THEN –outcome2
4. IF time = Condition2, on, AND risk = Condition8, on,
      THEN =outcome2
5. IF Age – Condition3, on,
      THEN =outcome2
6. IF reporter = Condition2, off, AND dependent = Condition2, off, AND time = Condition4, off,
      THEN –outcome4
7. IF occupation = Condition2, off,
      THEN =outcome1
8. IF time – Condition4, off, AND last_seen – Condition2, off, AND day – Condition4, off,
      THEN =outcome1
9. IF day = Condition1, off,
      THEN –outcome2
10. IF mn_problems = Condition6, on, AND longterm_stressors = Condition4, off, AND day = Condition3, on,
      THEN =outcome2
11. IF risk – Condition4, off, AND character – Condition1, off,
      THEN =outcome1
12. IF Age = Condition2, on,
      THEN –outcome2
13. IF season = Condition3, off, AND day = Condition1, on,
      THEN =outcome2
14. IF occupation – Condition6, on, AND longterm_stressors – Condition2, off, AND reporter – Condition1, off,
      THEN =outcome3
15. IF mn_problems = Condition2, on, AND longterm_stressors = Condition1, off,
      THEN –outcome3
16. IF occupation = Condition2, off, AND occupation = Condition7, on,
      THEN =outcome2
17. IF urban – Condition2, off, AND history – Condition2, on, AND season – Condition1, off,
      THEN =outcome2
18. IF appearance = Condition1, off, AND ph_outcome2 =Condition2, on, AND reporter = Condition4, on,
      THEN =outcome3

Fig. 6. A rule set example generated by GP algorithm [7]

The final category is compositional algorithm. The compositional algorithms neither focus on local models that mirror the behavior of individual units, nor treat the network as a black box, like pedagogical approaches. Representatives of this category include algorithms proposed by Omlin et al. [59] and Giles et al. [30]. This category has been designed for recurrent ANN model where the relationships between different outputs are used to find models of the underlying recurrent neural networks. Theses are typically in the form of finite-state machines that mimic the network to a satisfactory degree.

5 Recent Research Direction: Multi-Agent Systems

Agents are software entities that carry out some set of operations on behalf of a user or other programs. Agents usually have a set of beliefs, goals and rules to select plans and are capable of communicating messages concerning their beliefs and goals. Several agents can collectively and collaboratively form a multi-agent system to perform complex tasks that cannot be done by individual agents. For researchers who are working in the field of AI, the term agent generally means a software entity that use AI techniques to learn and adapt to its environment in order to perform a set of actions and attain the goal of a user or other programs. Intelligent Multi-Agent System (IMAS) frameworks usually have capability to deal with complex systems and support agent mental state representation, reasoning, negotiation, planning and communications. Recently intelligent agents and multi-agent systems have attracted much interest in data mining community. AI-based data mining methods such as genetic programming have been applied to many multi-agent learning tasks [33, 46]. Over the past few years GP has been successfully used for embedding intelligence in newly crated agents [35, 36].

There are numerous software agents in the market. Many of these software packages have proven to be very useful in variety of applications. Etzioni and Weld [23] provide a good reference for currently available intelligent software agents. One of the recent examples of software agent systems is Agent Academy (AA) [56]. The Agent Academy framework operates as an intelligent multi-agent system, which can train new agents or retrain its existing agents in a recursive mode. A user can issue a request to the agent factory (AF) for a new agent. In this system an untrained agent (UA) can have a minimal degree of intelligence that has been defined by the software designer. This agent then can be send to agent training (AT) module, where its world perception can be developed during a virtual interactive session with an agent master (AM). Perhaps the most important part of this framework is the part that store statistical data on prior agent behavior and experience. The data miner (DM) module then can use the stored information and mine the data, in order to generate useful knowledge about the application domain under the investigation. Knowledge extracted by the DM can be used by the AT module to train new agents or to enhance the behavior of existing agents.

5.1 An Example of Coupling Data Mining with Intelligent Agents

This section demonstrates a methodology for IMAS development in management of chronic diseases such as diabetes; which relies on the ability of DM to generate knowledge models for agents. The roles of all agents and their responsibilities and services are detailed as follows (see Fig. 7):

- *Personal agent* is one type of interface agents, which provides the user with a graphical interface to the multi-agent system and initiates a search or shows the results of a query to the user. This agent also known as *Diagnosis agent* can be used to analyses the situation and makes an accurate judgment for the patient. It can also be designed in a way to implement the method of therapy agent.
- *Monitoring agent* can be used to monitor the diabetic patients in real time and send out the monitored data to data processing agent.

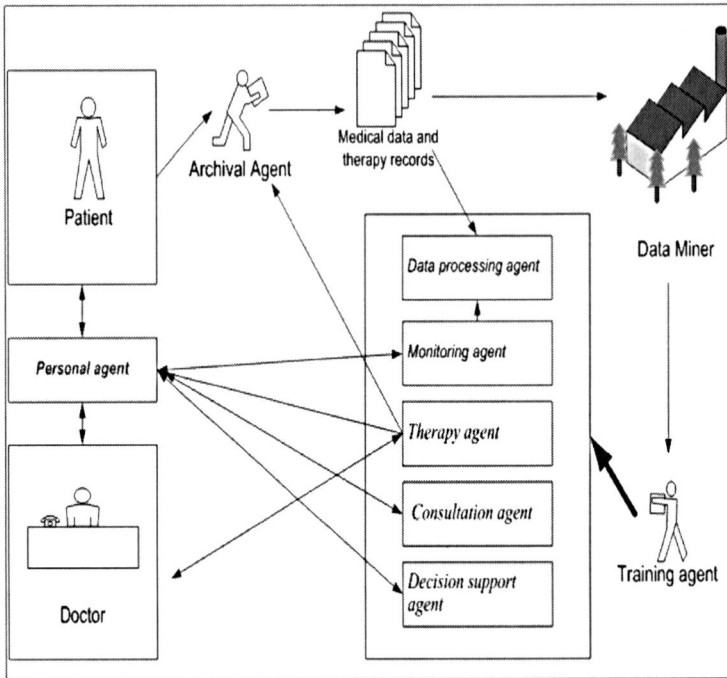

Fig. 7. Agent based architecture for designing an e-medicine system for the management of diabetes disease

- *Data processing agent* can provide statistic and integrate the monitored data.
- *Therapy agent* has the capability to determine a proper therapy method.
- *Consultation agent* can be used to provide consultation to the enquiry of patients and contacts with diagnosis or personal agent.
- *Decision support agent* has the capability to provide decision support and cooperation for personal agent.
- *Training agent* use knowledge extracted by the Data Miner in order to train new agents or enhance the behavior of agents already in running. The Data Miner is the core component of this IMAS prototype and can be used to provide the user with a number of DM algorithms. The DM algorithms are then validated before they can be store as a set of facts in knowledge based models of a rule engine. For Example, as described in [6], the implementation of the rules engine can be done through the Java Expert System Shell (JESS)[1].
- *Archival agent* integrates with the Medical database and has the capability to edit and store the patient record and therapy methods, and updates the database of the individual patients.

---

[1] Available at: http://herzberg.ca.sandia.gov.jess.

6 Summary

Research shows that the state of the art in computer-based clinical knowledge discovery and decision support research has matured. Over the years, computer-based clinical decision support systems have been used successfully in the clinical setting [6, 10, 13, 25, 47–52, 60, 65, 66]. Expert systems are now the common form of clinical decision support systems in routine clinical use. Many early expert systems that provided good user interface design, reasonably accurate results with clear justification and reasoning are still in routine use in health care systems, clinical laboratories and educational institutions.

More recently, researchers in data mining and knowledge-based systems have turned to AI-based data mining techniques such as neural networks and decision tree algorithms. These techniques can be either used as stand alone systems or deeply integrated into many existing expert systems. This chapter presented selected data mining and knowledge discovery techniques that can be used in clinical knowledge discovery and decision support systems.

The literature showed that such models are becoming a widely accepted technology in many medical sectors. Furthermore, combinations of techniques or specifically adapted training methods have great potential to be effectively used in predictive modelling tasks. Multiple techniques can be potentially integrated to improve the transparency of results, ease of use and the accuracy of medical prediction tools. Of these techniques some are being actively developed (notably hybrid neural network-decision trees), where the goal is to extract useful explanations from individual neural network classifiers.

There are still many challenges to be overcome. Each technique has its own advantages and disadvantages under different circumstances. However, it is clear that integrating AI-based data mining modules into computerised patient records and clinical findings can provide a great chance for improving the quality of care.

References

1. Aikens, J. S., Kunz, J. C., Shortliffe, E. H., and Fallat, R. J. (1983), "PUFF: An expert system for interpretation of pulmonary function data." *Computers and Biomedical Research* 16: 199–208.
2. Andrews, R. and Geva, S. (1994), "Rule extraction from a constrained error back propagation MLP." *Proceedings of the 5th Australian Conference on Neural Networks*: 9–12.
3. Andrews, R. and Geva, S. (1995), "Inserting and extracting knowledge from constrained error back propagation networks." *Proceedings of the 6th Australian Conference on Neural Networks*: 29–32.
4. Andrews, R., Diederich, J., and Tickle, A. B. (1995), "A survey and critique of techniques for extracting rules from trained artificial neural networks." *Knowledge Based Systems* 8: 373–89.

5. Barnett, G. O., Cimino, J. J., Hupp, J. A., and Hoffer, E. P. (1987), "DXplain: an evolving diagnostic decision-support system." *The Journal of the American Medical Association* 258 (1): 67–74.

6. Bates, D. W., Cohen, M., Leape, L. L., Overhage, J. M., Shabot, M. M., and Sheridan, T. (2001), "Reducing the frequency of errors in medicine using information technology." *The Journal of the American Medical Association* 8 (4): 299–308.

7. Blackmore, K. and Bossomaier, T. (2003), "Using a neural network and genetic algorithm to extract decision rules." *Proceeding of the 8th Australian and New Zealand Intelligent Information Systems Conference (ANZIIS)* 1: 187–91.

8. Boz, O. (2002), "Extracting decision trees from trained neural networks." *Proceedings of the 8th ACM SIGKDD International Conference on Knowledge Discovery and Data Mining*: 456–61.

9. Breiman, L., Friedman, J., Olshen, R., and Stone, C. (1984), *Classification and Regression Tree* (Wadsworth, Inc., NY).

10. Brownbridge, G., Evans, A., Fitter, M., and Platts, M. (1986), "An interactive computerized protocol for the management of hypertension: effects on the general practitioner's clinical behaviour." *Royal College of General Practitioners* 36: 198–202.

11. Buchanan, B. G. and Feigenbaum, E. A. (1978), "DENDRAL and META-DENDRAL: their applications dimensions." *Artificial Intelligence* 11: 5–24.

12. Chandrasekaran, B. (1986), "Generic tasks for knowledge-based reasoning:High-level building blocks for expert system design." *IEEE Expert*, 1 (3): 23–30.

13. Chase, C. R., Vacek, P. M., Shinozaki, T., Giard, A. M., and Ashikaga, T. (1983), "Medical information management: improving the transfer of research results to presurgical evaluation." *Medical Care* 21: 410–24.

14. Clancey, W. J. and Letsinger, R. (1981), "NEOMYCIN: reconfiguring a rule-based expert system for application to teaching." *Proceedings of the Seventh Intl Joint Conf on AI*: 829–835.

15. Clancey, W. J. and Shortliffe, E. H. (1984), "Readings in Medical Artificial Intelligence – The First Decade" (Addison-Wesley, MA).

16. Craven, M. (1996), "Extracting Comprehensible Models from Trained Neural Networks." Department of Computer Sciences, University of Wisconsin, PhD thesis, Madison.

17. Craven, M. W. and Shavlik, J. W. (1994), "Using sampling and queries to extract rules from trained neural networks." *Proceedings of the 11th International Conference on Machine Learning*: 37–45.

18. Craven, M. W. and Shavlik, J. W. (1996a), "Extracting tree-structured representations from trained networks." *Advances in Neural Information Processing Systems* 8: 24–30.

19. Craven, M. W. and Shavlik, J. W. (1996b), "Extracting tree-structured representations of trained networks." *Proceedings of the Advances in Neural Information Processing Systems* 8: 24–30.

20. Craven, M. and Shavlik, J. (1999), "Rule Extraction: Where Do We Go from Here?" *University of Wisconsin Machine Learning Research Group Working Paper, 99-1*.

21. d'Avila Garcez, A. S., Broda, K., and Gabbay, D. M. (2001), "Symbolic knowledge extraction from trained neural networks: a sound approach." *Artificial Intelligence* 125 (1–2): 155–207.

22. Domingos, P. (1998), "Knowledge discovery via multiple models." *Intelligent Data Analysis* 2: 187–202.
23. Etzioni, O. and Weld, D. S. (1995), "Intelligent agents on the internet - fact, fiction, and forecast." *IEEE Expert – Intelligent Internet Services* (4): 44–49.
24. Fu, L. (1994), "Rule generation from neural networks." *IEEE Transactions on Systems, Man, and Cybernetics* 28 (8): 1114–24.
25. Fuchs, J., Heller, I., Topilsky, M., and Inbar, M. (1999), "CaDet, A computer-based clinical decision support system for early cancer detection." *Cancer Detection and Prevention* 1999 23 (1): 78–87.
26. Gallant, S. (1988a), "Connectionist expert systems." *Communications of the ACM*, 31 (2): 152–169.
27. Gallant, S. I. (1988b), "Connectionist expert systems." *Communications of the ACM*, 31 (2): 152–169.
28. Gallant, S. I. and Hayashi, Y. (1991), "A neural network expert system with confidence measurements," in *Processing and Management of Uncertainty in Knowledge-Based Systems, IPMU'90*, B. Bouchon-Meunier, R. R. Yager, and L. A. Zadeh, Eds., (Springer, Berlin Heidelberg New York), 562–567.
29. Giarratano, J. C. and Riley, G. D. (1998), "Expert Systems: Principles and Programming." (Third Edition, Course Technology Publishing Company), 624 pages.
30. Giles, C. L., Miller, C. B., Chen, D., Chen, H., Sun, G. Z., and Lee, Y. C. (1992), "Learning and extracting finite state automata with second-order recurrent neural networks." *Neural Computation* 4 (3): 393–405.
31. Goldberg, D. (1989), *Genetic Algorithms in Search, Optimization, and Machine Learning* (Addison-Wesley, Reading, MA).
32. Golea, M. (1996), "On the complexity of rule extraction from neural networks and network querying." *Proceedings of the Rule Extraction From Trained Artificial Neural Networks Work shop. Society for the study of Artificial Intelligence and Simulation of Behavior Workshop Series (AISB'96)*: 51–59.
33. Haynes, T., Wainwright, R., and Sen, S., (1995), "Evolving a Team." *Working Notes of the AAAI-95 Fall Symposium on Genetic Programming, AAAI Press.*
34. Holland, J. (1986), "Escaping brittleness.," in *Machine Learning*, vol. 2, R. Michalski, C. J., and S. Mitchell, Eds. (Morgan Kaufmann, Los Altos, CA), 593–623.
35. Iba, H. (1998), "Evolutionary learning of communicating agents 108(1–4), 1998", *Information Sciences* 108: 1–4.
36. Iba, H. a. and Kimura, M., Robot (1996), "Programs generated by genetic programming." *Advanced Institute of Science and Technology, IS-RR-96-0001I, in Genetic Programming.*
37. Johansson, U., König, R., and Niklasson, L. (2003), "Rule extraction from trained neural networks using genetic programming." *13th International Conference on Artificial Neural Networks*: Supplementary Proceedings 13–16.
38. Johansson, U. and Niklasson, L. (2001), "Predicting the impact of advertising - a neural network approach." *Proceedings of the International Joint Conference on Neural Networks*: 1799–1804.
39. Johansson, U. and Niklasson, L. (2002), "Neural networks - from prediction to explanation." *Proceedings of the International Conference Artificial Intelligence and Applications*: 93–98.

40. Katz, S., A.S.K., Lowe, N., and Quijano, R. C. (1994), "Neural net-bootstrap hybrid methods for prediction of complications in patients implanted with artificial heart valves." *Heart Valve Dis.* 3: 49–52.
41. Keedwell, E., Narayanan, A., and Savic, D. (2000), "Creating rules from trained neural networks using genetic algorithms", *Int. J. Comput. Syst. Signals(IJCSS)* 1: 30–42.
42. Koutroumbas, K., Paliouras, G., Karkaletsis, V., and Spyropoulos, C. D. (2001), "Comparison of computational learning methods on a diagnostic cytological application." *Proceedings of the European Symposium on Intelligent Technologies, Hybrid Systems and their implementation on Smart Adaptive Systems (EUNITE)*: 500–508.
43. Krishnan, R. (1997), "A systematic method for decompositional rule extraction from neural networks." *Proceedings of the NIPS'96 Workshop on Rule Extraction from Trained Artificial Neural Networks*: 38–45.
44. Ledley, R. S. and Lusted, L. B. (1959), "Reasoning foundations of medical diagnosis." *Science* 130: 9–21.
45. Lindsay, R. K., Buchanan, B. G., Feigenbaum, E. A., and Lederberg, V. (1980), "Application of Artificial Intelligence for Chemistry : The DENDRAL Project" (New York: McGraw-Hill),
46. Luke, S. A. and Spector, L. (1996), "Evolving teamwork and coordination with genetic programming." *Genetic Programming 96*, MIT Press, Cambridge, MA.
47. McDonald, C. J. (1976a), "Use of a computer to detect and respond to clinical events: its effect on clinician behavior." *Annals of Internal Medicine* 84: 1627.
48. McDonald, C. J. (1976b), "Protocol-based computer reminders, the quality of care and the non-perfectability of man." *N Engl J Med* 295: 1351–5.
49. McDonald, C. J., Wilson, G. A., and McCabe, G. P. (1980), "Physician response to computer reminders." *The Journal of the American Medical Association (JAMA)* 244: 1579–81.
50. McDonald, C. J., Hui, S. L., Smith, D. M., Tierney, W. M., Cohen, S. J., and Weinberger, M. E. A. (1984), "Reminders to physicians from an introspective computer medical record. A two-year randomized trial." *Ann. Intern. Med.* 100: 130–8.
51. McDowell, I., Newell, C., and Rosser, W. (1986), "Comparison of three methods of recalling patients for influenza vaccination." *Canadian Medical Association* 135: 991–7.
52. McDowell, I., Newell, C., and Rosser, W. (1989), "Computerized reminders to encourage cervical screening in family practice." *The Journal of Family Practice* 28: 420–4.
53. Medsker, L. R. (1994), *Hybrid Neural Network and Expert Systems.*, (Kluwer Academic Publishers, Boston).
54. Miller, R. A., Masarie, F. E., and Myers, J. D. (1986), "Quick medical reference (QMR) for diagnostic assistance." *MD Computing* 3 (5): 34–8.
55. Mitchell, T. (1996), *An introduction to Genetic Algorithms* (MIT Press, Cambridge, MA).
56. Mitkas, P. A. (2002), "Agent academy: a data mining framework for training intelligent agents." European Projects Presentation Day of the 13th Eur. Conf. on Machine Learning & 6th Eur. Conf. on Principles and Practice of Knowledge Discovery in Databases (ECML/PKDD 2002), Helsinki, Finland.
57. Morio, S. (1989), "An expert system for early detection of cancer of the breast." *Computers in Biology and Medicine* 9 (5): 295–306.

58. Musen, M. A. (1999), "Stanford medical informatics: uncommon research, common goals." *MD Computer* 16 (1): 47–50.
59. Omlin, C. W., Giles, C. L., and Miller, C. B. (1992), "Heuristics for the extraction of rules from discrete time recurrent neural networks." *Proceedings of the International Joint Conference on Neural Networks* 1: 33–38.
60. Petrucci, K., Petrucci, P., Canfield, K., McCormick, K. A., Kjerulff, K., and Parks, P. (1991), "Evaluation of UNIS: Urological Nursing Information Systems", *Proceedings of the Fifteenth Annual Symposium on Computer Applications in Medical Care*.: 43–7.
61. Quinlan, J. R. (1987), "Generating production rules from decision trees." *In Proceedings of the Tenth International Joint Conference on Artificial Intelligence*: 30107.
62. Quinlan, J. R. (1993), *C4.5: programs for machine learning* (Morgan Kaufmann, San Francisco, CA).
63. Quinlan, J. (1996), "Bagging, boosting, and C4.5." *Proceedings of the Thirteenth National Conference on Artificial Intelligence*: 725–30.
64. Quinlan, J. R. (2000), "Data mining tools See5 and C5.0." available online from http://www.rulequest.com/see5-info.html.
65. Rogers, J. L., Haring, O. M., and Goetz, J. P. (1984), "Changes in patient attitudes following the implementation of a medical information system." *QRB* 10: 65–74.
66. Rosser, W. W., Hutchison, B. G., McDowell, I., and Newell, C. (1992), "Use of reminders to increase compliance with tetanus booster vaccination." *Canadian Medical Association* 146: 911–7.
67. Saito, K. and Nakano, R. (1988), "Medical diagnostic expert system based on PDP model." *Proceedings of the IEEE Int. Conf. Neural Networks* 1: 255–62.
68. Sestito, S. and Dillon, T. S. (1991), "The use of sub-symbolic methods for the automation of knowledge acquisition for expert systems." *Proceedings of the Eleventh International Conference Expert Systems and their Applications. Avignon* 1: 317–28.
69. Sestito, S. and Dillon, T. S. (1994), *Automated Knowledge Acquisition.*, (Prentice Hall, Australia).
70. Setiono, R. (1997), "Extracting rules from neural networks by pruning and hidden-unit splitting." *Neural Computation* 9 (1): 205–25.
71. Shortliffe, E. H. (1976), *Computer-Based Medical Consultation: MYCIN.*, (American Elsevier, New York. NY).
72. Shortliffe, E. H., Axline, S. G., Buchanan, B. G., Merigan, T. C., and Cohen, S. N. (1973), "An artificial intelligence program to advise physicians regarding antimicrobial therapy." *Computer Biomedical Research.* 6 (6): 544560.
73. Sotos, J. (1990), "MYCIN and NEOMYCIN: two approaches to generating explanations in rule based expert systems." *Aviation, Space, and Environmental Medicine* 61 (950–4):
74. Szolovits, P. (1982), "Artificial Intelligence in Medicine.," in *AAAS Selected Symposia Series*, Westview Press, Ed., Colorado).
75. Taha, I. and Ghosh, J. (1996), *Symbolic Interpretation of Artificial Neural Networks. Technical Report TR-9701-106, University of Texas, 1996.* Technical Report TR-9701-106, University of Texas, Texas).
76. Thrun, S. B. (1994), "Extracting provably correct rules from artificial neural networks." *Technical Report IAI-TR-93-5*: University of Bonn, Germany.

77. Tickle, A. B., Orlowski, M., and Diederich, J. (1996), "DEDEC: a methodology for extracting rules from trained artificial neural networks." *Proceedings of the AISB'96 Workshop on Rule Extraction from Trained Neural*: 90–102.
78. Tickle, A. B., Andrews, R., Golea, M., and Diederich, J. (1998), "The truth will come to light: directions and challenges in extracting the knowledge embedded within trained artificial neural networks." *IEEE Transactions on Neural Networks* 9 (6): 105767.
79. Towell, G. and Shavlik, J. (1993), "The extraction of refined rules from knowledge based neural networks." *Machine learning* 13 (1): 71–101.
80. Van Melle, W. (1980), "A Domain independent system that aids in constructing knowledge based consultation programs." Computer Science Department, Doctoral thesis, Stanford University.
81. Van Melle, W., Shortliffe, E. H., and Buchanan, B. G. (1981), "EMYCIN: A Domain Independent System that Aids in Constructing Knowledge-Based Consultation Programs." *In Pergamon-Infotech Report on Machine Intelligence*: 249263.
82. Walther, E., Eriksson, H., and Musen, M. A. (1992), "Plug and Play: Construction of task-specific expert-system shells using sharable context ontologies." *AAAI Workshop on Knowledge Representation Aspects of Knowledge Acquisition, San Jose, CA*: 191–198.
83. Wang, C. H. and Tseng, S. S. (1990), "A brain tumour diagnostic system with automatic learning abilities." *Proceedings of the 3rd Annual IEEE Symposium on Computer Based Medical Systems.*: 313–20.
84. Waterman, D. A. (1985), "A Guide to Expert Systems..", (Addison-Wesley The knowledge Series In Knowledge Engineering, Boston, USA), 419 pages.
85. Witten, I. H. and Frank, E. (2000), *Data Mining: Practical Machine Learing Tools and Techniques with Java Implementations.*, (Morgan Kaufmann, San Diego, CA).
86. Yu, V. L., Fagan, L. M., and Wraith, S. M., et al. (1979), "Antimicrobial selection by a computer: a blinded evaluation by infectious diseases experts." *The Journal of the American Medical Association* 242 (12): 1279–82.
87. Zhou, Z. H., Chen, S. F., and Chen, Z. Q. (2000), "A statistics based approach for extracting priority rules from trained neural networks." *Proceedings of the IEEE-INNS-ENNS International Joint Conference on Neural Networks* 3: 401–06.

Part II

Applications

Evolving Trading Rules

Adam Ghandar[1], Zbigniew Michalewicz[2], Martin Schmidt[3],
Thuy-Duong Tô[4], and Ralf Zurbrugg[5]

[1] School of Computer Science, University of Adelaide, Adelaide, SA 5005,
Australia, adam.ghandar@adelaide.edu.au
[2] School of Computer Science, University of Adelaide, Adelaide, SA 5005,
Australia; also at Institute of Computer Science, Polish Academy of Sciences, ul.
Ordona 21, 01-237 Warsaw, and Polish-Japanese Institute of Information
Technology, ul. Koszykowa 86, 02-008 Warsaw, Poland,
zbyszek@cs.adelaide.edu.au
[3] SolveIT Software Pty Ltd, Level 12, 90 Kings William, Adelaide 5000, Australia,
martin.schmidt@solveitsoftware.com
[4] School of Commerce, University of Adelaide, Adelaide, SA 5005, Australia,
td.to@adelaide.edu.au
[5] School of Commerce, University of Adelaide, Adelaide, SA 5005, Australia,
ralf.zurbrugg@adelaide.edu.au

Summary. This chapter describes a computational intelligence system for portfolio
management and provides a comparison of the relative performance of portfolios
of managed by the system using stocks selected from the ASX (Australian Stock
Exchange). The core of the system is the development of trading rules to guide
portfolio management. The rules the system develops are adapted to dynamic market
conditions.

An integrated process for stock selection and portfolio management enables a
search specification that produces highly adaptive and dynamic rule bases. Rule base
solutions are represented using fuzzy logic and an evolutionary process facilitates a
search for high performing fuzzy rule bases. Performance is defined using a novel
evaluation function involving simulated trading on a recent historical data window.

The system is readily extensible in terms of the information set used to construct
rules, however to produce the results given in this chapter information derived only
from price and volume history of stock prices was used. The approach is essentially
referred to as technical analysis – as opposed to using information from outside the
market such as fundamental accounting and macroeconomic data. A set of possible
technical indicators commonly used by traders forms the basis for rule construction.

The fuzzy rule base representation enables intuitive natural language interpretation of trading signals and implies a search space of possible rules that corresponds
to trading rules a human trader could construct. An example of a typical technical
trading rule such as "buy when the price of a stock X's price becomes higher than
the single moving average of the stock X's price for the last, say, 20 days" (indicating a possible upward trend) could be encoded using a fuzzy logic rule such as
"if single moving average buy signal is High then rating is 1"; conversely we could

A. Ghandar et al.: *Evolving Trading Rules*, Studies in Computational Intelligence (SCI) **92**,
95–119 (2008)
www.springerlink.com

have a trading rule such as "sell stocks with high volatility when the portfolio value is relatively low" encoded by a fuzzy rule: "if Price Change is High and Portfolio Value is Extremely Low then rating is 0.1".

The fuzzy rule bases undergo an evolutionary process. An initial population of rule bases (genotypes) is selected at random and may be seeded with some rule bases that correspond to accepted technical trading strategies. Further generations are then evolved by evaluating the rule bases comprising the initial population and then taking offspring of the best performing rule bases from the previous generation.

Fuzzy rule base performance is evaluated through analysis of the results of applying a particular rule base to simulated trading. In the simulated scenario an initial capital is allocated to construct an initial portfolio at the beginning of the simulation period. This initial portfolio is managed over the rest of a data window.

The empirical results from testing the system on historical data show that the system can produce quite impressive results over the test period with respect to a range of portfolio evaluation tools. Particularly given that we impose both costs to trading and restrictions on how trades can occur and limit the information set used to information derived only from price and volume data.

1 Overview

This chapter describes a computational intelligence system for learning trading rules and provides a comparison of the relative performance of portfolios managed by the system in the Australian Stock Exchange. Using only price and volume data the system determines rules to buy and sell stocks periodically. The rules adapt as market conditions change over time.

The system combines portfolio construction and management activities to build and manage a portfolio of stocks over time. The integrated process for stock selection and portfolio management allows for a search specification resulting in a highly adaptive and dynamic rule base system. Solutions are represented using fuzzy logic rule bases and a search for the best rule bases is implemented using an evolutionary process. The system produces rule bases that are adapted to market conditions using feedback from performance by adapting the length of the training data window and the action of a repair operator used in the evolutionary search.

The result section describes the application of portfolio evaluation tools to give a detailed assessment of the system's performance. A portfolio managed by the system is compared to the ASX200 index and a portfolio constructed using a non-adaptive rule base.

The rest of this chapter is organized as follows: Sect. 2 provides background information. Section 3 explains the approach to constructing the computational intelligence system in detail. Section 4 gives an evaluation of the systems performance and describes the methods for financial portfolio performance evaluation. Section 5 provides concluding remarks.

2 Background

There is a relatively large body of work describing various applications of nature-inspired algorithms to financial modeling; in this section we survey some of this work and give a context to the system described in this chapter.

A selection of this work includes: [21] presents an index forecasting approach; [9] applies an ANN to currency exchange rate prediction by anticipating the direction of price change using signal processing methods for series with high noise and small sample sizes; [4] describes a neural evolutionary approach to find models of correlation between financial derivatives; [3] discusses assessing credit risk and predicting using ANNs; and [2] discusses a neural network for option pricing. More recently, numerous applications of evolutionary computation have been published: [15] describes a dynamic stock selection system in which a model optimized using evolutionary computation determines optimal portfolio weights given trade recommendations; a genetic programming approach for combining trading rules in autonomous agents so that the rules compliment each other is given in [18]; [19] presents a linear genetic programming system for trading that uses intraday data; grammatical evolution for evolving human readable trading rules is extensively discussed in [5]; finally an application of genetic programming for discovering trading rules that are applicable in the short term is given in [16].

One of the possible financial applications has been in the area of developing trading rules to signal when investors should buy or sell financial instruments. Research in this area has received greater attention over recent years as an appreciation for the ease by which computational algorithms can develop complex trading strategies are further realized. Research such as described in [1, 13] highlight the possibilities for evolutionary computation to provide trading strategies, based on pattern recognition, to profit from equity market trading. Published research in academic finance journals primarily focuses on examining how well genetic algorithms can develop specific trading rules using historical prices to test, ex post, their profitability. A significant benefit of genetic programming in trading rule based systems is the grammatical phenotype structure which enables expression of rules combining several input in a form that is able to be readily understood and applied, this feature shared with the fuzzy rule base representation used in this paper.

This type of research is also directly related to the study of market efficiency. In an efficient capital market it would not be possible for traders to make a profit from past data as all relevant information for pricing a security today would be incorporated in today's price. Therefore, many finance papers (see [12], [14], and [8] for example) inter-relate the issue of market efficiency with the ability for genetic algorithms to literally "beat the market". Results are somewhat mixed. Although there is general consensus that financial markets do sometimes exhibit periods where certain trading rules work (see [6]), it is hard to find clear evidence that a single trading rule can function over an extended period of time. This is probably due to the fact that financial markets are ever-evolving, and in fact given the number of technical analysts that are employed in all the major financial trading institutions, when a trading rule is found to work it would not take long before it is exploited until it no longer yields a significant profit. It is therefore possibly more interesting to see if trading rules can be constructed that also continually evolve as the markets change. An adaptive trading strategy seems to be more promising than static approaches.

The system described in this chapter forms trading rules using price and volume history of stock prices and adapts the rules to changing market conditions. The approach is essentially referred to as technical analysis – rather than using fundamental accounting and macroeconomic data to determine which stocks to buy or sell, trading rules are developed solely applying historical data series from the previous trades of these stocks. The representation and evaluation functions used in this system however are able to be readily extended to include or test more extensive information sets.

3 System Design

This section describes the design of the computational intelligence system in detail. The system produces and applies trading rules taking into account evolving market conditions. The trading rules are represented using fuzzy logic rule bases and an evolutionary process facilitates the search for the best fuzzy logic rule bases with respect to the training data. In the following subsections we describe the fuzzy rules, the linguistic variables used to construct these rules, the evolutionary process, the evaluation function and methods which implement increased adaptation ability of the rules to changing market conditions.

3.1 Fuzzy Rule Base Representation

Fuzzy representation enables intuitive natural language interpretation of trading signals and implies a search space of possible rules that corresponds to trading rules a human trader could construct. An example of a typical technical trading rule such as "buy when the price of a stock X's price becomes higher than the single moving average of the stock X's price for the last 20 days" (indicating a possible upward trend) could be encoded using a fuzzy logic rule such as "If *Single Moving Average Buy Signal* is *High* then *rating* is 1"; conversely we could have a trading rule such as "sell stocks with high price movement when the portfolio value is relatively low" encoded by a fuzzy rule: "If *Price Change* is *High* and *Portfolio Value* is *Extremely Low* then *rating* is 0.1".

Each fuzzy rule base consists of a set of "if–then" rules where the "If" part specifies properties of technical indicators and the "then" part specifies a rating with ten discrete levels given a stock with these properties. The rule inputs are termed linguistic variables in the fuzzy logic component. Clearly, at least one linguistic variable must be defined to construct rules. Detailed definitions of the linguistic variables used in the system are given below. To construct the rules used to obtain the results presented in this chapter we used $V = 33$. The construction of the linguistic variables is described in Sect. 4. The output is interpreted as a rating of the strength of a buy recommendation given fulfillment of the If part. It is possible for the If part of a rule to refer to any combination of the technical indicators the system uses to give one output rating. A rule base may contain at least one and no more than $O = 30$ rules.

The value of each linguistic variable is described by one of a possible seven fuzzy membership sets. These are defined describing the relative magnitude of a particular observation: *Extremely Low (EL)*, *Very Low (VL)*, *Low L (L)*, *Medium (M)*, *High (H)*, *Very High (VH)*, and *Extremely High (EH)*. Membership functions map crisp data observations to degrees of membership of these fuzzy sets.

Figure 1 shows a visualization of the membership functions for a linguistic variable. For the triangular membership functions, the mapping from an observation to a degree of membership for each membership function is fully defined by specifying a minimum, center and maximum value where the min and max values refer to the lowest and highest linguistic variable observations at the edges of the triangle that belong to the membership set to the least degree and the center belongs to the membership set to the highest degree (the top of the triangle). The ramp membership functions are specified by a minimum and maximum value, any values greater than or equal to the maximum are classified as having full membership.

To construct the membership functions for each linguistic variable a series of historical data observations (of the variable) is sorted from lowest to highest

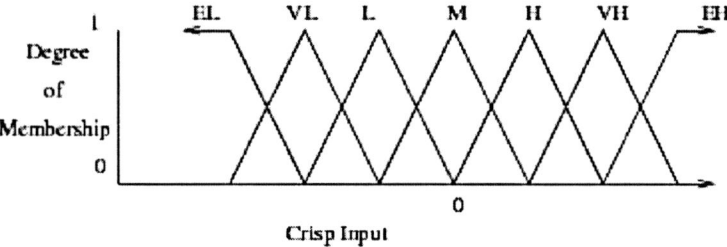

Fig. 1. Membership functions of the Fuzzy Sets *Extremely Low, ... Extremely High* for a linguistic variable. The middle membership functions (*Very Low* to *Very High*) are triangular and the two extreme membership functions are ramp membership functions. Note that the size of each function with respect to the crisp input that is mapped to a positive membership depends on the distribution of the data observations

and duplicates are removed. This series is then divided into M ordered sets of equal size where each set corresponds to a membership set, the sets overlap. The extreme low and high membership sets take as their maximum the lowest and highest data observations of the sets that contain the lowest and highest observations. For the middle membership functions the lowest and highest members of each set are the minimum and maximum values that belong to the set and the center is found by taking the mean of all the observations that belong to the set. When new data is input to the system this procedure is repeated.

Any "if" part may include up to V linguistic variables. The output for each rule gives one of B different ratings; there can be up to $O = 30$ rules in each rule base.

To estimate the total number of possible phenotypes, we may first estimate the number of possible phenotypes which consist of a single rule. Note that if a single rule has one linguistic variable present, then there are $M \times V \times B$ such phenotypes. If a single rule has two linguistic variables present, then there are $M^2 \times V \times B$ such phenotypes (M possible combinations of 2 linguistic variables out of available V, M^2 possible values for a pair of linguistic variables, B possible outcomes). In general, the number of possible phenotypes for a single rule (with one, two, ..., V linguistic variables present) is:

$$p = 10 \times \sum_{i=1}^{V} 7^i \times \binom{V}{i}.$$

The total number of phenotypes with k rules can be estimated as p^k.

An example of a phenotype rule base that could be produced by the system is given below. It consists of three rules:

- If *Single Moving Average Buy Signal* is *Extremely Low* then *rating* = 0.9.
- If *Price Change* is *High* and *Double Moving Average Sell* is *Very High* then *rating* = 0.4.
- If *On Balance Volume Indicator* is *Extremely High* and *Single Moving Average Buy Signal* is *Medium* and *Portfolio Value* is *Medium* then *rating* = 0.5.

Internally, each rule is represented using a sequence of slots. The columns in Fig. 2 have the following meanings. Column 1 contains a Boolean value to indicate whether the rule is active. Columns 2 through to 10 represent the rule inputs (each corresponds to a linguistic variable) and contain (a) a Boolean value indicating whether or not the linguistic variable is active, and (b) a number from 1 to 7 representing a membership function for the variable (1 corresponds to *extremely low* and 7 to *extremely high*). Finally, column 11 indicates the rule output rating and contains a single floating point value from the set $\{0.1, 0.2, \ldots, 1.0\}$). The internal representation for a rule base is simply a 30×11 matrix (note that columns 2–10 contain a Boolean and an integer).

	V_1	V_2	V_3	V_4	V_5	V_6	V_7	...	V_{33}	
B	B I	B I	B I	B I	B I	B I	B I	...	B I	F
B	B I	B I	B I	B I	B I	B I	B I	...	B I	F
B	B I	B I	B I	B I	B I	B I	B I	...	B I	F
B	B I	B I	B I	B I	B I	B I	B I	...	B I	F
B	B I	B I	B I	B I	B I	B I	B I	...	B I	F

Fig. 2. Internal rule base representation for a rule base with 5 rules ($O = 5$) and 33 linguistic variables ($V = 33$). B indicates a boolean value: $B \in \{T, F\}$; I an integer: $I \in \{1, 2, 3, 4, 5, 6, 7\}$; and F a float: $F \in \{0.1, 0.2, 0.3, 0.4, 0.5, 0.6, 0.7, 0.8, 0.9, 1.0\}$

	V_1	V_2	V_3	V_4	V_5	V_6	V_7	...	V_{33}	
T	F 2	T 1	F 7	F 1	F 6	F 1	F 7	...	F 1	0.9
F	F 4	T 2	F 2	T 3	T 4	F 5	F 1	...	F 2	0.3
F	T 1	F 3	F 2	F 2	F 2	T 4	F 4	...	F 4	0.7
F	T 4	F 5	F 1	F 4	F 2	T 5	F 2	...	F 5	0.4
F	F 6	T 3	F 3	T 3	F 7	F 3	T 7	...	F 3	0.5

Fig. 3. Example of the internal rule base representation. The first column indicates if a rule is used or not, the last column indicates the output rating give full satisfaction of that rule. The central columns (V_1 to V_{33}) show whether a linguistic variable is used and the membership function required for that variable in the rule. The order of the middle columns is important and corresponds to the particular linguistic variables

The number of possible genotypes for rule bases is

$$2^O \times 2^{V \times O} \times 7^{V \times O} \times 10^O,$$

as there are $2^{V \times O}$ possible truth assignments for the Boolean variable in column 1 of the matrix, $2^{V \times O}$ possible truth assignments for all Boolean variables in columns 2–33 of the matrix and $7^{V \times O}$ possible assignments for integer variables in columns 2–10 of the matrix, and 10° possible assignments for the variable in the 11-th column of the matrix as there are 10 possible output ratings.

As an example see Fig. 3. This rule base consists of a single active rule (in the first row) indicated by the true value in the first column and this rule has one active linguistic variable – V_2. This rule would be interpreted in the pheonotype representation (with reference to Table 1 as: If *Price Change* is *Extremely High* then *rating* = 0.9).

Linguistic Variables

In this section we describe the linguistic variables used in the fuzzy rules as described above. The linguistic variables are based on technical indicators used by finance practitioners that are calculated using close price and volume data and in some cases Index and interest rate data.

Table 1 lists the definitions of the linguistic variables used in the system, Table 2 lists conditions that must be met for the corresponding linguistic variable formulae in Table 1 to apply. The abbreviations used in the tables

Table 1. Linguistic variables used in the system and calculation

No.	Name	Formula	Restrictions
1	Price Change 1		$\delta = 20$
2	Price Change 2	$\ln\left(\frac{p_t}{p_{t-\delta}}\right)$	$\delta = 50$
3	Price Change 3		$\delta = 100$
3	SMA Buy	$\frac{p_t}{ma_t}$	$len_{ma} \in \{10, 20, 30\}$
4	SMA Sell	$\frac{ma_t}{p_t}$	$len_{ma} \in \{10, 20, 30\}$
5	DMA Buy 1		$len_{ma2} \in \{10, 20, 30\}$ $len_{ma2} \in \{40, 50, 60\}$
6	DMA Buy 2	$\frac{ma1_t}{ma2_t}$	$len_{ma1} \in \{60, 70, \ldots, 120\}$ $len_{ma2} \in \{130, 140, \ldots, 240\}$
7	DMA Buy 3		$len_{ma1} \in \{60, 70, \ldots, 120\}$ $len_{ma2} \in \{130, 140, \ldots, 240\}$
8	DMA Sell 1		$len_{ma2} \in \{10, 20, 30\}$ $len_{ma2} \in \{40, 50, 60\}$
9	DMA Sell 2	$\frac{ma2_t}{ma1_t}$	$len_{ma1} \in \{60, 70, \ldots, 120\}$ $len_{ma2} \in \{130, 140, \ldots, 240\}$
10	DMA Sell 3		$len_{ma1} \in \{60, 70, \ldots, 120\}$ $len_{ma2} \in \{130, 140, \ldots, 240\}$
11	PPO 1		$len_{ma2} \in \{10, 20, 30\}$ $len_{ma2} \in \{40, 50, 60\}$
12	PPO 2	$\frac{ma1_t - ma2_t}{ma1_t} \times 100$	$len_{ma1} \in \{60, 70, \ldots, 120\}$ $len_{ma2} \in \{130, 140, \ldots, 240\}$
13	PPO 3		$len_{ma1} \in \{60, 70, \ldots, 120\}$ $len_{ma2} \in \{130, 140, \ldots, 240\}$
14	OBV Buy	$\frac{\left(p_t - \max[p_{t-1}, \ldots p_{t-n}]\right)}{p_t} + \frac{\left(\max[obv_{t-1}, \ldots obv_{t-n}] - obv_t\right)}{obv_t}$	
15	OBV Sell	$\frac{\left(\min[p_{t-1}, \ldots, p_{t-n}] - p_t\right)}{p_t} + \frac{\left(obv_t - \min[obv_{t-1}, \ldots, obv_{t-n}]\right)}{obv_t}$	
20	SD 1		$\delta = 20$
21	SD 2	$sd(\ln(\frac{P_t}{P_{t-1}}), \ldots, \ln(\frac{P_{t-\delta}}{P_{t-\delta-1}}))$	$\delta = 50$
22	SD 3		$\delta = 100$
23	RSI	$RSI = 100 - \frac{100}{1+RS}$ $RS = \frac{total gains \div n}{total losses \div n}$	

Table 1. (continued)

No.	Name	Formula	Restrictions
24	MFI	$MFI = 100 - \frac{100}{1+MR}$ $MR = \sum \frac{MF^+}{MF^-}$ $MF^+ = p_i \times v_t, where\, p_i > p_{i-1}, and$ $MF^- = p_i \times v_t, where\, p_i < p_{i-1}$	
25	Bol 1	$Bol = \frac{p_t - ma_t}{2 \times sd(P_t, \ldots, P_{t-\delta})}$	$\delta = 20$
26	Bol 2		$\delta = 50$
27	Vol. DMA Buy 1		$len_{vma1} = 5,$
28	Vol. DMA Buy 2	$\frac{vma1_t}{vma2_t}$	$len_{vma2} = 20$ $len_{vma1} = 20,$ $len_{vma2} = 100$
29	Vol. DMA Sell 1		$len_{vma1} = 5,$
30	Vol. DMA Sell 2	$\frac{vma2_t}{vma1_t}$	$len_{vma2} = 20$ $len_{vma1} = 20,$ $len_{vma1} = 100$
30	PVO 1		$len_{ma1} = 5,$
31	PVO 2	$\frac{ma1_t - ma2_t}{SMt} \times 100$	$len_{ma1} = 20$ $len_{ma1} = 20,$ $len_{ma1} = 100$
32	DMI	see [20]	
33	%R	$\%R = \frac{p_t - min[p_{t-1}, \ldots, p_{t}-10]}{max[p_{t-1}, \ldots, p_{t}-10] - min[p_{t-1}, \ldots, p_{t}-10]}$	

The period length of the moving average variables is subject to evolution and may take values within the restrictions given in the third column

Table 2. Conditions for the linguistic variables in Table 1 to be defined

No.	Conditions
4	$(p_{t-\epsilon} < sma_{t-\epsilon})$ and $(p_t > sma_t)$
5	$(p_{t-\epsilon} > sma_{t-\epsilon})$ and $(p_t < sma_t)$
6–9	$(ma1_{t-\epsilon} < ma2_{t-\epsilon})$ and $(ma1_t > ma2_t)$
10–13	$(ma1_{t-\epsilon} > ma2_{t-\epsilon})$ and $(ma1_t < ma2_t)$
14	$(p_t > \max[p_{t-1}, p_{t-2} \ldots p_{t-n}])$ and $(obv_t \leq \max[obv_{t-1}, \ldots, obv_{t-n}])$
15	$(p_t < \min[p_{t-1}, p_{t-2} \ldots p_{t-n}])$ and $(obv_t \geq \min[obv_{t-1}, \ldots, obv_{t-n}])$
23,24	$(vma1_{t-\epsilon} < vma2_{t-\epsilon})$ and $(vma1_t > vma2_t)$
25,26	$(vma1_{t-\epsilon} > vma2_{t-\epsilon})$ and $(vma1_t < vma2_t)$

ϵ is an interval in days to control the period some of the variables take values, it was fixed to 5 days in all tests described in this paper

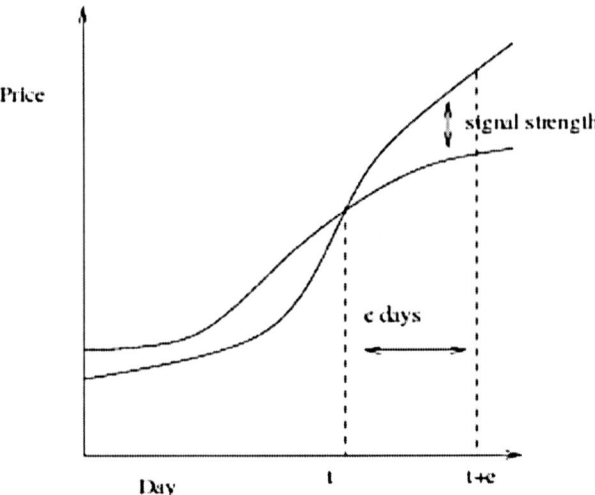

Fig. 4. Moving average signal design. The period "e" is the period during which the indicator emits a positive signal, and "signal strength" is the magnitude of the signal

have the following expansions: SMA, single moving average; DMA, double moving average; PPO, price percentage oscillator; OBV, on balance volume indicator; SD, standard deviation; RSI, relative strength index; MFI, money flow index; Bol, bollinger band; Vol. DMA, volume double moving average; PVO, percentage volume oscillator; DMI, directional movement index; %R, percent R.

For the moving average variables the value ϵ refers to the period in days between the signal trigger (when the shorter moving average becomes lower or higher than the longer) and the day t for a signal to occur, this is clarified in Fig. 4. In the OBV linguistic variable the value obv_t for each day t is calculated from historical data using the algorithm:

Initially at $t = 0$ $obv_0 = v_0$, then for each subsequent day t of historical data observations obv_t is calculated as follows:

- if $p_t > p_{t-20}$ then $obv_t = obv_{t-1} + v_t$,
- if $p_t < p_{t-20}$ then $obv_t = obv_{t-1} - v_t$,
- if $p_t > p_{t-20}$ then $obv_t = obv_{t-1}$,

3.2 Evolutionary Process

The fuzzy rule bases undergo an evolutionary process. An initial population of rule bases (genotypes) are selected at random and may be seeded with some rule bases that correspond to accepted technical trading strategies.

The evolutionary algorithm used in our asset allocation system is summarized by the following sequence of steps:

1. Initialize population P of n solutions (each solution RB_i is a rule base):

$$P = \langle RB_1, RB_2, \ldots, RB_n \rangle,$$

2. Evaluate each solution: calculate $eval(RB_i)$ for $i = 1, \ldots, n$,
3. Identify the best solution found so far ($best$),
4. Alter the population by applying a few variation operators (tournament selection of size 2 is used),
5. Apply a repair operator to each offspring; this operator controls diversity of offspring with respect to the best solution $best_{previous}$ from the previous generation (elitism is not used),
6. Repeat steps 2–5 successively for N generations,
7. The best solution after N generations represents the final solution.

Three variation operators (one mutation and two crossovers) and one repair operator are used in the process, the probabilities of applying a particular operator are evolved. We discuss each operator in turn below.

The mutation operator works by possibly modifying each gene of a single parent rule base in the process of producing an offspring. The type of gene remains the same: for instance a Boolean value cannot become an integer used to represent a membership function nor a decimal used to represent an output rating. If a gene is Boolean it is flipped. Otherwise if it is an integer or float, one of three events occur with equal probabilities:

1. The corresponding gene in the parent is incremented or decremented (equal probability for either) by a small amount, δ, to derive the offspring gene: for floats $\delta = 0.1$ and for integers $\delta = 1$. Since integers represent membership sets the change corresponds to a shift of one degree of membership (for example from low to very low).
2. The gene in the offspring is assigned a new value at random. For an integer gene the new value is selected from the domain $1, 2, \ldots 7$ and for a float from the domain $0.1, 0.2, \ldots, 1.0$.
3. The corresponding gene in the parent is passed unaltered to the offspring.

The two crossover operators combine genes from two parents to produce a single offspring. The first one, uniform crossover, assigns each gene in the offspring the value of a gene selected from one of the parents (the parent that provides the gene value is selected with equal probability). The second crossover operator assigns the rows of the offspring matrix by selecting – with equal probability – rows from both parents. In other words, the effect of this operator is to build a new rule base by choosing complete rules from each parent.

The last operator used in the system is a repair operator. It is used to maintain stability between generations. It is a binary operator with two rule bases as arguments and its effect is to modify the first genotype in such a way that it is no more than p percent different from the second genotype, no more than p percent different from the second genotype, which is best before we started the most recent adaptation. The number p is a parameter of the method. We found this parameter to be very important in controlling the type of rules generated, in Sect. 4 the values we used for p are given.

3.3 Evaluation of a Fuzzy Rule Base

The evaluation process comprises of three stages: in the first stage individual stocks are evaluated according to a rule base; in the second stage, the overall rule base's performance is evaluated. The return on investment (ROI) is adjusted in the final stage of the evaluation process. Each of these three stages are discussed in this section.

Rating of Individual Stocks

In this section the procedure to assign a rating to stocks with respect to a rule base is explained. For any stock X a rating $RB(X)$ is defined. This mapping from stock data to stock rating will be described using an example. Consider a rule base as follows:

1. If *Single Moving Average Buy Signal* is *High* then *rating* $= 0.7$.
2. If *Price Change is High* and *Volume Change* is *Very High* then *rating* $= 0.4$.

On a particular day t the following observations are made of technical indicators for stock X:

1. *Volume Change* $= 0.5$
2. *Single Moving Average Buy Signal* $= 0.95$
3. *Price Change* $= 0.2$

Initially each rule in the rule base is processed separately. The first and only "If" part of the first rule is: If *Single Moving Average Buy Signal* is *High*. It was observed that for stock X on day t the *Single Moving Average Buy Signal* had value 0.95. The membership function for *High* is defined by its min, center and max which are, in this case, 0.12, 0.97 and 3.88 respectively. Using (1), below, a membership function defined by these values maps the observed value 0.95 to a degree of membership of 0.97 in *High* or 97% *High*, a visualization of this procedure is given in Fig. 5.

$$m(x) = \begin{cases} \frac{x-min}{center-min}, & \text{if } min \le x \le center \\ 1 & , \text{ if } x = center \\ \frac{x-max}{center-max}, & \text{if } center \le x \le max \\ 0 & , \text{ otherwise} \end{cases} \quad (1)$$

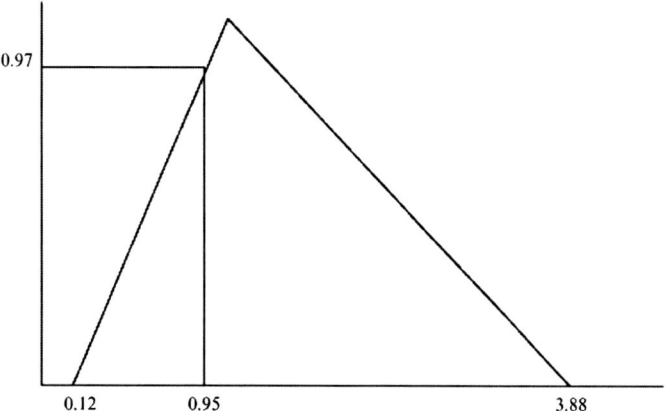

Fig. 5. Finding the degree of membership of observed single moving average buy signal 0.95 for stock X is "High" with a degree of 0.97

We now move to the output rating part of the first rule: then *rating* = 0.7. As the rule fulfilled the "If" part of the rule to the degree of 0.97 we adjust the output rating: $0.97 \times 0.7 = 0.679$.

The system looks at each rule in turn, the second rule in this example has two inputs: If *Price Change* is *High* and *Volume Change* is *Very High*. Each conjunction is processed in turn in the same way as the single conjunction in first rule. By this procedure, it is determined that the observation *Price Change* = 0.2 implies membership in the fuzzy set *High Price Change* = 0.5; and that *Volume Change* = 0.5 implies membership in *Very High Volume Change* = 1.0. These two values are combined by multiplying the membership degrees for each part resulting in a combined membership degree for the second rule: $0.5 \times 1 = 0.5$. The output rating is then adjusted using this combined membership degree: $0.5 \times 0.4 = 0.2$.

Finally an output rating with respect to the whole rule base is calculated. The final rating combines the sub-ratings from each rule to give a rating for stock X given some input data, we term this rating $RB(X)$. $RB(X)$ is defined as the center of mass of the sub-ratings from each rule in a rule-base:

$$RB(X) = \frac{\sum o_i}{\sum r_i}, \qquad (2)$$

where o_i is the output of rule i for stock X, and r_i is the rating of rule i.

Recall that for the first rule the sub-rating was 0.679 and for the second it was 0.200. Therefore the combined rating is: $(0.679 + 0.2) \div (0.7 + 0.4) = 0.799$.

Evaluation of Rule Base Performance

The procedure described in the previous section is applied to each stock in
the market. This enables the definition of a ranking of stocks in a market M
ordered by rating:

$$R(M) = \langle X_{i1}, X_{i2}, \ldots, X_{in} \rangle, \tag{3}$$

where $M = \{X_1, X_2, \ldots, X_n\}$ and $RB(X_{ik}) \geq RB(X_{ik+1})$.

The performance of a rule base is measured by analysis of simulated
trading on a historical data window. A decoder Fig. 6 defines the interpretation
of the ranking to make decisions for portfolio construction. In the simulated
scenario an initial capital is used to construct a portfolio on day 1 of the
simulation period, this portfolio is then rebalanced over the rest of the period
depending on the signals generated by the rule base under evaluation.

In the system, a portfolio P_t is defined as a vector of holdings of stocks in
$M = (X_1, \ldots, X_n)$ at time t:

$$P = [a_1 X_{i1}, \ldots, a_k X_{im}], \tag{4}$$

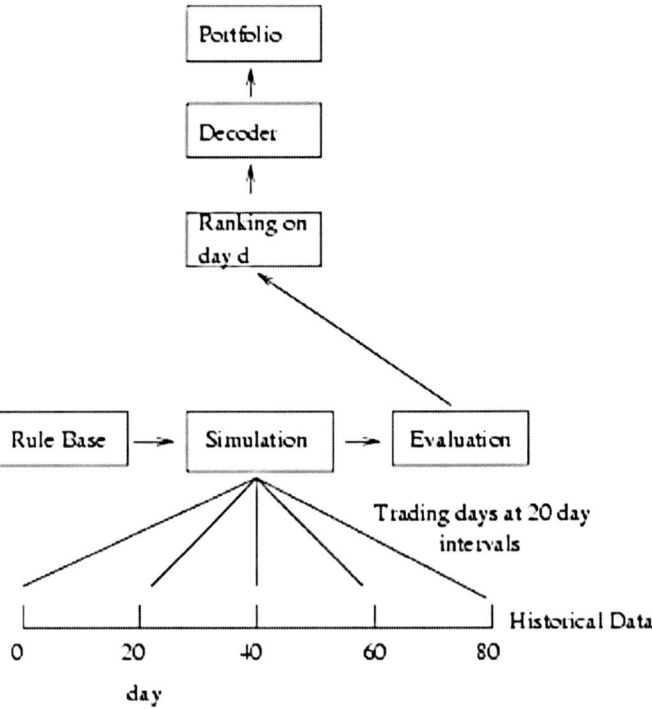

Fig. 6. The decoder takes a ranking and recommends a portfolio

where $a_1, \ldots, a_k$ are natural numbers, $\{X_{i1}, \ldots, X_{im}\} \subseteq M$, and $Value(P, t) = \sum a_j \times price(X_{ij})$, for $j = 1, \ldots, m$.

The decoder interprets the ranking in terms of two parameters: *buy best stocks percentage* (BBS) and *sell Worst Stocks Percentage* (SWS). BBS is the percentage of stocks to select from top of the ranking and SWS is a percentage to sell from the bottom of the ranking. At set intervals $dist = 20$ during the simulation period stocks are sold according the ranking on the trading day according to SWS and then bought according to BBS. In this way the initial portfolio P_1 is updated to get a new portfolio P_2 and so on until the end of the simulation period is reached.

The measure used for evaluation of portfolio performance is Return on Investment (ROI) over the simulation period (see (5)).

$$ROI = \frac{\ln(V_{t1}) - \ln(V_{t0})}{t_1 - t_0}, \tag{5}$$

where V = Portfolio Value, t_1 = End Time and t_0 = Start Time.

The result of the simulation is the return on investment over the whole simulation period for $RB_X : ROI(RB_X)$. To compare RB_X to another rule base RB_Y it is the case that if $ROI(RB_X) > ROI(RB_Y)$ then RB_X is better than RB_Y. The basic ROI criteria is supplemented by consideration of a few additional characteristics of performance, this takes place in the final evaluation and are described in the following section.

Final Evaluation

The ROI is the basic measure for rule base evaluation. However to further guide the search for optimal performing rule bases ROI is adjusted using penalty functions. The final evaluation value equals the ROI from the simulation minus penalties. There are two penalties applied to modify ROI, and they are:

1. Portfolio loss penalty
2. Ockham's razor penalty

Let us discuss each penalty in turn starting with the portfolio loss penalty.

The portfolio loss penalty implements a capacity for reducing risk. In simulation we measure the portfolio gain or loss on each trading day (see Sect. 3.3). Solutions that result in a reduction of portfolio value are penalized if they result in losses on any trading day even if at the end of the simulation period the return was high. The penalty becomes progressively higher for large losses.

The loss penalty is calculated using the formula:

$$Penalty_1 = \sum_{t=d,2d,\ldots,\frac{n}{d}d} m(t),$$

where d is the interval between trading days and $m(t)$ is calculated as follows,

$$m(t) = \begin{cases} 0.01, & \text{if } P_t - P_{t-d} \leq -5 \\ 0.1, & \text{if } -5 \leq P_t - P_{t-d} \leq -10 \\ 10, & \text{if } P_t - P_{t-d} \leq -10 \end{cases}$$

The second penalty, Ockham's razor, is calculated as follows:

$$Penalty_2 = r \times k_1,$$

where r is the number of active rules and k_1 is a constant which was set to 0.1 for the results given here. Ockam's razor reduces the fitness of solutions with many rules. The magnitude of k_1 controls a balance where the return produced using a candidate rule base must be justified by the number of rules. For example, setting the value to 0.1 causes the evolutionary process to produce rules where each rule must contribute at least 0.1 to the fitness. A rule base with fewer rules that results in similar return on investment to one with a greater number of rules is preferred. Rule bases with fewer rules are also less likely to be over-fitted to the training data.

The penalties are added together to get an overall value for a rule base and this value is deducted from the ROI for that rule base. Figure 7 gives an overview of the process used to determine a fitness value – the penalized ROI. Adjusting the ROI using the penalty functions promotes the discovery of rule bases which when applied give high return with less risk and generalize well outside training data.

3.4 Adaptation

In this section methods to cause rule bases to adapt to market conditions are discussed. Approaches towards this are through selection of data windows and by controlling the search. During the search process the performance of rule bases is evaluated based on data as described in the previous section. We first discuss the methods to select and alter the historical data used to evaluate the rules and then describe a method used to influence the ability of the

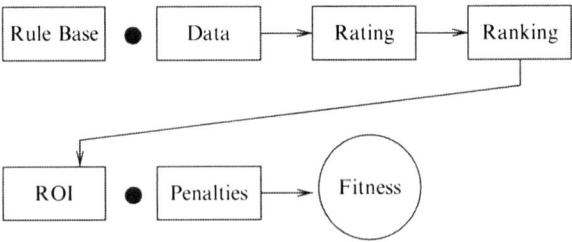

Fig. 7. Complete evaluation process

evolutionary search to use information previously discovered across training windows.

Selection of Training Data Window

The selection of the training data window controls the period of training data used when generating rule bases. Three methods for selecting a data window are able to be used by the system:

1. Static window
2. Sliding window
3. Variable length sliding window

The static window method uses a single initial period to evolve a rule base and then the rules from this period are used for all future trading. The sliding window uses a recent historical time windows for evaluation optimizing the rule base to recent periods.

The system also allows the length of the period of training data used for each sliding window to vary according to performance measured using a Sharpe ratio [17] which relates benchmark returns to the return of a real portfolio found from applying the rule base and decreases the window length when performance is worse than the benchmark. The results given in this chapter used the variable length sliding window method and the window length was controlled using the formula:

$$windowlength = e^{k \times ln(sharpe)} \times maxLength,$$

where k is constant set to control the sensitivity of the window change which was set to 2, sharpe is a Sharpe ratio $sharpe = \frac{(r_p - r_f)}{\sigma_p}$ and $maxLength$ is the maximum window length. The new window date is never set earlier than the previous window start date. In the case that the window length is calculated as earlier than this, the previous start date is used.

Adapting the Action of the Repair Operator

The repair operator is used to maintain stability between generations. It is a binary operator with two rule bases as arguments and its effect is to modify the first genotype in such a way that it is no more than p percent different from a second genotype. On initialization the second genotype is the solution from the previous window and during the search it is the best solution found at the current generation.

The parameter p is adjusted depending on the performance of a real portfolio in relation to the index which serves as a benchmark. It is reduced when performance is worse than the benchmark and increased if the real portfolio is out performing the benchmark. The rationale is to focus the search close to solutions while they give good performance and to broaden the search when this performance decays. p is increased or decreased according to the formula:

$$p = \begin{cases} sharpe \times 0.5 \times (1+k), & \text{if} & P_t^{real} < P_{t-d}^{real}, \\ sharpe \times 0.5 \times (1-k), & \text{if} & P_t^{real} > P_{t-d}^{real}, \\ 0, & \text{otherwise}, \end{cases}$$

where the sharpe ratio $sharpe = \frac{(r_p - r_f)}{\sigma_p}$ and k is a constant set in the configuration file to control the sensitivity of p to variation in portfolio performance. The sharpe ratio combines measurement of both risk and return with respect to the benchmark to make the adjustment of p (controlling the level of restriction on the search) dependant not only on the difference between returns between the current and previous month, but also relative to benchmark returns which are expected to be achievable.

4 Performance of the System

The system described in the previous sections is applied to form portfolios from stocks traded on the Australian Stock Exchange (hereafter ASX). Every month during the period of August 2001 to December 2006 two different portfolios are formed. The first one is created using a full adaptive evolutionary process, referred to as the "Adaptive EA" portfolio. The second one is created using a static rule base, a rule base is generated for the first window and then used for the rest of the simulation (see Sect. 3.4). This is called the "Static EA" portfolio and serves as a comparison benchmark for the advantages provided by using an adaptive rule base. Finally we also compare the performance of our EA portfolios to the ASX index, which reflects the performance of the market as a whole.

Figure 8 shows the portfolio values over the whole trading period, where the starting points have all been standardized to 1,000. Both of the EA portfolios perform very well, clearly dominating the index. The total increase in portfolio value over the whole investment period is measured by "holding period return" reported in Table 3. The whole market during the investment period improves slightly with the Index's holding period return of only 123% over a period of more than five years. During the same time, the Static EA portfolio value increases by 310% and the Adaptive EA portfolio value increases by 474%. Adapting the system fully to the market conditions results in a higher value of the Adaptive EA portfolio than that of the Static EA portfolio by 1.5 times. The annualized return, using either arithmetic or geometric method, is very impressive for the EA portfolios given the market conditions. However, it should be noted that returns of the EA portfolios are more volatile than the market. Further analysis is required to confirm whether investors are rewarded sufficient returns for the risk they bear.

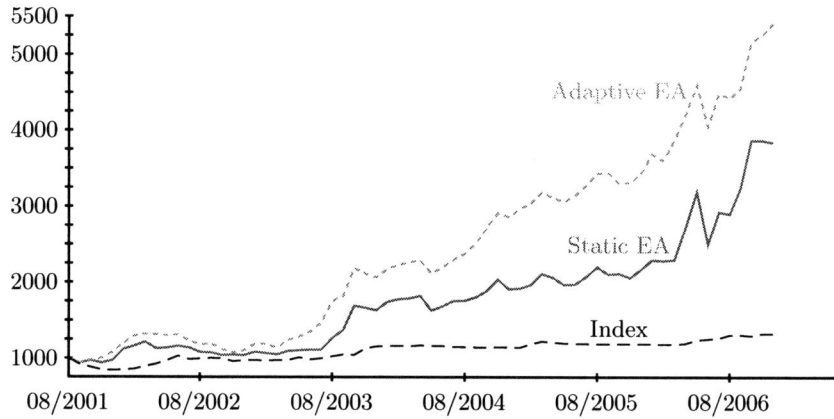

Fig. 8. Portfolio values from 08/2001 to 12/2006. Each portfolio starts at a value of 1,000 in 08/2001

Table 3. Portfolio Returns

Panel 1. Raw portfolio value

	ASX Index(%)	Static EA(%)	Adaptive EA(%)
Holding Period Returns	123.34	310.06	473.99
Annualized Arithmetic Returns	14.31	28.59	34.06
Annualized Geometric Returns	13.67	25.48	32.08
Annualized Volatility	11.39	25.51	20.45

Panel 2. Excess portfolio value

	ASX Index(%)	Static EA(%)	Adaptive EA(%)
Holding Period Returns	117.99	304.7	468.63
Annualized Arithmetic Returns	8.96	23.25	28.71
Annualized Geometric Returns	8.32	20.13	26.72
Annualized Volatility	11.36	25.49	20.45

Panel 2 of Table 3 reports statistics for excess portfolio values, ie. portfolio values above the risk free return. It can be argued that over time, even if investors hold risk free assets, such as Treasury securities, they are rewarded with financial returns. Therefore it is more relevant to investors to check the performance of the excess returns. Due to the increase in risk free returns over time, the excess returns of all portfolios are slightly lower than the raw returns. However, the relationship between different portfolios does not change qualitatively.

The monthly returns of each portfolio are illustrated in Fig. 9 while Table 4 further reports some characteristics of the return distributions. More than half of the time the Adaptive EA portfolio has a monthly return greater than 2.8%, whereas a half of the time the Static EA portfolio and the Index Portfolio

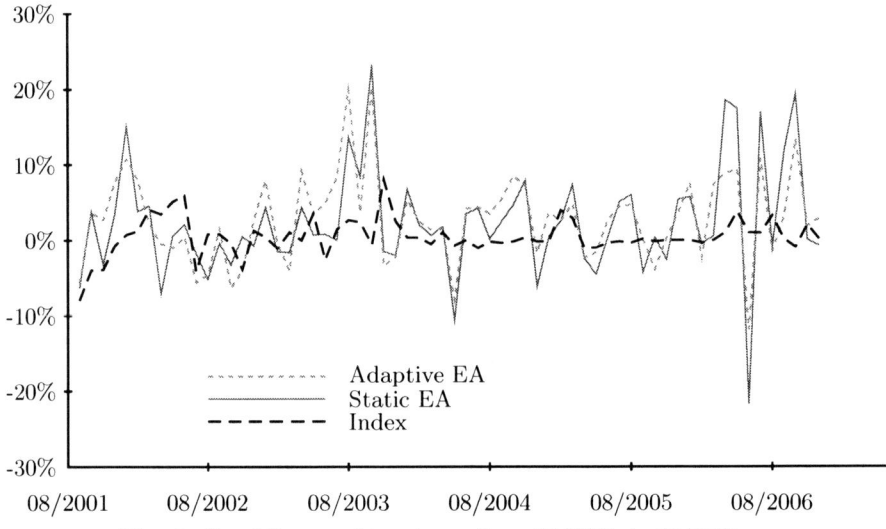

Fig. 9. Portfolio monthly returns from 08/2001 to 12/2006

Table 4. Portfolio return characteristics

	ASX Index	Static EA	Adaptive EA
Average monthly return	0.470%	2.383%	2.838%
Median monthly return	0.078%	0.685%	2.791%
Largest positive return	8.071%	23.257%	20.367%
Largest negative return	−8.008%	−21.769%	−11.930%
Skewness	−0.0829	0.3759	0.4356
Kurtosis	2.7062	2.1657	1.2516
Frequency of gain greater than 10%	0.000%	12.500%	7.813%
Frequency of loss greater than 10%	0.000%	3.125%	1.563%

have a return smaller than 2.4 and 0.8% respectively. Compared to the Index portfolio, the EA portfolios occasionally do experience higher loss. The largest negative returns for the two EA portfolios are 21.8% and 11.9%, whereas the largest loss for an index portfolio is only 8%. The index portfolio does not experience any loss greater than 10% in any given month, compared to the occurrence frequency of 3.15% and 1.56% for the two EA portfolios. This result may be indicative of the penalty function not sufficiently discarding the choice of stocks that are more likely to experience a large decline. However, both of the EA portfolios exhibit positive skewness, which is what investors prefer. A positive skewness portfolio has a high probability of large returns. It should be noted that the EA portfolios also have lower kurtosis, which implies less regular and smaller swing away from the mean in investor returns. Overall, the Adaptive EA portfolio has better return potential compared to the Static EA portfolio while maintaining a lower level of risk.

We have noted that our EA portfolios have much better return potential than the Index, and at the same time are more volatile. The Sharpe ratio [17] measures how much excess returns (portfolio return r_p above the risk free rate r_f) investors are awarded for each unit of volatility, ie.

$$\text{Sharpe} = \frac{r_p - r_f}{\sigma_p}.$$

As can be seen in Table 5, the Sharpe ratio for the Index is only 0.79, whereas that for the Static EA portfolio is 0.91. Adaptive EA portfolio has the best Sharpe ratio of 1.4, nearly double the reward investors receive for holding a passive index portfolio. The improvement in return potential for a fully adaptive system has been well above some additional level of risk for investors.

The Sharpe ratio focuses on portfolio volatility which measures total risk of the portfolio. Modern portfolio theory further decomposes volatility into systematic risk and unsystematic risk. The systematic risk component reflects how the changes in market conditions affect portfolio values, whereas the unsystematic risk component is unique to each portfolio. The constraints under which we form portfolios, such as the maximum stocks in each country or each sector, are in fact constraints to build well diversified portfolios. A well diversified portfolio should have return awarded to compensate for the systematic risk component only. Denote $r_{m,t}$ the returns at time t of the market, the systematic risk β_p of portfolio p is determined by the Capital Asset Pricing Model (CAPM) equation:

$$r_{p,t} - r_{f,t} = \alpha_p + \beta_p(r_{m,t} - r_{f,t}) + e_{i,t}. \tag{6}$$

Since the excess return $r_{p,t} - r_{f,t}$ of any portfolio should be fully explained by its level of systematic risk β_p and the market risk premium $r_{m,t} - r_{f,t}$, in an efficient market the alpha value of the portfolio, α_p, should be zero. If it is not and in fact there is a positive value then the portfolio is outperforming relative to its level of systematic risk and the performance of benchmark index. The higher the alpha value, the better the portfolio is to hold. Both of the EA portfolios have positive alpha, indicating a superior performance (see Table 5). The Adaptive EA portfolio has a value of alpha 1.5 times larger than that of the Static EA portfolio.

The robustness of alpha values can also be measured through the information performance rank, sometimes also known as the appraisal ratio. Essentially, it evaluates the active stock-picking skills of the strategy, once unsystematic risk generated from the investment process is accounted for. As we are comparing each of our portfolio's with the ASX Index, the information ratio is calculated as:

$$\text{Annualized Information Ratio} = \frac{\sqrt{T}\alpha}{\sigma_e}, \tag{7}$$

Table 5. Standard portfolio performance measures

	ASX Index	Static EA	Adaptive EA
Sharpe ratio	0.789	0.912	1.404
Alpha	NA	0.124	0.184
Information Performance Rank	NA	0.572	1.160
Selectivity	NA	0.124	0.184
Net Selectivity	NA	0.031	0.126

where T is the period multiple to annualize the ratio and σ_e is the standard error of (6). Grinold and Kahn, [11], have argued good information ratios should be between 0.5 and 1, with 1 being excellent. Goodwin [10] examined over 200 professional equity and fixed income managers over a ten year period and found that although the median information ratio was positive, it never exceeded 0.5. The information ratio of the Static EA portfolio is 0.57, indicating a very good performance. The Adaptive EA portfolio has an information rank of 1.16, double that of the Static EA portfolio. The information rank very close to 1 is indicative of a very strong and consistent performance.

Finally Fama's Net Selectivity measure [7] provides a slightly more refined method to analyze overall performance for an actively managed fund. Overall performance, measured as the excess returns of the portfolio over the risk-free rate, can be decomposed into the level of risk-taking behavior of the strategy and security selection skill. This security selection skill, or Selectivity, can be measured as a function of the actual return of the portfolio minus the return that the benchmark portfolio would earn if it had the same level of systematic risk. Both of the EA portfolios have a positive Selectivity, and the Adaptive EA portfolio still have much stronger performance than the Static EA one.

The Selectivity value, however, can be broken down still further to calculate Net Selectivity. Given that a portfolio's strategy may not be limited to simply track the benchmark portfolio, which would be the case for our portfolios under examination, it is also necessary to take into account the fact that the portfolios are not fully diversified, relative to the chosen benchmark. In fact, the number of stocks in our EA portfolios is much smaller than in the ASX index. To account for this, net selectivity is the value of selectivity that the strategy adds to the portfolio minus the added return required to justify the loss of diversification from the portfolio moving away from the benchmark. This effectively means any returns that the portfolio earns above the risk free rate must be adjusted for *both* the returns that the benchmark portfolio would earn if it had the same level of systematic risk and the same level of total risk to the benchmark. After accounting for the difference in total risk, the Static EA portfolio still maintain a positive Net Selectivity, but quite marginal. On the other hand, the Adaptive EA portfolio has a very substantial positive Net Selectivity of 0.13, again indicating a very strong performance.

To track the performance of our portfolios over time, rolling Alphas and rolling Information Ratios are calculated. Each month new Alpha and

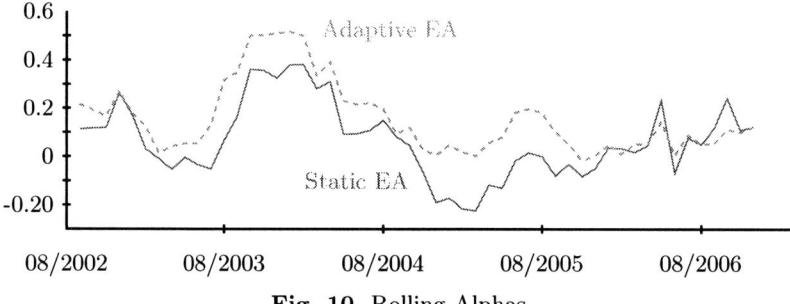

Fig. 10. Rolling Alphas

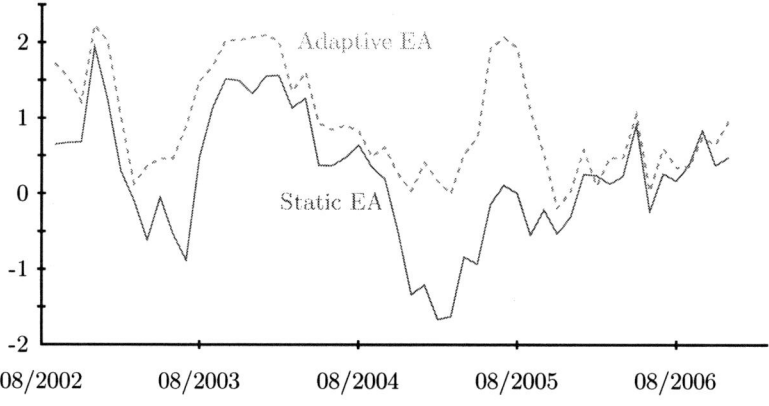

Fig. 11. Rolling Information Ranks

Information Ratio are calculated based on the data for the previous year. Figures 10 and 11 graph these two series. Even though both of the EA portfolios have positive overall Alphas and larger than 0.5 Information Ranks as shown in Table 5, only the Adaptive EA portfolio is able to maintain a consistent level of strong performance over time. This clearly shows the advantages of adapting the portfolio to changing market conditions.

5 Concluding Remarks

This chapter has provided a discussion of the framework in which an evolutionary algorithm manages a portfolio of selected stocks chosen using only price and volume information of individual company shares. Fuzzy logic rules are designed to optimize towards a specific fitness function, which is a simple return on investment measure. A specific financial penalty function is also incorporated to penalize solutions that select a portfolio of stocks that experiences significant losses. Effectively, it is a penalty for downside risk.

The computational intelligence system described in this chapter is tested on historical data of stocks traded on the Australian Stock Exchange during the period of August 2001 to December 2006. The empirical results show that portfolios constructed using evolutionary algorithm beat the market index by all standard portfolio performance measures. The clear dominance of the Adaptive EA portfolio over the Static EA portfolio is a result of the system being dynamic and adaptive to changing market conditions. Given that we impose both costs to trading and restrictions on how trades can occur, the EA portfolio performance is a relatively impressive result. This is even more particularly true when considering that only price and volume information is used to generate trading signals.

References

1. F. Allen and R. Karjalainen. Using genetic algorithms to find technical trading rules. *Journal of Financial Economics*, 51:245–271, 1999.
2. P.C. Andreou, C. Charalambous, and S.H. Martzoukos. Robust artificial neural networks for pricing of european options. *Computation Economics*, 27(2–3):329–351, 2006.
3. A. Atiya. Bankruptcy prediction for credit risk using neural networks: a survey and new results. *IEEE Transactions on Neural Networks*, 12(4):929–935, 2001.
4. A. Azzini and A.G.B. Tettamanzi. A neural evolutionary approach to financial modeling. In *GECCO '06: Proceedings of the 8th annual conference on Genetic and evolutionary computation*, pages 1605–1612, New York, NY, USA, 2006. ACM Press.
5. A. Brabazon and M. O'Neill. *Biologically Inspired Algorithms for Financial Modelling (Natural Computing Series)*. Springer, New York, Secaucus, NJ, USA, 2006.
6. W. Brock, J. Lakonishok, and B. LeBaron. Simple technical trading rules and the stochastic properties of stock returns. *Journal of Finance*, 47:1731–1764, 1992.
7. E.F. Fama. Components of Investment Performance. *Journal of Finance*, 27(3): 551–567, 1972.
8. C. Fyfe, J. Marney, and H. Tarbert. Technical analysis versus market efficiency a genetic programming approach. *Applied Financial Economics*, 9:183–191, 1999.
9. C.L. Giles, S. Lawrence, and A.C. Tsoi. Noisy time series prediction using recurrent neural networks and grammatical inference. *Machine Learning*, 44(1–2): 161–183, 2001.
10. T.H. Goodwin. The Information Ratio. *Financial Analysts Journal*, 54(4):34–43, 1998.
11. R.C. Grinold and R.N. Kahn. *Active Portfolio Management* (2nd edn.) McGraw-Hill: New York, 2000.
12. P. Hsu and C. Kuan. Reexamining the profitability of technical analysis with data snooping checks. *Journal of Financial Econometrics*, 3(4):606–628, 2005.
13. A.W. Lo, H. Mamaysky, and J. Wang. Foundations of technical analysis: Computational algorithms, statistical inference, and empirical implementation. *Journal of Finance*, 55(4):1705–1765, 2000.

14. C. Neely, P. Weller, and R. Dittmar. Is technical anlaysis in the foreign exchange market profitable? a genetic programming approach. *Journal of Financial and Quantitative Analysis*, 32(4):405–426, 1997.

15. O. Jangmin, J. Lee, J.W. Lee, and B.-T. Zhang. Dynamic asset allocation for stock trading optimized by evolutionary computation. *IEICE-Transaction Information System*, E88-D(6):1217–1223, 2005.

16. J.-Y. Potvin, P. Soriano, and M. Valle. Generating trading rules on the stock markets with genetic programming. *Computers and Operation Research*, 31(7):1033–1047, 2004.

17. W.F. Sharpe. Mutual fund performance. *Journal of Business*, 39(1):119–138, 1966.

18. H. Subramanian, S. Ramamoorthy, P. Stone, and J. Benjamin. Kuipers. Designing safe, profitable automated stock trading agents using evolutionary algorithms. In *GECCO '06: Proceedings of the 8th annual conference on Genetic and evolutionary computation*, pages 1777–1784, New York, NY, USA, 2006. ACM Press.

19. N. Svangard, S. Lloyd, P. Nordin, and C. Wihlborg. Evolving short-term trading strategies using genetic programming. In D.B. Fogel, M.A. El-Sharkawi, X. Yao, G. Greenwood, H. Iba, P. Marrow, and M. Shackleton, (eds), *Proceedings of the 2002 Congress on Evolutionary Computation CEC2002*, pages 2006–2010. IEEE, 2002.

20. J.W. Wilder. *New Concepts in Technical Trading Systems*. Trend Research, 1978.

21. S.-I. Wu and R.-P. Lu. Combining artificial neural networks and statistics for stock-market forecasting. In *CSC '93: Proceedings of the 1993 ACM Conference on Computer Science*, pages 257–264, New York, NY, USA, 1993. ACM Press.

A Hybrid Genetic Algorithm for the Protein Folding Problem Using the 2D-HP Lattice Model

Heitor S. Lopes and Marcos P. Scapin

Bioinformatics Laboratory, Federal University of Technology – Paraná, Av. 7 de setembro, 3165 80230-901 Curitiba, Brazil, hslopes@utfpr.edu.br

Summary. Proteins perform vital functions in all living beings and their function is determined by their tri-dimensional structure. Predicting how a protein will fold into its native structure is of great importance, and is one of the challenges of modern Computational Biology. Due to the complexity of the problem, simplified computational models were proposed, such as the 2-dimensional Hydrophobic-Polar (2D-HP) model. We proposed a hybrid genetic algorithm (GA) for the problem, with the following features: a fitness function based on the concept of radius of gyration, two biologically inspired genetic operators, and directed local search for improvement of solutions. Computational experiments include a detailed parameter analysis, tests with a data set of nine synthetic amino acid chains (from 20 to 85 residues long), and six real-world proteins translated to the 2D-HP model (from 96 to 330 residues long). The performance of the GA for the first data set was better than other approaches in the recent literature. For the second data set, the analysis considered both the quality of the folding and the processing time. Overall results were very promising and suggest the suitability of the hybrid GA.

1 Introduction

Proteins perform vital functions in all living beings. The function of a protein is determined after being folded into a specific 3D structure under certain physiological conditions. This structure is called its native conformation. Therefore, understanding how proteins fold is of great importance for Biochemistry and Biology. Proteins are formed by amino acid chains that can be roughly classified as hydrophobic (aversion to water) or hydrophilic (affinity to water).

It is commonly accepted that the main driving force of the formation of the structure of a protein is the internal interaction of hydrophobic residues. This means that the amino acids with hydrophobic side-chains in a protein sequence tend to be grouped together to form a hydrophobic core. The hydrophilic amino acids tend to be pushed away from the core and interact with surrounding solvent molecules.

H.S. Lopes and M.P. Scapin: *A Hybrid Genetic Algorithm for the Protein Folding Problem Using the 2D-HP Lattice Model*, Studies in Computational Intelligence (SCI) **92**, 121–140 (2008)
www.springerlink.com

According to Anfinsen's experiments performed in the 1960s, the 3D structure of a protein can be predicted by the analysis of its amino acid sequence. The task of predicting a protein tertiary structure is called the Protein Structure Prediction Problem (PSP), and the way it folds into its native conformation is known as Protein Folding Problem (PFP). Both are current problems in the modern Computational Biology that has drawn attention of Computer Science experts.

The exhaustive search of the conformational space of a protein is not possible with the current computational technology, even for small proteins. Therefore, simple lattice models have been proposed to decrease the complexity of the problem. However, even so, the PFP is NP-complete and, therefore, intractable for most real-world instances [1]. Consequently, heuristic optimization methods seem to be the most reasonable algorithmic choice to solve this problem. Among the many approaches to this problem, evolutionary computation methods have played an important role, since they are reputed as efficient search and optimization methods. Genetic Algorithms (GAs) do not guarantee that the global optimum will be found, but, during the evolution, it is powerful enough to find quasi-optimal solutions [2]. Therefore, many researchers have already applied GAs to the PFP [3–7]. GAs have achieved the most promising results in this area, although other methods, such as differential evolution [8] and ant colony optimization [9] have also been used.

This chapter is structured as follows: first we present a brief definition of the 2D hydrophobic-polar (2D-HP) model, followed by a detailed description of the hybrid genetic algorithm for the protein structure prediction problem using the 2D-HP model. Also, we describe the improved features of the algorithm designed to improve its predictability. Next, a complete parameter analysis is done aiming to understand their influence on the performance of the GA. We evaluated the algorithm with a benchmark of five synthetic amino acid chains, resulting in a better performance when compared with results in the recent literature. Exploring further, we also present the results of the application of the algorithm to five real-world proteins. Finally, results are discussed thoroughly and conclusions are done.

2 The 2D-HP Model

The Hydrophobic-Polar (HP) model was introduced by [10] and it is the most known and studied discrete model for protein tertiary structure prediction. This model is based on the concept that the major contribution to the free energy of the native conformation is due to interactions among hydrophobic residues. Such residues tend to form a core in the protein structure while being surrounded by the hydrophilic residues in such a way that the core is less susceptible to the environmental influence.

The HP model classifies the 20 standard amino acids in just two types: hydrophobic (H for non-polar) or hydrophilic (P for polar). This decision is taken for its simplicity, although some amino acids cannot be clearly classified as being of one of the two types [11]. Therefore, a protein is a string of characters defined over a binary alphabet $\{H, P\}$.

As it is a lattice model, the chain is embedded in a 2D square lattice and the allowed movements of the chain are restricted by the lattice. At each point, the chain can turn 90° left or right, or continue ahead. In a legal (valid) conformation, the adjacent residues in the sequence must be adjacent in the lattice and each lattice point can be occupied by only one residue.

The free energy of a conformation is inversely proportional to the number of hydrophobic non-local bonds (or H–H contact) where a H–H contact takes place if two hydrophobic residues occupy adjacent grid points in the lattice but are not consecutive in the sequence. Each such interaction contributes with −1 to the energy value. This yields two basic characteristics of real proteins: the protein fold must be compact and the hydrophobic residues are buried inside to form low-energy conformations [12]. The problem may be also considered as the maximization of the hydrophobic non-local bonds, since this is the same as the minimization of the free energy of a conformation in this model. Figure 1 shows an example of an 18 amino acids-long chain embedded in a square lattice. Black and white dots represent the hydrophobic and polar amino acids, respectively, and the square dot is the first element of the chain. The chain is folded in such a way that six hydrophobic non-local bonds are formed (shown as dotted lines).

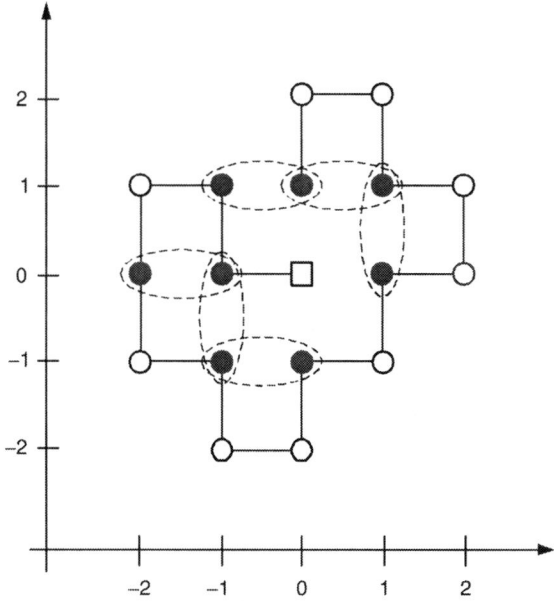

Fig. 1. Example of an amino acids chain folded in a square lattice

In addition to its simplicity, the folding process with the model has behavioral similarities with the real process of folding [13]. Notwithstanding, from the computational point of view, the problem of finding the native structure in using the 2D-HP model is proved to be NP-complete [1, 14].

3 Methodology

Genetic algorithms (GAs) comprise a class of heuristic methods inspired in the natural evolution of living beings and in the Darwinian principle of the survival of the fittest. GAs have been used in many computational applications including machine learning, combinatorial optimization, data mining and others. This approach is particularly interesting when dealing with NP-complete problems, whose search space is intractable by enumerative methods, or/and problems whose evaluation of the objective function is too complex, ill-defined or strongly constrained [15]. Besides the usual features of a GA, we proposed several enhancement strategies in an attempt to improve its performance for the problem.

3.1 Chromosome Encoding

It is well-known that the dynamics and the effectiveness of a GA are strongly influenced by the way individuals are represented. Individuals represent possible solutions for a problem and, in GAs, such solutions are represented indirectly. Individuals in a GA encode a genotypical representation, while the real-world solution is the phenotypical representation, decoded from the genotype. The three most usual ways to represent a protein structure using the HP model are [16]:

(a) Cartesian coordinates: This representation is straightforward but it is too sensible to translation and rotation, in such a way that identical structures can have completely different coordinates.
(b) Internal coordinates: In this representation an amino acid position is related to the previous one, and so there are two possibilities: absolute or relative coordinates. Using absolute coordinates, a movement is defined according to the grid axes, while using relative coordinates a movement is defined according to the previous one.
(c) Distance geometry: This representation describes a structure using a matrix of all distances between every pair of points.

Most GA-based approaches so far use internal coordinates because the connectivity of the chain is implicit in the encoding, i.e., every consecutive amino acid in the chain must be adjacent in the grid. Besides, a study presented by [17] showed empirically that the use of relative coordinates in the chromosome encoding achieved better results in comparison with absolute coordinates. Therefore, in our implementation we also used relative coordinates. In a 2D space, there are only three possible moves: (R)ight, (L)eft and (F)orward. Thus, an individual encoding is defined over the alphabet $\{R, L, F\}$. Considering a N-residue long chain, the individual will have $N - 1$ genes, representing the moves required to form a specific conformation.

When applying GA to solve constrained problems, depending on the representation used, invalid individuals may appear as result of the application of operators or in the initial random population. In principle, this problem can be handled in three different ways: eliminating invalid individuals, fixing them or allowing them to survive. The first approach is simple and straightforward, but possible useful genetic material can be lost and not recovered later. The second approach is interesting but usually it is computationally intensive. In the last approach, invalid individuals are allowed to survive in the population but their fitness value is decreased proportionally to the constraints violation. This penalty strategy is useful for preserving potentially interesting genetic material for further generations.

Despite the advantages of the representation chosen, it allows two or more amino acids to occupy the same position in the lattice. This fact, known as collision, leads to an illegal conformation. Therefore, the penalty strategy was used to penalize individuals that results in such unfeasible folds. These individuals can appear during evolution as result of the application of genetic operators and a penalty is added to the fitness function for every lattice point at which there is more than one amino acid. Another important reason to allow individuals that decode to unreal folds during evolution is that the shortest path from one compact legal conformation to another legal conformation may be very short if illegal conformations are allowed, when compared to the shortest path if only legal ones exist [17].

3.2 Initial Population

According to [4], the encoding using relative internal coordinates exhibits the problem that the initial population (randomly initialized) tends to have increasing number of collisions as the length of the protein increases. Therefore, the GA spends much time working with the illegal conformations before good results can be obtained. To circumvent this problem, we proposed a different technique to build the initial population, taken from another type of evolutionary algorithms, called Genetic Programming [18]. This technique, named ramped-half-and-half, does not ensure that individuals in the initial population will be free from collisions, but it tends to minimize them generating a greater diversity of conformations.

To implement this technique, the population is divided into two parts generated differently. The proportion of each part is a user-defined parameter as a percentage of the population size. The first part of the population is generated in a totally random manner, as usual. The second part is generated considering each individual as totally unfolded and applying several random mutations that varies between three and the number of genes in the chromosome. This is done in a way that this part contains a proportional number of individuals for each value, ranging from three to the number of genes in the chromosome. The minimum of three mutations was chosen because of the fact that conformations with 0, 1 and 2 mutations will have few hydrophobic contacts, and such individuals will bring very little contribution to the evolution.

The initial population created in this way will have a large diversity, a necessary condition to evolution. Besides, we empirically observed that the evolution process would be helped if a considerable number of individuals having few mutations (that is, they have large unfolded parts) is present.

3.3 Fitness Evaluation

To evaluate an individual, it is necessary to decode its genotypical representation (chromosome) defined over the alphabet $\{R, L, F\}$ to obtain the corresponding Cartesian coordinates (phenotypical representation). This set of coordinates describes how the residues are disposed in the lattice, that is, it represents a 2D conformation. After, the conformation is evaluated by a fitness function that gives a numerical value representing the goodness of the solution for the folding problem. The fitness function proposed is the product of three terms, as shown in (1):

$$fitness = HnLB \times RadiusH \times RadiusP, \tag{1}$$

where $HnLB$ is the number of H–H non-local bonds (related to the free energy of the conformation), and $RadiusH$ and $RadiusP$ are terms computed using the radius of gyration of the hydrophobic and hydrophilic amino acids, respectively. These terms are explained in the next sections.

Hydrophobic Non-Local Bonds

It is believed that the hydrophobic non-local bonds are the main force that drives the protein folding process. Since the problem is treated as the maximization of the H–H contacts, for every hydrophobic non-local bond, the energy function is added by 1. The complete energy function is shown in (2):

$$HnLB = \left(\sum_{i<j} e_{v_i v_j} . \Delta(r_i - rj) \right) - (NC.PW), \tag{2}$$

where NC is the number of collisions of the folding; PW is the penalty weight; and $\Delta(r_i - r_j) = 1$ if residues r_i and r_j form a non-local contact and $\Delta(r_i - r_j) = 0$, otherwise. Depending on the contact types between residues, the energy $e_{v_i v_j}$ will be e_{HH}, e_{HP} or e_{PP}, corresponding to H–H, H–P or

P–P contacts, respectively. In our implementation, e_{HH} was set to 1 and the remaining, to 0.

Recalling that we are using a penalty strategy for illegal conformations, the penalty term $(NC.PW)$ is subtracted directly from the energy function. This term comprises the number of grid points which are occupied by more than one amino acid, multiplied by the penalty weight that is set according to the chain length: the longer the chain, the higher is the penalty weight. In preliminary experiments with chains of different lengths we observed that for chains around 20 amino acids, $PW = 2$ was adequate. For chains of length around 50, $PW = 3$ lead to better results, while for chains around 80 amino acids, $PW = 4$ was more effective. Therefore, a straight line was interpolated using these points, and (3) shows how PW is adjusted as function of the number of amino acids (NA) of the chain

$$PW = (0.033 \times NA) + 1.33. \tag{3}$$

Radius of Gyration

An important issue to be taken into account, while attempting to predict the structure of a protein, is related to the energy hypersurface, i.e., how the free energy is distributed considering all possible conformations. In fact, it is not only the size of the energy landscape that matters, but also its shape. The energy landscape is protein-dependent, that is, for each protein, a different energy landscape exists. Therefore, the energy hypersurface has the dimension corresponding to the number of amino acids in the chain and can have many local minima.

The original HP model uses only the number of hydrophobic non-local bonds to evaluate a solution. Due to the nature of this discrete model, it creates large plateaus in the energy landscape on which local search cannot find a descent direction, and where it effectively performs a random search. This fact was also confirmed in our preliminary experiments and motivated the creation a modified fitness function. Using a discrete energy function (such as that of (2)), this landscape is much more difficult to be searched. In order to make the corresponding fitness landscape smoother, we propose the use of a new concept, the Radius of gyration (Rg) in the fitness function. The use of Rg in the fitness function can help the GA to escape from the plateaus. Furthermore, Rg increases the compactness of solutions: given two conformations with the same number of H–H contacts, Rg allows the fitness function to reward the most compact one, and thus makes the evaluation closer to reality. Radius of gyration is a physical concept from classical mechanics and is defined as the radial distance from a given axis at which the mass of a body could be concentrated without altering the rotational inertia of the body about that axis [19].

Bringing the Rg concept to the folding problem, it is a measure that indicates how compact a set of chained amino acids is: the more compact a conformation, the smaller its radius of gyration. Considering only the hydrophobic amino acids, Rg will measure how compact is the core formed by these residues. Equation (4) shows how Rg is computed for hydrophobic residues:

$$RgH = \sqrt{\frac{\sum_{i=1}^{NH}[(x_i - \bar{X})^2 + (y_i - \bar{Y})^2]}{NH}} , \qquad (4)$$

where x_i and y_i are the Cartesian coordinates of the i-th hydrophobic amino acid, $\bar{X}$ and $\bar{Y}$ are the mean values of all hydrophobic and x_i and y_i, respectively, and NH is the number of hydrophobic amino acids in the chain. Recalling again that the problem is dealt as maximization, the final term that contributes to the objective function is given in (5):

$$RadiusH = MaxRadiusH - RgH, \qquad (5)$$

where $MaxRadiusH$ is the calculated radius of gyration of the chain totally unfolded, supposing that this is the maximum value that can be reached.

This term of the fitness function regarding the hydrophilic radius of gyration (RgP) has the opposite purpose as $RadiusH$. That is, it fosters hydrophilic residues to lie far away from the hydrophobic core. The hydrophilic radius of gyration is computed in the same way as in (4), considering only hydrophilic residues, and without being subtracted from $MaxRadiusP$ (as in (5)). Using the RgP value, the formal definition of the term that contributes to the fitness function is shown in (6), and it will be always between 0 and 1:

$$RadiusP = \begin{cases} 1 & , \quad if \quad (RgP - RgH) \geq 0 \\ \frac{1}{1-(RgP-RgH)}, & otherwise \end{cases}. \qquad (6)$$

A positive difference, that is, $(RgP - RgH) > 0$, means that the hydrophobic residues are buried inside the conformation while the hydrophilic ones are outside. Such situation is desired and, in this case, $RadiusP$ has no influence in the fitness function, since the two first terms will be multiplied by 1. However, if the opposite is true, meaning that the hydrophobic residues are more spread than the hydrophilic residues, what is not desired, this conformation will be penalized by decreasing the fitness function value.

Overall, the use of radius of gyration concept in the fitness function makes the energy hypersurface smoother, allowing the GA to do a more efficient search.

3.4 Genetic Operators and Selection Method

Genetic operators are used to create new individuals in the population, by modifying existing ones. First, it is necessary a method for selecting individuals from the current population according to a criterion previously established. The selection method is not a genetic operator itself, but a procedure that must be executed before their application. In this implementation we used the tournament selection method. First, *tourneysize* individuals are randomly selected from the current population to take part of a tournament. Parameter *tourneysize* is usually taken as a percentage of the population size. Next, the best individual of this group is selected. This procedure is performed whenever genetic operators request individuals.

The use of problem-tailored genetic operators is commonly found in the literature, especially when the genetic algorithm is used for hard optimization problems. This is the case of the PFP, where one can find specific operators, biologically inspired or not (see, for instance, [20–22]). Apart from other characteristics, special genetic operators play a very important role in the PFP and, possibly, this is one of the main features that differs between implementations.

The first operator applied during the generation of a new population is the crossover operator. According to [2], this operator plays an important role for the PSP, since a piece of structure that has been useful for a given solution may be also useful for others. Two types of crossover were used in this work: 1- and 2-point crossover and both are applied with the same probability during the evolution.

Mutation is another operator commonly used in evolutionary computation. In this work, two types of mutation are used. The first is the simple mutation where each gene is tested, according to the mutation probability, to verify whether or not the actual value of the gene will be changed. The second type, called Improved-Mutation, also changes a gene according to the same mutation probability. However, just after the application of this operator, the individual is evaluated by the fitness function. If the fitness increases, the operation is successful, otherwise, the operation is undone and the process continues with the remaining genes. In most times, this special type of mutation improves the fitness of an individual and, in the worst case, the fitness will be unchanged. During evolution, the choice between the two types of mutation is controlled by a user-defined parameter of the GA.

In preliminary experiments, we observed that using only the basic genetic operators did not lead the GA to achieve good results. Thus, new and more specialized operators were developed in order to aid the evolution process to obtain better results.

The first specialized operator called U-Fold because it tends to fold the conformation into an U shape so as to maximize the number of H–H contacts between the two faces of the sequence that are laid face to face. Firstly, this operator searches all the individual for the longest straight segment. The length of this segment must be, at least, 1/10 of the total protein length. After selecting the longest straight segment, all the possible folding points for that segment is checked in order to choose the best way to fold this segment such that the number of H–H contacts will be maximized.

Another specially designed operator is Make-Loops. It was implemented with the objective of increasing the number of contacts in an individual by grouping hydrophobic residues that were found in a straight line in the conformation. The same as in U-Fold, it searches for the longest straight segment inside the conformation but, in this case, its minimum length was set to be four residues to allow its correct application. After finding a valid segment, the operator finds the first hydrophobic residue and fixes its position in the lattice. Then, the next hydrophobic residue is searched, but it needs to be, at least, three residues far from the first. Besides, in order to make a loop, it is necessary that an even number of intermediary residues exist between the two hydrophobic ones, due to the lattice that is being used. After fixing the two positions, the operator generates a loop by positioning the intermediary residues for both sides in order to find the conformation that best contribute for the fitness of the individual. It may happen that both conformations are worse than the original individual, making the loops to be undone. This procedure is repeated until the end of the selected segment.

It is easy to realize that both operators U-Fold and Make-Loops can be useful only when there are long straight parts in the chain being folded. This happens only at the beginning of the evolution process. Therefore, such operators work only in the initial generations of the GA, and are disabled as they became useless. The U-Fold and the Make-Loops operators are biologically inspired, since they aim at simulating the construction of stable secondary structures found in real folded proteins, such as β-sheets and α-helices [23].

A third operator used during the evolution is named Partial-Optimization, and is based on a concept taken from algorithms for combinatorial optimization problems, such as the TSP (Travelling Salesman Problem), and it is known as 2-opt. Here, we implemented a generalized version of the concept first proposed by [24]. The idea of this operator is to select randomly two non-consecutive residues of the protein chain and make their positions fixed in the lattice. Then, all possible conformations are evaluated, keeping the connectivity of the chain in the fixed points and changing the intermediate residues in the lattice. After this local search, the best conformation is kept. In the evolutionary computation area, algorithms that use some kind of local search strategy are considered as hybrid. Although there is no consensus on this issue, we prefer to consider the proposed genetic algorithm as hybrid to make clear this important feature.

The Partial-Optimization operator must be used with parsimony since, depending on the length of the internal segment between the two selected residues, this operation may become computationally expensive. It is easy to verify that the number of possibilities increases exponentially as the number of intermediate residues increases. This operator is applied to all the population during the evolution, according to a user-defined probability. The length of the intermediary segment ($POsize$) is also an user-defined parameter of the GA.

In Fig. 2, it is presented an example of the application of the Partial-Optimization operator to the sequence PHHPPPPHHPPHPP. In the figure, black and white dots represent, respectively, hydrophobic and polar amino acids. Arrows indicate the points selected to be kept fixed. The left part of

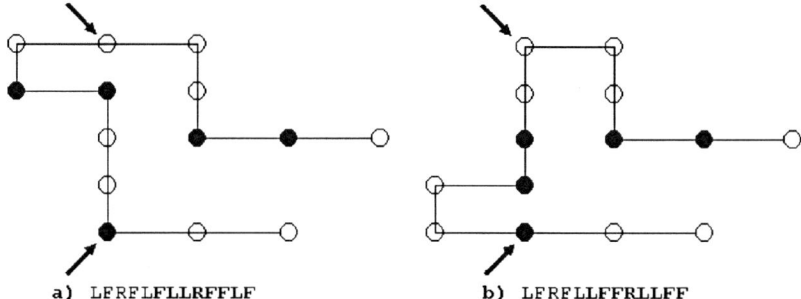

a) LFRFLFLLRFFLF   b) LFRFLLFFRLLFF

Fig. 2. Example of the application of the Partial-Optimization operator

the figure shows the original conformation with no H–H contacts while in the right, after the application of the operator, the new conformation presents two H–H contacts.

3.5 Improvement Strategies

Preliminary tests have shown that both the proposed fitness function and the special genetic operators have helped the GA to achieve good performance. Even so, it is impossible to assure that the GA will not get trapped in a local maximum. When reaching a local maximum, depending on the running parameters, the genetic diversity of the population can be lost rapidly. This phenomenon, known as convergence, has the undesirable effect of preventing further improvement. In this situation, the only possible evolution is due to mutation, by chance. Therefore, it is useless to go on running the algorithm. Instead of stopping and restarting the algorithm from the scratch, we devised a new strategy, called Decimation-and-Hot-Boot (DHB), described as follows.

Throughout generations an elitist strategy is used jointly with the tournament selection method. This strategy always copies the best-fitted individual of a generation to the next one. If an even-better individual is not found after many generations, this is a clear signal that the population has possibly converged and, consequently, the evolution has stagnated. In this work, the number of generations with no improvement necessary to trigger DHB was fixed to 10% of the total number of generations. At this point, 50% of the population is deleted (decimated) and a number of individuals is generated according to the method mentioned in Sect. 3.2.

Applying the DHB strategy makes the population to be restarted with a large genetic diversity giving chance for the evolutionary process to continue. At the restart, it must be taken into account the fact that all the newly created individuals probably will have very low fitness values compared to those individuals that survived decimation. Obviously, it is necessary to change running parameters temporarily so as to decrease the selective pressure (towards the best individual) in such a way that all individuals can have the opportunity to evolve. This is accomplished by increasing the frequency of application of the Improved-Mutation and decreasing the tournament size. The net effect of these changes is a fast improvement in the average fitness of the population in few generations. Hopefully, this strategy can contribute to find better individuals than before. The decimation strategy is applied whenever the no-improvement counter reaches the preset value and the maximum number of generations was not reached.

While generating and evaluating the last population, the Improved-Mutation and Partial-Optimization operators are to all the individuals with 100% probability. Although expensive, this local search operation is done once, as a last chance to improve further the best solution found by GA.

4 Computational Experiments and Results

The implementation of the previously described GA features, resulted in a software system called "HGAPF"(Hybrid Genetic Algorithm for Protein Folding), which was tested using several amino acid chains to verify its efficiency. The system was developed using Borland Delphi 7, under Microsoft Windows XP server running in a PC desktop with Athlon XP-2.4 processor and 512 Mbytes of RAM.

4.1 Parameters' Adjustment

Before running HGAPF with the benchmarks, a number of experiments was done for fine-tuning the GA parameters. For these experiments we used a real-world protein, known as 1qql, extracted from the Protein Data Bank (PDB) [25] and converted to the HP model (see Sect. 4.3 for more details). The Improved-Mutation operator is not used in the fine-tuning experiments, except when explicitly mentioned. The following standard parameters (and ranges) of the GA were tested: population size – $popsize$ (200 and 500), number of generations – gen (100, 200 and 300), tourney size – $tourneysize$ (3 and 5% of the population), probability of crossover – p_{cross} (70 and 90%), probability of mutation – p_{mut} (2, 5 and 8%). For each of these 72 experiments, 100 independent runs were done with different random seeds. Both the maximum number of H–H bonds ($maxHH$) and the average number of H–H bonds ($avgHH$) were considered. The two better combinations of parameters differ only in the number of generations, giving slightly different results. The standard parameters were then set to: $popsize = 500$, $gen = 300$, $tourneysize = 3\%$, $p_{cross} = 90\%$ and $p_{mut} = 2\%$. Using these parameters, the HGAPF took an average of 35.3 s for running.

Using the standard parameters, the Improved-Mutation operator was tested using the following probabilities of being selected among the two mutation operators ($p_{selectIM}$: 10, 30, 50, 70, 90 and 100%). This probability should not be confounded with p_{mut}, which is the probability of using a mutation operator, while $p_{selectIM}$ is the probability of selecting Improved-Mutation instead of the regular mutation operator. The results of 100 independent runs for each experiment showed that, independently of the value of $p_{selectIM}$, the use of Improved-Mutation always improve results. However, increased values for $p_{selectIM}$ lead to a faster convergence of the GA. Therefore, $p_{selectIM}$ was set to 10%.

Next, the probability of applying the U-Fold (p_{UF}) and the probability of applying the Make-Loops (p_{ML}) operators were tested, both in the range from 10% to 100%. Again, the standard parameters were used with the Improved-Mutation operator turned off. Besides $maxHH$ and $avgHH$, the average number of generations in which each operator was used was also observed. For each of the 20 experiments, again 100 independent runs were done. The best results were found when both probabilities were set to 10%, which maximized, at the same time $maxHH$ and the average number of generations the operator was used. The increment in the processing time for these operators was not significant.

The local search operator (Partial-Optimization) was tested, regarding its probability of application (p_{PO}) and the length of the intermediary segment ($POsize$) to be optimized. The range for these two parameters were: 1–8% and 5–8, respectively. The combinations of parameters lead to 32 experiments, done in the same way as before. Considering that the processing time tends to grow exponentially as the size of such parameters grow (in special, $POsize$), not only $avgHH$ should be taken into account, but also the processing time. Experimentally, we observed that when p_{PO} and $POsize$ were set both to 8%, the average processing time (using the previously mentioned standard parameters) raises to 660 s, when compared with 35.3 s with no Partial-Optimization. To meet both performance and processing time, we used the concept of Pareto optimality, commonly used in multiobjective problems [26]. Out from the Pareto front, we select $p_{PO} = 8\%$ and $POsize = 7$ amino acids as the operating point with better trade-off between the two conflicting objectives. At this point, the average processing time is only 122.1 s with a small decrement in performance, when compared with the combination of parameters that gives the better results.

4.2 Experiments with Synthetic Proteins

The first benchmark used is a set of synthetic amino acid chains, which optimal folding is known. Table 1 shows details of these chains that were used in several works in recent literature. In this table, NA and $maxHH$ stand, respectively, for the number of amino acids in the chain and the maximum number of H–H bonds, according to the 2D-HP model. This benchmark was presented by [27] (except the last instance that was introduced by [12]). The $maxHH$ values in the table are those originally given by authors, although for chains with 60 and 85 amino acids other results can be found in the literature.

Due to the stochastic nature of GA, all experiments described in this and the next sections were run for 100 times with different random seeds, and

Table 1. Benchmark of synthetic amino acids chains used in the experiments

NA	Amino acids chain	$maxHH$
20	$(HP)^2PH^2PHP^2HPH^2P^2HPH$	9
24	$H^2(P^2H)^7H$	9
25	$P^2HP^2(H^2P^4)^3H^2$	8
36	$P^3H^2P^2H^2P^5H^7P^2H^2P^4H^2P^2HP^2$	14
48	$P^2H(P^2H^2)^2P^5H^{10}P^6(H^2P^2)^2HP^2H^5$	23
50	$H^2(PH)^3PH^4PH(P^3H)^2P^4H(P^3H)^2PHPH^4(HP)^3H^2$	21
60	$P^2H^3PH^8P^3H^{10}PHP^3H^{12}P^4H^6PH^2PHP$	35
64	$H^{12}(PH)^2(P^2H^2)^2P^2HP^2H^2PPH^2P^2HP^2(H^2P^2)^2(HP)^2H^{12}$	42
85	$H^4P^4H^{12}P^6(H^{12}P^3)^3HP^2(H^2P^2)^2HPH$	52

Table 2. Comparison of results

NA	[27]	[12]		[28]	[22]	[29]	HGAPF	
	Best	Best	Mean	Best	Best	Best	Best	Mean
20	9	9(100)	9	9	9	9	9(74)	8.74
24	9	–	–	9	9	9	9(63)	8.61
25	8	–	–	8	8	8	8(69)	7.69
36	14	14(8)	12.40	14	14	14	14(6)	12.44
48	22	23(1)	18.50	23	22	23	23(2)	20.06
50	21	–	–	21	21	21	21(6)	18.72
60	34	–	–	36	35	35	35(6)	32.65
64	37	37(1)	29.30	42	42	39	40(5)	34.58
85	–	46(1)	40.80	53	50	52	51(2)	45.80

the averages are reported. Table 2 presents the results obtained by HGAPF together with those from [3] and [12] who also used a GA approach. More comparison of results is done with [28] that used PERM (Pruned Enriched Rosenbluth Method), [22] that used a variation of tabu search (GTabu), and [29] who employed an evolutionary Monte Carlo algorithm. Numbers within parenthesis indicate how many times the best score was found. It is important to highlight that the results from [3] represent the best individual taken out of five runs while those from HGAPF and [12] were run for 100 times. There is no information about repetibility for the other works. Comparing the results obtained for the four shortest chains and the 50 amino acids-long chain, it can be noticed that all algorithms have reached the maximum value of H–H bonds, despite some differences in the average. For the 48-residue chain, [27] and [22] did not find the best conformation while the others did. Considering the chain with 60 amino acids, [27] presented its result (34 contacts) as being the optimal, all the other algorithms found better conformations with 35 hydrophobic non-local bonds and PERM found another one with 36. For the two longest chains (64 and 85 amino acids), it can be observed that HGAPF found conformations better than the other GA implementations and,

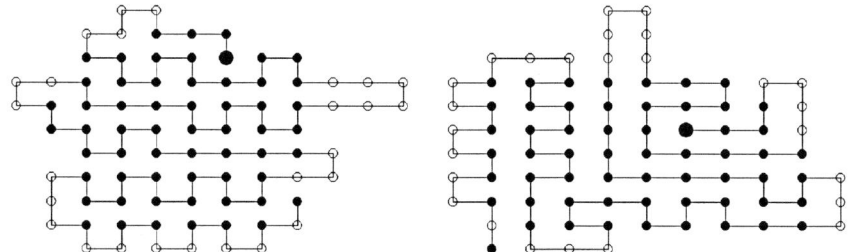

Fig. 3. Two best conformations for the 85 amino acids-long chain

sometimes, better or worse than [22] and [29], but very close to the maximum known, found by [28].

At this point it is important to recall that the difference of a single H–H bond from a conformation to another indicates a great improvement obtained by the algorithm. Consequently, jumping from the closest local minimum to the global minimum can be considered a great achievement. For instance, the two best results found by HGAPF for the 85-residue chain are presented in Fig. 3. In the figure we can notice that both conformations are quite different, but have 51 hydrophobic non-local bonds. This example confirms the assertion mentioned before that the fitness hypersurface has many local maxima of comparable amplitudes. This fact, by itself, makes the task of finding the global optimum of the PFP very hard, for any algorithm.

4.3 Experiments with Real-World Proteins

A second set of experiments to evaluate the performance HGAPF was done using real-world proteins drawn from the Protein Data Bank – PDB [25]. PDB is a reference for Molecular Biology and all proteins in the database had their tertiary structure precisely determined by means of X-ray spectroscopy or similar methods. Therefore, data from PDB are considered reliable and the conformations of proteins are supposed to represent their real native states.

A PDB file has detailed information about structure, but it is necessary to "translate" a protein sequence, that uses the 20 standard amino acids, to a sequence of H's and P's, in order to use the HP model. A translation matrix, based on the amino acids chemical features, is used to define the hydrophilic (polar) or hydrophobic nature of standard amino acids. There are some divergences between authors (see, for instance, [23]) and the PDB, concerning to this translation matrix. Therefore, we decided to use the PDB definition of hydrophobic and hydrophilic amino acids, and we used this approach to construct the translation matrix for our experiments. Some real small proteins, ranging from 96 to 330 amino acids, were selected at random for building this benchmark, and their translation to the HP model, as well de PDB identification code, are presented in Table 3.

Table 3. Results for real-world proteins

PDB	NA	Protein sequence	$maxHH$
1mli	96	$H^{10}(HP)^3PH^4PH^3P^2HPH^2P^2HP^3H^3PH^3PH^6(PH)^2$	46
		$(HP)^2(PH)^2H^2PH^3(PH)^2HPH^2(PH)^3H(HP)^2P^2H^2P^4$	
256b	106	$(HP)^2P(PH)^2HP^3H^4(PH^2P)^2(H^2P)^2H^5PH^4P(HP)^2H^3$	48
		$P^2H^2P^2HP^2(HP)^2P^2HPH^2P^3H^4PH^2P(PH^2)^2P^2H^5P$	
		$H^4(PH)^2(H^2P^2)^2HPP$	
1hmd	113	$H^2PHP^3H^3(PH)^3H^5P^3H^5PH^5(P^2H)^2PH^2P^2HP^2H^7$	61
		$P^4H^2P^4H^5(PH^5P)^2H^2(PH)^2H^4(PH^3)^2H^2PH^3PH^3$	
2ccy	128	$(P^3H)^2H^3P^2H^3PH^3P^2H^2PH^9(PH)^3(HP^2)^2H(H^7P)^2$	72
		$H^2P^2HPH^2P^2H^3(PH)^3HPH^7P^2H^{10}P^2H^4PH^{12}P^2H^2P^2$	
1d9i	288	$HPH^3P^2H^2PH^2PHP^2H^2P(P^2H^2)^2(PH)^2(HP)^2(PH)^2$	126
		$H^2P^2HPH^4PHP^4H^5PH^2P^3H^{11}P(PH)^3HP^3H^3PH^4P^2$	
		$(HP)^2P(PH)^2HPH^2PH^5P^2(HP)^2P^2HP^3H^8PH^3P^2H^3P$	
		$H^3P^4H^3PH^2P(H^5P)^2H^2P(H^4P)^2P(PH^3)^2(PH)^2HP^3H^3$	
		$P^2(HP)^2H^2P^2H^7P^3H(HP)^2H^2PHP^2H^2PH^4P^2HP^3H^2$	
		$PH^4(PH^2)^2P^3H^{10}PH^5PH^3P^2H^2$	
1epr	330	$(PH^2)^2H^2(PHP)^2PH^4(PH)^2H^2(PH)^2(HP)^3H^2P^3H^4P^3$	160
		$H^3P(PH)^3H^3P(PH)^2H^5PH^4(PH)^2(HP)^2P^3HPH^3$	
		$(PH^2)^2H^6PH^2P^2H^4(P^3H^2)^2PH^7(PH^2)^2(P^2H)^2H^3P^2H^3$	
		$PHP^3H^4PH^5(PH^3)^2H(PH^6)^2(PH^2)^2H(PH^2)^2(H^4P)^2$	
		$(HP)^2H^3PH^4(H^4P)^3HPH^3P^4H^6(PH)^2H^2P^2H^6(PH)^2H^2$	
		$(PH)^2HPH^2(PH)^2HP^3H^5P^3H^2(H^3P)^2H^{10}PH^4PH^5PH$	

Since the maximum number of hydrophobic non-local bonds for these sequences is not known a priori, and they were not synthetically designed, it is not possible to evaluate precisely how good the solutions found are. Notwithstanding, a graphical analysis of the folding (not shown here) reveals that the best foldings reported here are not the global optimum for these sequences. However, hopefully these chains and results will be important for forthcoming researchers to improve their heuristic methods. Results are also shown in Table 3, where $maxHH$ stands for the maximum number of H–H contacts found by HGAPF with its standard parameters.

5 Discussion and Conclusions

This paper described an hybrid genetic algorithm for the protein folding problem using the 2D hydrophobic-polar (HP) model.

The use of the concept of radius of gyration in the fitness function is one of the main contributions of this approach. It takes smoothness to the fitness landscape, allowing better solutions to be found by forcing the hydrophobic residues to be placed in the core of the conformation. Using this fitness function improvement, two conformations with the same number of H–H bonds can be adequately discriminated.

Another important enhancement is the Partial-Optimization operator, which performs a local search in a part of the individual and, most times, improves the overall conformation. Despite its utility, it tends to be computationally expensive as the number of intermediate residues increases, and should be used with parsimony.

The Improved-Mutation operator also added significant improvement to the GA, working together with the regular genetic operators. This special operator is especially useful in the last generation when it is applied to all individuals of the population hopefully improving further the solutions found. The U-Fold and Make-Loops operators have also contributed to the interesting results obtained.

The Decimation-and-Hot-Boot strategy was necessary because of the many local maxima in the fitness landscape. This strategy allows the algorithm to restart without loosing useful information (represented by the best individual found up to that moment). DHB does not improve individuals, but gives conditions to the GA to recover from the effects of a premature convergence, possibly allowing a more efficient exploration of the search space.

Genetic algorithms have been applied to many problems and, as a matter of fact, the more complex and complicated a problem is, the more sophisticated a GA have to be. As a consequence, a GA implementation will have many parameters to be tuned for the specific class of problems. Therefore, such kind of application will require a previous analysis of parameters' sensitivity. Our analysis of the GA parameters aimed at suggesting a set of values to the parameters. Those values are expected to make the GA perform well for many instances of the problem. Several tests were done to evaluate the influence of each parameter over the evolution process. For the parameters related to improved mutation and partial optimization operators, a different analysis was done because they are very time costly. The analysis of the Pareto front identifies not a single value, but a set of values that are equally good, considering both conflicting objectives: performance maximization and processing time minimization. Users will choose specific sets of parameter values according to specific situations, deciding what is more important at the moment, better results or a shorter processing time. It is worth to recall that the more sophisticated a GA, the more difficult is to adjust its running parameters for a given problem. This is where methods for self-adjustment of parameters [30] may be a valuable aid to the user.

It is important to emphasize the results obtained over the case studies. The first data set was taken from the literature and comprises synthetically designed amino acids chains with different lengths whose optimal folding is supposed to be known. Most algorithms in the literature perform well for short chains. However, for the longer ones, different local maxima were found, suggesting that there is room for further testing with longer chains. Our proposed GA performed better or, at least, similarly to other algorithms found in the literature.

The second data set was composed by six real-world proteins selected from PDB and translated to the HP model according to the PDB definitions. After some non-exhaustive runs, the best conformation found for each sequence was presented. The maximum number of H–H bonds presented for these proteins is certainly far from the optimal unknown value, but, even so, they represent good foldings. A graphical analysis of the conformations found reveals that, our GA was able to group most of the hydrophobic residues conveniently in the core of the protein, pushing hydrophilic residues away to peripheral regions. We can also notice that in real proteins there are some hydrophilic residues isolated in the sequence, what makes this kind of sequence very complicated to be folded. As we mentioned before, due to the nature of the fitness landscape, small improvements (i.e., increasing the number of H–H bonds by few units) in a given conformation are very hard to obtain.

Let us consider that the folding process takes several steps, from an unfolded sequence to its minimal energy conformation. It is not difficult to imagine that, the more folded the chain, the more difficult is to find one or more changes in the structure that can lead to a better conformation. Therefore, from the computational point of view, the folding is a problem of increasing difficulty. Hence, any heuristic algorithm devised to solve this problem must be powerful enough to cope with this situation. The hybrid GA approach proposed in this work has some limitations, especially when dealing with long chains. This fact suggests the necessity of even more intelligent operators that can analyze regions of the conformation and make context-dependent changes. Possibly, such improvements will lead to computationally intensive local search strategies, and this is an issue to be addressed in a future work.

The 2D-HP model is very popular, but it is far away from reality. More realistic models are needed, despite its computational complexity. In the same way, heuristic methods will play an important role to offer solutions for the protein folding problem. Also, more powerful computational resources certainly will be required for dealing with real-world instances of the problem. This is a current trend in the area, since researchers are moving from simple workstations to hardware-based reconfigurable computing [31] and grid computing [32]. Overall, results of this work encourage the continuity of the research, since there is still no efficient method for solving large instances of the PFP. Particularly, future developments will be towards a more complex lattice model and improved genetic operators.

Acknowledgements

The authors gratefully acknowledge the Brazilian National Research Council (CNPq) for the financial support to H.S. Lopes (grant 305720/04-0) and M.P. Scapin (grant 131355/04-0).

References

1. Berger B, Leight T (1998) Protein folding in the hydrophobic–hydrophilic (HP) model is NP-complete. J Comput Biol 5:27–40
2. Unger R, Moult J (1993) On the applicability of genetic algorithms to protein folding. In: Proc. of IEEE 26th Hawaii Int. Conf. on System Sciences, vol. I, pp. 715–725
3. Unger R, Moult J (1993) A genetic algorithm for three dimensional protein folding simulations. In: Proc. of the 5th Annual Int. Conf. on Genetic Algorithms, pp. 581–588
4. Patton AL, Punch III WF, Goodman ED (1995) A standard GA approach to native protein conformation prediction. In: Proc. of the 6th Int. Conf. on Genetic Algorithms, pp. 574–581
5. Pedersen JT, Moult J (1997) Protein folding simulations with genetic algorithms and a detailed molecular description. J Mol Biol 269:240–259
6. Krasnogor N, Pelta D, Lopez PEM, Canal E (1998) Genetic algorithm for the protein folding problem: a critical view. In: Proc. of Engineering of Intelligent Systems, pp. 353–360
7. Day RO, Lamont GB, Pachter R (2003) Protein structure prediction by applying an evolutionary algorithm. In: Proc. of Int. Parallel and Distributed Processing Symp., pp. 155–162
8. Bitello R, Lopes HS (2006) A differential evolution approach for protein folding. In: Proc. IEEE Symp. on Computational Intelligence in Bioinformatics and Computational Biology, pp. 1–5
9. Shmygelska A, Hoos HH (2005) An ant colony optimisation algorithm for the 2D and 3D hydrophbic polar protein folding problem. BMC Bioinformatics 6:30–52
10. Dill KA (1985) Theory for the folding and stability of globular proteins. Biochemistry 24:1501–1509
11. Jiang T, Cui Q, Shi G, Ma S (2003) Protein folding simulations of the hydrophobic–hydrophilic model by combining tabu search with genetic algorithms. J Chem Phys 119:4592–4596
12. König R, Dandekar T (1999) Improving genetic algorithms for protein folding simulations by systematic crossover. Biosystems 50:17–25
13. Dill KA, Bromberg S, Yue K, Fiebig KM, Yee DP, Thomas PD, Chan HS (1995) Principles of protein folding – a perspective from simple exact models. Protein Sci 4:561–602
14. Crescenzi P, Goldman D, Papadimitriou C, Piccolboni A, Yannakakis M (1998) On the complexity of protein folding. J Comput Biol 5:423–465
15. Michalewicz Z (1996) Genetic Algorithms + Data Structures = Evolution Programs, 3rd edition. Springer, Berlin
16. Piccolboni A, Mauri G (1998) Application of evolutionary algorithms to protein folding prediction. Lect Notes Comput Sci 1363:123–136
17. Krasnogor N, Hart WE, Smith J, Pelta DA (1999) Protein structure prediction with evolutionary algorithms. In: Proc. Int. Genetic and Evolutionary Computation Conf., pp. 1596–1601
18. Koza JR (1992) Genetic Programming: On the Programming of Computers by Means of Natural Selection. MIT, Cambridge
19. Beer FP, Johnston ER (1980) Vector Mechanics for Engineers – Statics. McGraw Hill, New York

20. Cox GA, Mortimer-Jones TV, Taylor RP, Johnston RL (2004) Development and optimisation of a novel genetic algorithm for studying model protein folding. Theor Chem Acc 112:163–178
21. Hoque MT, Chetty M, Dooley LS (2006) A guided genetic algorithm for protein folding prediction using 3D hydrophobic–hydrophilic model. In: Proc. IEEE Congr. on Evolutionary Computation, pp. 2339–2346
22. Lesh N, Mitzenmacher M, Whitesides S (2003) A complete and effective move set for simple protein folding. In: Proc. 7th Ann. Int. Conf. Research in Computational Molecular Biology, pp. 188–195
23. Lehninger AL, Nelson DL, Cox MM (1998) Principles of Biochemistry, 2nd edition. Worth, New York
24. Croes GA (1958) A method for solving traveling salesman problems. Oper Res 6:791–812
25. Berman HM, Westbrook J, Feng Z, Gilliland G, Bhat TN, Weissig H, Shindyalov IN, Bourne PE (2000) The protein data bank. Nucleic Acids Res 28:235–242
26. Deb K (2000) Multi-Objective Optimization Using Evolutionary Algorithms. Wiley, Chichester
27. Unger R, Moult J (1993) Genetic algorithms for protein folding simulations. J Mol Biol 231:75–81
28. Chikenji G, Kikuchi M, Iba Y (1999) Multi-self-overlap ensemble for protein folding: ground state search and thermodynamics. Phys Rev Lett 83:1886–1889
29. Liang F, Wong WH (2001) Evolutionary Monte Carlo for protein folding simulations. J Chem Phys 115:3374–3380
30. Maruo MH, Lopes HS, Delgado MRBS (2005) Self-adapting evolutionary parameters: encoding aspects for combinatorial optimization problems. Lect Notes Comput Sci 3448:154–165
31. Armstrong Jr. NB, Lopes HS, Lima CRE (2007) Reconfigurable Computing for Accelerating Protein Folding Simulations. Lect Notes Comput Sci 4419:314–32
32. Tantar A-A, Melab N, Talbi E-G, Parent B, Horvath D (2007) A parallel hybrid genetic algorithm for protein structure prediction on the grid. Future Gen Comput Syst 23(3):398–409

Optimal Management of Agricultural Systems

D.G. Mayer[1], W.A.H. Rossing[2], P. deVoil[3], J.C.J. Groot[2], M.J. McPhee[4], and J.W. Oltjen[5]

[1] Animal Research Institute, Department of Primary Industries & Fisheries, Brisbane, Australia
[2] Biological Farming Systems Group, Wageningen University and Research Centre, Wageningen, The Netherlands
[3] Agricultural Production Systems Research Unit, Department of Primary Industries and Fisheries, Toowoomba, Australia
[4] Beef Industry Centre of Excellence, NSW Department of Primary Industries, Armidale, Australia
[5] Department of Animal Science, University of California, Davis, USA

Summary. To remain competitive, many agricultural systems are now being run along business lines. Systems methodologies are being incorporated, and here evolutionary computation is a valuable tool for identifying more profitable or sustainable solutions. However, agricultural models typically pose some of the more challenging problems for optimisation. This chapter outlines these problems, and then presents a series of three case studies demonstrating how they can be overcome in practice. Firstly, increasingly complex models of Australian livestock enterprises show that evolutionary computation is the only viable optimisation method for these large and difficult problems. On-going research is taking a notably efficient and robust variant, differential evolution, out into real-world systems. Next, models of cropping systems in Australia demonstrate the challenge of dealing with competing objectives, namely maximising farm profit whilst minimising resource degradation. Pareto methods are used to illustrate this trade-off, and these results have proved to be most useful for farm managers in this industry. Finally, land-use planning in the Netherlands demonstrates the size and spatial complexity of real-world problems. Here, GIS-based optimisation techniques are integrated with Pareto methods, producing better solutions which were acceptable to the competing organizations. These three studies all show that evolutionary computation remains the only feasible method for the optimisation of large, complex agricultural problems. An extra benefit is that the resultant population of candidate solutions illustrates trade-offs, and this leads to more informed discussions and better education of the industry decision-makers.

1 Introduction

Across the globe, the traditional concept of 'the family farm' is increasingly being replaced by more business-orientated enterprises. Many agricultural enterprises are being taken over by large corporations. Along with the

D.G. Mayer et al.: *Optimal Management of Agricultural Systems*, Studies in Computational Intelligence (SCI) **92**, 141–163 (2008)
www.springerlink.com

pressures of increasing costs and decreasing returns, these competitive trends are leading to an increasingly complex decision environment, in which modern-day agricultural systems have to become more efficient. Those not meeting these challenges tend to fall by the wayside.

In this scramble to remain competitive, agricultural businesses are increasingly adopting new technologies, both in terms of hard infrastructure and "soft systems". The latter include the incorporation of producers' knowledge, historical farm records, and meteorological and other data into a systems model of the enterprise. These models are useful for learning and exploratory "what–if" simulations, and logically lead to formal optimisation exercises, as exemplified in the case studies of this chapter. In these optimisations of agricultural systems, evolutionary computation is more and more proving to be the leading practical methodology.

Concisely, these optimisation exercises require (a) a model of the system of interest, (b) an objective function to be optimised, such as the enterprise's gross margin or profit, (c) modeled input parameters, such as coded management options, which drive the model and are progressively improved during the optimisation process, and (d) an optimisation algorithm for this process. An alternate important use of optimisation is in model tuning, where the internal model parameters are tuned to minimise the deviation between the model's predictions and real-world data.

Unfortunately for practitioners, agricultural models generally pose some of the more difficult problems for optimisation methods. The "curse of dimensionality" often applies as many of these models have very large search-spaces. Whilst some of the modeled management options may be discrete (such as choice of pasture type or breed of animal), many more are continuous (levels of fertilizer or irrigation, day of the year for implementation), which also contributes to the complexity of the search-spaces. The response function being optimised typically has a non-smooth surface, including sharp cliffs and discontinuities when the system is over-utilised and collapses both biologically and economically. Multiple local optima are likely, where quite-different alternate management strategies can have similar outcomes in economic terms, and these can very often confound the search and lead to suboptimal convergence. Epistasis (the interacting effects of the input variables) is usually very pronounced in these systems.

This range of problems means that only the most efficient optimisation methods are likely to succeed with "real-world" agricultural models. In particular, the somewhat dated gradient and deterministic direct-search methods are poorly suited for this task. Of the more modern algorithms, "tabu search" and "ant-colony" methods appear well suited to combinatorial optimisation problems that are effectively Euclidean in nature, but not to the optimisation of multidimensional models. Similarly, simulated annealing has proven to be thorough and reliable, but is too slow and generally inefficient to be of practical use with larger modelling problems. This leaves evolutionary computation, which encompasses genetic algorithms and evolution strategies, as the most practical methodology.

Literature examples of evolutionary computation in practice abound. Appendix 1 of Mayer [17] lists 35 published studies on the optimisation of agricultural systems. This trend has continued since, with "most" optimisation studies of agricultural systems now using evolutionary algorithms. In the wider practical optimisation literature, differential evolution in particular is becoming quite widely used, with the methodology and applications for this being well covered in Price et al. [26].

Three case studies of the application of evolutionary computation to real-world agricultural systems follow. These studies range from a farm-level model with relatively few management options and a single economic objective, through to landscape studies with multidimensional objective functions. They describe the types of problems currently under investigation around the world, and illustrate how evolutionary computation is being used as a practical methodology to provide solutions.

2 Australian Livestock Enterprises

The modelling of agricultural livestock enterprises across Australia has been actively researched for over 20 years. During this time, we have developed three livestock systems models. Initially, dairy farms of south-east Queensland were studied, and then an increasingly complex series of tropical beef property models was developed, and tuned to data across northern Australia. More recently, we have been modifying the Davis Growth Model from the USA to make phenotypic predictions of beef cattle, and developing a generic model for beef herds in states across southern Australian. In parallel with the development of these models, comparisons of the available optimisation methods have been conducted in sequence. These have repeatedly demonstrated the superiority of a range of evolutionary computation algorithms. With the simpler models, a number of optimisation methods reliably found the global optimum of these systems, although genetic algorithms were amongst the more efficient. As the size and complexity of the models rose, the only practical method of optimisation was via the most efficient versions of these evolutionary algorithms.

2.1 Dairy Farm Model

A "typical" model of an agricultural system is our farm-level dairy model [20], as shown in Fig. 1. Pasture and milk production are simulated, with model parameters tuned to data from on-farm research. For a single-year run of the model, there are 16 independent continuous and categorical management options to implement, and (at the specified level of precision) this equates to a search-space of the order of 10^{23} possible combinations of these management options.

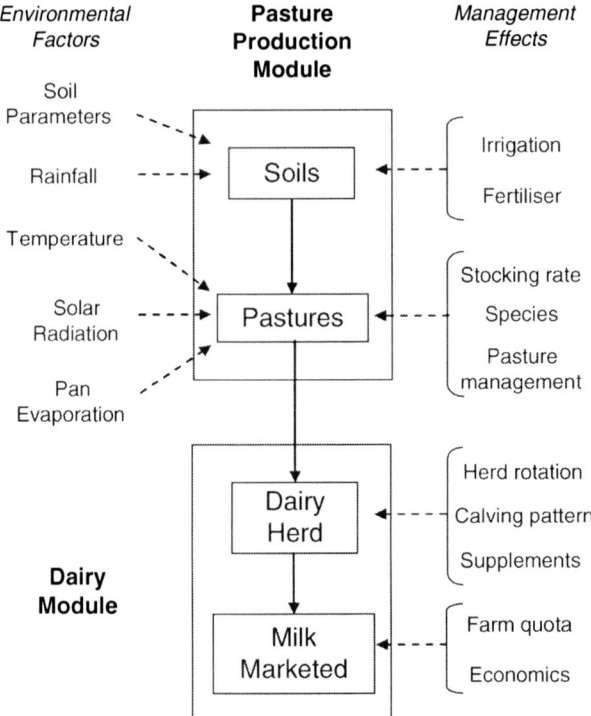

Fig. 1. System schematic for dairy farm model

Initial optimisations of this model [18] using hill-climbing (quasi-Newton) and direct-search (Nelder–Mead simplex) library routines were disappointing. The random replicates terminated at different management combinations, with a wide range of reported "optimal" gross margins. More comprehensive comparisons [19] confirmed these somewhat dated methods as unsuitable for model optimisation. Here, simulated annealing proved the most reliable method, always arriving at the system's global optimum. This comparative study also included Genesis [7], a binary genetic algorithm.

Using the default options for most of the Genesis operational parameters, a factorial structure of 48 replicates was trialed – two population sizes by three cross-over rates by two mutation rates by two generation gaps by two selection types. Whilst these options did tend to interact somewhat, in overall terms this genetic algorithm performed quite well. On average, the reported optima were 0.32% short of the global optimum, but each of these was achieved with one-tenth the computation time of simulated annealing. In practical terms, the magnitude of this apparent error is not likely to be of concern, as there will always be some degree of mismatch between the model's predicted performance and the real world. The model correctly identified the near-optimal region, which generated hypotheses for testing and proving with follow-up on-farm experimentation.

2.2 Beef Enterprises (Cohort Level)

Throughout the 1990s, the important breeding and trading options for whole-property herds were modeled, using animal cohorts as the base unit. This series of increasingly-complex models [21,22] was coded onto Sun workstations for the repeated rapid runs required by optimisation. The initial version was based on and validated against DYNAMA [11], a commercially-available PC herd dynamics model. This package is widely used by producers and agribusinesses throughout northern Australia, for budgeting, planning, educational and "what–if" purposes.

For the investigative optimisations of the Sun version, the proven methods of simulated annealing and the binary genetic algorithm (Genesis) were again used. In addition, a new evolutionary algorithm, Genial [33], was trialed on the more complex version of the herd dynamics model. This package has the advantage of real-value coding, and it has been suggested that the "natural alignment" of these to their real-world counterparts, along with the expanded range of genetic operators, will make this more efficient than binary coding. Furthermore, optimisations using Genial can be run in either "genetic algorithm" mode (where each of the management decisions to be optimised has one "allele"), or as an "evolution strategy" (where each has two separate alleles, the first being the option value, and the second its variance). Table 1 lists the relative performance of these methods.

It is obvious from this table that simulated annealing is no longer the best option – on both the restricted and full models it does not get acceptably close to the global optimum, whilst also taking considerably longer to converge. Similarly, the evolution strategy is not the best choice for the full model – it appears that the additional "overhead" of taking the option variances along in the process (and hence doubling the required number of "alleles") has penalized its relative performance.

This leaves the genetic algorithms as the preferred methods. On the full model, both converged (on average) to acceptable solutions, taking around

Table 1. Average percentage short of the modeled global optimum (and, in square brackets, the average number of model evaluations to achieve convergence, in millions), for optimisation methods applied to the "cohort-level" herd models

Model (dimensions, search-space)	Simulated annealing (ASA)	Binary GA (Genesis)	Evolution strategy (Genial)	Real-value GA (Genial)
Restricted (20, 10^{50})	0.88% [0.82]	0.24% [0.14]		
Full (40, 10^{100})	1.36% [3.44]	0.04% [2.43]	0.26% [10.42]	0.05% [3.46]

the same level of computation to achieve this. The binary version appeared marginally more efficient, however this may have been due to the nature of this particular model. Quite a number of the trading options (buying and selling proportions for each cohort of animal) had their optimal values on the boundary (of zero). Under a binary coding it is likely that these options eventually all "jumped" from "very little" to "zero", these being the adjacent binary-coded options. For the real-value coding, however, each of these options would have to "creep" onto the boundary, thus taking somewhat longer to converge. For more general models, particularly with a higher proportion of continuous (non-categorical) management options, the natural alignment of these with real-value coding should be more efficient.

However, the intended application of these optimal regimes to real-world beef properties identified certain deficiencies with the model. Whilst the cohort-level DYNAMA model has been widely used in the industry, it cannot handle animal differences, as all animals within an age and sex cohort are assumed identical. This is not the case in the real world. For example, heifers which conceive and produce a calf in the first year of mating tend to have lower weights than those which fail, and thus a lower probability of conception in the following year. This and other identified shortcomings of the cohort model resulted in the rewriting of the model.

2.3 Beef Enterprises (Animal Level)

The revised model, on an "individual animal" basis, tracks the animal liveweights, conception and mortality probabilities over a 10 year horizon [23]. This takes about 30 s of runtime on the Sun workstations for one model evaluation, so only around 10^5 evaluations are possible in each month of computation. This is far short of the 10^6 or 10^7 model evaluations as previously used to achieve convergence. Also, this expanded model now has 70 independent management options, so forms a 70-dimensional problem with a search-space in the order of 10^{120}, and with strong interactions between these options expected. Hence only the most efficient evolutionary algorithm can be considered for optimisation. The real-value coding of Genial was utilized, as only one of these 70 management options was categorical with the rest being continuous.

Initially, a factorial structure of four "exploratory" replicates was run, for up to 6 weeks each. Whilst these optimisation runs were obviously still short of convergence, some valuable comparisons were possible [23]. Firstly, the population size of 200 was superior to 500, indicating that (for this 70-dimensional problem) a factor of around 3 did carry sufficient genetic material, with the higher ratio of 7 merely slowing the rate of climb. Secondly, a "shotgun" combination of 12 mutation parameters (different types and probabilities) was trialed, as it was expected that these types would contribute successfully at different stages during the optimisation. However, this did not eventuate, with the more standard low-level Gaussian-creep mutation applied throughout proving superior.

About this time, our collaborators in the Cooperative Research Centre for Cattle and Beef Quality identified "differential evolution" (DE) as a particularly simple and efficient variant of evolutionary computation [29]. Initial trials against Genial, using approximately-equal operational parameters, showed that it performed similarly in the early and middle stages of optimisations of the individual-animal beef herd model. However, in the latter stages the exploratory nature of DE continued to find better management strategies, with more-optimal solutions continually being discovered whilst Genial had "converged" to suboptimal solutions, of the order of 5% lower [24]. Thus DE was adopted as the "optimiser of choice". Following some parameter tuning, one DE optimisation was run for 9 months, and approached 10^6 model evaluations. Its optimum (which had not increased in the last 10% of runtime) was then adopted as the global optimum of this model.

The overall conclusion from this series of optimisation studies (using Genesis, Genial and DE) is that most combinations of operational parameters do "work" in practice. Whilst there are some differences between the performance of different combinations, these tend to be only slight, and even the "worst" evolutionary algorithms are usually better than alternate optimisers. Regarding efficiency of the optimisations, the population size appears the most important operational parameter. It is generally accepted that "too small" populations carry insufficient genetic material to work well, but "too large" populations are hampered by the excessive overhead of these numbers. In our experience, the optimal population size appears between 1.5 and 2.0 times the number of parameters being optimised (i.e., dimensionality of the problem). The best ratio will depend on the problem being studied, so perhaps the 'safer' value of 2.0 should be recommended.

Whilst the key operations of recombination and mutation have been shown to be synergistic, the exact types and probability levels used are less critical. In practice, so long as "some" of these operators are used in conjunction, the evolutionary algorithm should perform well. Elitism (retention of the best or best few individuals) should be used, to guard against losing the optimum. The other operational parameters, such as selection and replacement method, tend to only have minor effects in our problems.

2.4 Davis Growth Model: Fat Deposition in Beef Cattle

In the beef production and feedlot finishing systems across southern Australia, fat deposition holds a great deal of interest. This is because intramuscular fat (i.e., marbling) increases the value of carcasses, while the other three fat depots (intermuscular, subcutaneous and visceral) are considered waste products at time of slaughter, and excessive fat trim is penalized. Conversely, carcasses containing less than about 28–30% fat are heavily discounted, so the market contains conflicting objectives [28]. An animal model that encompasses the underlying principals of biochemistry, nutrition and genetics has the potential to improve our understanding of the mechanisms of fat deposition.

Simulation of beef growth is an essential tool for maintaining a profitable beef industry, as growth rates (by different breeds, and from different feed types) largely determine the patterns of fat deposition. The Davis Growth Model encompasses the concepts of cellular hyperplasia and hypertrophy. Developed in the USA, it has been adopted to assist the beef industry make phenotypic predictions [15].

The phenotypic prediction program of the Cooperative Research Centre for Beef Genetic Technologies is using this model to predict key carcass parameters. A case study of over 500 records was conducted to assess its accuracy. Negotiations with key industry contacts are continuing, and at this time further work is still required on (a) developing these concepts with the producer groups, and (b) improving the Davis Growth Model and adapting it to Australian conditions.

Our "optimal" evolutionary algorithm, namely differential evolution, has been incorporated with the first application of the Davis Growth Model, which is a feedlot scenario. Here, an actual feedlot intake of 306 steers from seven breeds was grown out on a mixed-grain diet for between 98 and 117 days. Simulations predicted animal performance and carcass characteristics and value for between 60 and 140 days, and preliminary model runs have already found better management solutions. Against the standard "minimal input" management scenario, segregation of the animals according to initial liveweight improved the gross margin per animal by 2.2%. The first optimisation (using differential evolution) was terminated before it had converged to the global solution, but at this point it had already increased the gross margin by 2.8%. Key industry personnel are currently evaluating this "more-optimal" allocation of the intake animals to different pens, and interpreting the interacting effects of days on feed, economics, and the animals' initial liveweights, fat depths, frame sizes (largely determined by breed), and condition scores.

3 Cropping Systems in Australia

APSIM (Agricultural Production Systems sIMulator: [14]) is a crop modeling platform used widely in the international arena. Its applicable use ranges from crop genetics [10] and paddocks [32] to farm and landscape processes [5]. In this example, we address system management at the paddock and farm level.

The Agricultural Production Systems Research Unit (APSRU) is a joint venture between Australian state and national research organizations and universities, and is the instigator behind APSIMs development. APSRU has a long history of participatory research programs and, as a result, a large proportion of the research agenda is driven by external bodies. In projects related to Farming Systems management, the questions from the farmer and farm extension community to the APSRU research community come in two general forms: "Can I improve my system?"; and "What is the impact of changing {price, climate or resource availability} on my system?". Here, in consultation with a farm business, we construct a farm model that is able to answer both forms. The model building process involves iterative stakeholder engagement, with learning outcomes for all parties [2].

Outcomes from this process are rarely presented as an "optimal solution". Indeed, the most important outcome is not any "result" but a continuing dialog between parties. As such, the model becomes not a decision support tool, but a discussion support tool [25]. While not the "final word", optimisation and search techniques are an important part of the model development process, and this section describes our use of evolutionary computation techniques in this larger context.

3.1 The Problem

Dryland cropping in the northern Australian grain growing region is characterised by highly opportunistic production of a range of cereal, pulse, oilseed, forage and fibre crops. A subtropical climate allows both summer and winter crops to be grown, with yields largely determined by crop water supply from both in-season rainfall and water stored in the soil prior to planting. However, seasonal rainfall in this region is highly variable, such that the prospects for any one crop are risky [10]. To reduce dependence on rainfall, fallowing the paddock between crops is a recommended technique to accumulate moisture for subsequent crops. This, however, has consequences: financial returns are less frequent (1 crop instead of 2); and resource degradation through increased soil erosion (via reduced soil surface cover) and solute leaching (via increased soil drainage) is possible. This issue of cropping vs. fallowing demonstrates the trade-offs that exist between economic and ecological objectives.

In this environment, each planting opportunity (a rainfall event sufficient to germinate a crop) presents the farm manager with the question of whether to plant a particular crop and, if so, how much of the farm to sow to that crop. In addition to the above tradeoffs concerning water supply, many agronomic factors also influence this decision. The chance that a crop will fail (due to drought or other stress) changes through each season. For example, an early planted summer crop has a higher likelihood of being exposed to extreme temperatures during its reproductive phase than a late planted crop, with a consequent reduction of grain yield. As well, if a particular crop is planted continuously in the same paddock, crop health will be reduced via disease and pathogens. Willingness to plant in risky conditions is seen as a measure of "aggressiveness", whilst conservative practices are seen to avoid risk.

3.2 Implementation: The Farming System Model

By means of several on-farm interviews with farm managers and extension officers, we developed a model of a 2,000 ha farm near Emerald, Central Queensland, Australia (23.53 °S, 148.16 °E, 179 m elevation, 637 mm annual rainfall). The business operates a no-till cropping system comprising three major soil types: a low plant available water capacity (PAWC) soil (120 mm PAWC) in 20% of the land area; a medium (150 mm PAWC) soil in 50% of the land area; and a high (180 mm PAWC) soil in 30%. The farm is divided into ten paddocks of 200 ha each. In this study, we only consider the farm's cropping enterprises (Fig. 2), which constitutes 80–95% of grain production including sorghum, wheat, chickpea and maize. In general 1/3 of the cropping area is dedicated to winter crops, though this usually decreases during wet years. Depending on soil water, double cropping is considered though summer cropping is predominant.

The available machinery of the farm determines work rates (e.g., a planting rate of 50 ha/day), and incurs an operating cost for every activity. Capital structure (loan repayments and machinery replacement) is modeled, though in

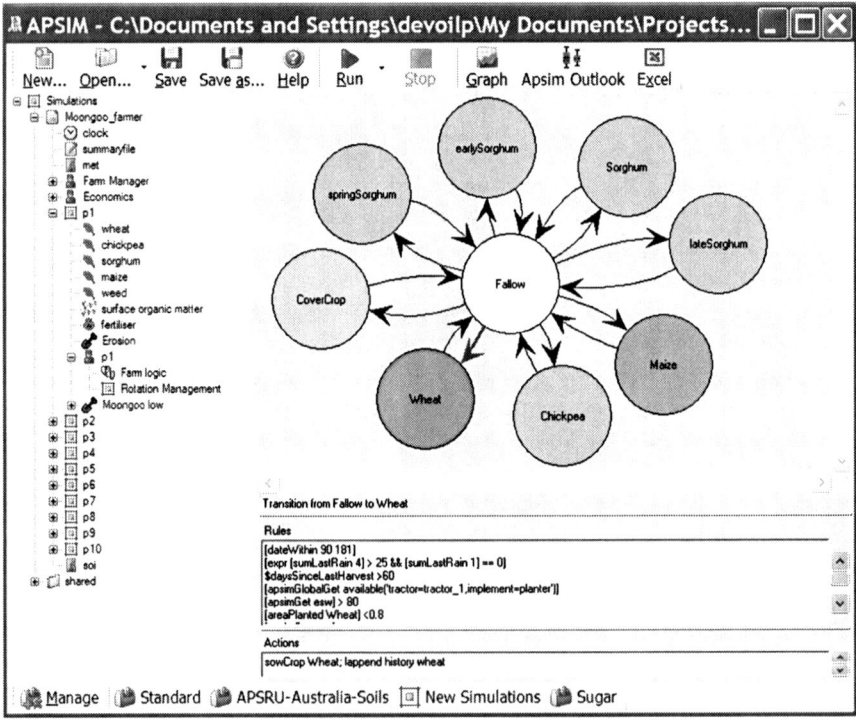

Fig. 2. An APSIM user interface showing the state transition network representing farm management. "Rules" allow transition between states, and "Actions" are taken when a transition takes place

this scenario we are not varying machinery investments. No-till systems have a higher initial capital investment and a consequent reduction in operating costs. Farm scale economics are monitored, thus the current cash status can be used as an input to management decisions.

The management of the farming system is modeled as a set of state and transition networks, with each network representing a particular paddock. Each paddock has a current state (e.g., fallow, under a particular crop) and "rules" that allow transition to adjacent states. These rules represent both feasibility (machinery is available, correct planting season) and strategy (crop type and area planted). They are usually expressed as a Boolean value (true for feasible, false otherwise), but these can also be real values, with higher values representing the desirability of that action. Each day, the model examines all paths leading away from the current state to adjacent states, and if the product of all rules associated with a path is non-zero, it becomes a candidate for action. The highest ranking path is taken, and the process repeats until nothing more can be done for that day.

Finding the critical values (how much to plant under what conditions) for these rules is the optimisation problem. But first we must decide what is "optimal". Outputs from the simulation run include production measures (individual crop yields), economic measures (annual operating return, chance of making a loss), efficiency measures (crop and whole farm water use efficiency) and environmental measures (e.g., deep drainage, runoff, and erosion). Each of these outputs becomes an "objective function" to be minimised or maximised.

The traditional approach to formulating such a problem is to assign costs to the non-monetary objectives and maximise the single profit objective – an approach which (a) hides the fact that there are tradeoffs to be made, and (b) promotes the falsity that such equivalences can be made a priori. Pareto methods [6] can be used to rank the multidimensional objective function of each individual in the population, replacing the lumped objective function value. The ranking procedure starts by finding the set of individuals that are Pareto-optimal in the current population, to which the highest rank is assigned. This set is called the first Pareto stratum. The individuals that are Pareto optimal in the remaining population are assigned the next highest rank. These individuals comprise the second Pareto stratum. This process continues until all the individuals are ranked. Figuratively speaking, the subsequent non-dominated fronts are peeled off step by step, as illustrated in Fig. 3. A complication can arise with Pareto methods when a population "collapses" to the same value in one or more dimensions. Here, the algorithm must re-rank these individuals (via recursion) over the remaining dimensions.

3.3 Implementation: The EC Algorithm

Genial [33] was initially adopted as the evolutionary computation algorithm, and we added both Pareto ranking techniques, and parallel evaluation on

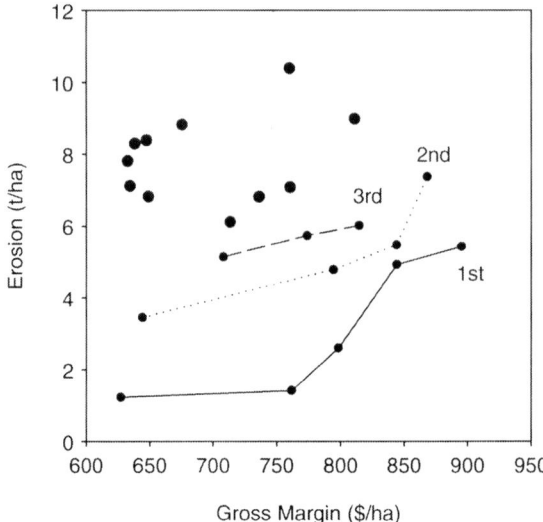

Fig. 3. Pareto ranking - the top 3 Pareto-optimal solution frontiers are joined by lines. Objectives are to maximise gross margin ($/ha) and minimise erosion (t/ha)

idle desktop PCs after hours [4]. Later, the simplifications [24] offered by differential evolution (DE: [29]) were noted, and DE has been in exclusive use since.

In simple terms, we start with the DE algorithm, replacing the objective function value with the Pareto rank of each individual in the population. We use parallel methods to evaluate new individuals within the EC population. These "individuals" are the results of APSIM simulation runs.

The APSIM "engine" (as distinct from its various user interfaces) is a command-line application that runs under both windows and UNIX. As such, controlling each model run is simply a matter of writing a simulation control/parameter file, running the application, and collecting outputs. Due to its slow runtime (typically 5 min for each of these simulations), we use a cluster of workstations to evaluate each simulation in parallel. The cluster is a shared resource managed by the Condor [13] batch processing system. We have written a communications wrapper that sits between the EC and the remote execution hosts, encoding an "individual" from the EC pool into a simulation control file, submitting the job to the system, waiting for termination, and finally reading output from each simulation back to objective function values within the EC.

This parallel EC structure is described as a "standard model" [26], where the master EC process resides on one machine, and objective function evaluation takes place on slave processors. This choice was taken as the computational effort involved in this problem rests largely on objective function evaluation, by a ratio of many thousands to one.

It is apparent that the EC must be parameterised (specifically population size) to available resources. While the (scalar) rules of thumb that exist for estimating population size still apply, efficient use of parallel evaluation indicates that the pool size should be an even multiple of the number of processors available. However, in our shared resource environment, theoretical throughput is never achieved, due to the unpredictable use of the shared resource. Neither is the method particularly efficient in terms of resource use – the overheads of submitting jobs via a batch control system are tall in comparison to other parallel systems. However, parallel evaluation represents a tradeoff on the part of researchers – the cost of a dedicated computing resource is much higher than a shared resource.

3.4 Results and Discussion

As stated, this exercise is not looking for a "final word" from the optimiser. Rather, it is used to show that the farm model, and thus our model of farmers' behavior, operates in a consistent manner. In the iterative model building process, most parameters for the farm model are elucidated from farm interviews. Candidates for optimisation are chosen where uncertainty in the farm interview is evident, or a suspicion that some gradual change has occurred in the industry. It is apparent that the farmers' attitudes to risk we are modeling are strongly influenced by economic, technological and climate changes over time.

We examine the farm model under two climate scenarios that are relevant to the farmer's experience: a wet decade (1986–1995) and a dry decade (1996–2005). We are interested in threshold values for planting crops, as listed in Table 2, and how they differ between the two decades.

The DE algorithm is configured with a population of 150 individuals, using crossover and mutation rates of 50%. There are three objective functions (cash flow, risk and soil loss), and 16 variables as described in Table 2. It is run

Table 2. Extractible soil water (ESW, in mm) required for planting crops during both wet and dry decades, and proportion of the total farm area planted (AP) to each crop

Crop	ESW wet	ESW dry	AP wet	AP dry
Chickpea	88	104	0.4	0.7
Maize	92	107	0.4	0.6
Spring sorghum	88	108	0.4	0.6
Early sorghum	87	75	0.4	0.5
Sorghum	90	78	0.4	0.6
Late sorghum	91	89	0.5	0.5
Cover crop	89	95	0.5	0.6
Wheat	90	94	0.3	0.7

ESW is bounded between 60 and 120 mm, and proportion of area between 0 and 1

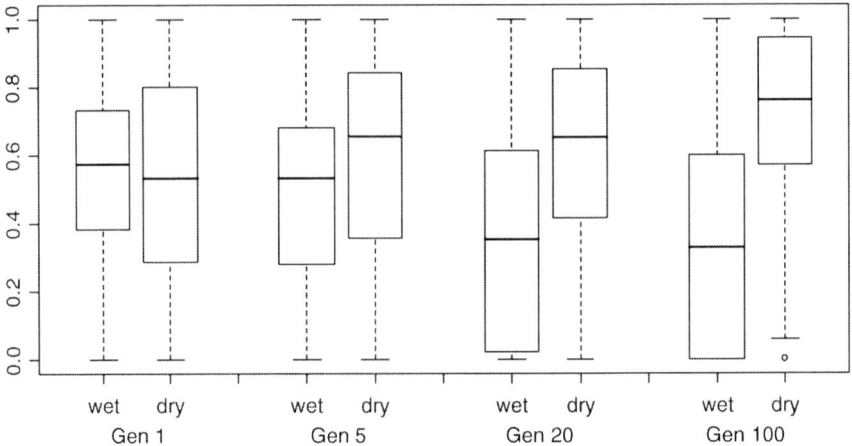

Fig. 4. Distributions of the proportion of the farm planted to wheat through two optimisation runs – one during a wet decade, the other dry

for 100 generations, though experience shows a stable population after 50 generations. The mass of data (100 generations of 150 evaluations of 10 years of 10 paddocks) precludes comprehensive data presentation here, but a sample of the EC's operation can be seen in Fig. 4, showing the distribution of one variable in both the wet and dry decades.

Table 2 shows the mean value of each variable in the top rank of the final DE population. While the threshold of water required and area planted for the main summer crops (sorghum) are about the same, we see that the area planted to winter crops (wheat & chickpea) is greater in the dry decade, though the extractible soil water threshold for chickpea is so high as to be very infrequent; and likewise for the early, high-risk summer crops (maize & spring sorghum). This same behavior was evident from the interview process – wetter decades were more favorable to summer crops, and drier decades to winter crops. The economic and environmental outputs were likewise consistent with interviews – wet decades were highly profitable, and that no-till systems are highly effective in reducing soil loss no matter what cropping intensity is applied.

4 Land-Use Planning in the Netherlands

Over the last two decades, attention in policy, land-use planning and research directed at intensively managed agricultural areas in North-Western Europe has shifted from production to provision of multiple services and functions by agriculture. Examples of multifunctional land-use aims are maintenance or improvement of landscape structure, sustainable management of renewable natural resources, preservation of biodiversity, and contribution to socioeconomic viability of rural areas. The required adjustments and innovations in landscapes and land-use systems can be characterized as complex, uncertain and value-laden issues, affecting various stakeholders. To support strategic thinking about land use options, we have been using systems approaches that integrate knowledge from different disciplines and at multiple scales. The resulting multiscale land-use or farming activities allocation problems include multiple and possibly conflicting objectives. Identifying just which trade-offs must be accepted in compromise solutions is an important outcome.

Pareto-based multiobjective optimisation methods using evolutionary computation can be useful. These methods enable straightforward and flexible model formulation including spatial interaction, while avoiding much of the mathematical complexity and rigidity that accompanies previously-used mathematical techniques such as linear or dynamic programming. Moreover, Pareto-based evolutionary computation enables simultaneous optimisation of multiple objectives without prior weighting or normalization, and also allows objectives to be expressed in their own units without the need of monetarisation of non-economic functions. The lack of guaranteed optimality associated with heuristic techniques is subordinate to these advantages, because our results feed into negotiation processes. Here, understanding the "leeway" provided by options dissimilar in land use but similar in performance levels may be more important than providing exact answers to questions that nobody actually poses.

A GIS-based optimisation methodology named Landscape IMAGES (Interactive Multigoal Agro-landscape Generation and Evaluation System) was developed [9]. This approach combines agronomic, economic and environmental indicators with biodiversity and landscape quality indicators. The following section shows the use of Landscape IMAGES to explore trade-offs and gives illustrations of applications.

4.1 Methodology

The exploration of the trade-offs between performance criteria or objectives can be formulated as a multiobjective design problem, which can be generally stated as follows:

$$\text{Max } F(x) = (F_1(x), \ldots, F_k(x))^T \tag{1}$$

$$x = (x_1, \ldots, x_n)^T \tag{2}$$

Subject to i constraints:

$$g_i(x) \le h_i \tag{3}$$

$F_1(x), \ldots, F_k(x)$ are the objective functions that are simultaneously optimised, and $(x_1, \ldots, x_n)$ are the decision variables that represent the activities allocated to the n spatial units. The decision variables can take on values from a predefined array $x \in S$, where S is the solution or parameter space. Constraints (3) can arise from the problem formulation, for instance by limitations on the inputs or outputs related to the activities.

The evolutionary strategy of differential evolution [29] was applied to a population of solutions to improve its average performance criteria generation by generation. During this iterative process, solutions are selected for each new generation on the basis of Pareto optimality [6] and reduction of crowding of solutions in sections of the solution space [3]. Because we are interested not only in the trade-off frontier, but also in the shape of the entire solution space constituted by the performance criteria, we developed a three-step approach: (1) determination of the extremes for individual performance criteria by single objective optimisation; (2) exploration of the trade-off frontiers between performance criteria by (2a) constrained single objective optimisation, followed by (2b) multiobjective optimisation; (3) homogeneous spreading of solutions within the solution space by maximising the Euclidian distance to neighboring solutions. Each step in this procedure uses results of the previous step as input.

The models were implemented in the Microsoft Visual Studio .NET development environment. An object-oriented architecture facilitates rapid extension of the framework with new optimisation problems. The model user can interact with the framework through a flexible and easy-to-use graphical user interface, namely the Model Explorer. This can be used to visualize input and output files, to adjust and view the model structure, and to show the generated landscapes.

4.2 Implementation

The Landscape IMAGES modeling approach was applied to a case study area located in the Northern Friesian Woodlands of The Netherlands. This region is characterized by a small scale landscape on predominantly sandy soils with dairy farming as the prevailing land-use activity. On some farms a limited proportion of up to 5% of the area is used for forage maize production, while the rest of the area is occupied by permanent grassland, rotationally grazed and mown. The fields with an average size of 2 ha are often surrounded by hedge rows. The selected case study area of 873 ha (and a subsection of 232 ha) is enclosed by (rail-)roads.

For the case study the landscape was compartmented into land units representing agricultural fields (polygons) and field borders (lines coinciding with polygon borders). Agricultural production activities were allocated to the fields, and field borders could be occupied by a hedgerow or remain unoccupied. An agro-ecological engineering approach was used to design production activities, which are defined as the cultivation of a crop or vegetation and/or management of a herd in a particular physical environment, completely specified by its inputs and outputs [31]. The inputs and outputs of the production activities were calculated from established empirical agro-ecological relations (see [9]). Here, we report two applications with objective functions from highly different domains.

In the first application, four objectives were evaluated – economic performance, nature value, landscape quality and nitrogen emissions:

(a) As an indicator for the economic performance of farms, gross margin was adopted. Returns from production per field were calculated directly from milk production and milk price. Costs per field were separated into costs related to grazing or mowing, fertilizer application and transport costs. Financial revenues from nature conservation packages were added to the value of the objective function for economic results. Revenues were assessed for each individual field on the basis of plant species abundance, and harvesting and fertilization regimes.

(b) Species abundance in grass swards and hedge rows was used as an indicator for nature conservation value and was estimated from an empirical relationship between nutrient availability and average species presence in grasslands.

(c) Landscape quality was related to variation in the landscape, calculated as the weighted sum of: (1) the variance of the species numbers for each field and its immediate surroundings; and (2) the half-openness of the landscape, represented by the squared deviation from 50% occupation of borders by hedges.

(d) Emission of nitrogen was calculated from the difference between uptake of N by grass and availability of N from natural soil fertility and fertilizer application.

In the second application, seven environmental functions were evaluated, representing the perspectives of different stakeholders involved in the landscape design process, and focusing on ecology (objective a), landscape quality (b, c), cultural history (d, e) and economic concerns for the short- (e, f) and long-term (g):

(a) Connectivity of the hedgerow structure is reflected as the length of the largest connected sub-graph [30], integrated over a range of dispersal distances. This was used as a measure of ecological quality and was maximised.

(b) Sight lines are defined as the distance between two consecutive hedgerows along the longitudinal axis of the fields. As the variation in sight lines determines the perception of the landscape as "half-open", this variation was maximised.

(c) Sight lines from road to road through the landscape are perceived to disturb the 'naturalness' of the landscape. These sight lines determine the landscape porosity, which was minimised.

(d) The reclamation history of the landscape caused a high ratio of hedgerows along the length of a field over hedgerows bordering the narrow side of a field, the L/T ratio. To maintain this characteristic the L/T-ratio was maximised.

(e) The removal of existing hedgerows may disrupt the historical characteristics of the landscape and is costly. Therefore, removal of hedgerows was
minimised.

(f) The addition of new hedgerows is also costly, therefore in the optimisation
the placement of new hedgerows was minimised.

(g) From the perspective of some farmers aiming to develop large-scale industrial farming systems, the presence of hedgerows forms a barrier to
maneuvering with machines and for enlargement of fields. Moreover, for
these farmers the hedgerows are an unwanted sink of labor for maintenance. To represent this business/economic perspective, minimising total
hedgerow length was included as objective.

In the application, the landscape of 2005 was compared to the current
(2007) landscape, which resulted after replanting of hedgerows based on a
plan developed by a local non-governmental organization (NGO) involved in
landscape management.

4.3 Results

In Fig. 5a the estimated solution space for the first application is described
in terms of two objectives – gross margin and nature value. Solution sets in
both applications cover a large range of possible landscape configurations in

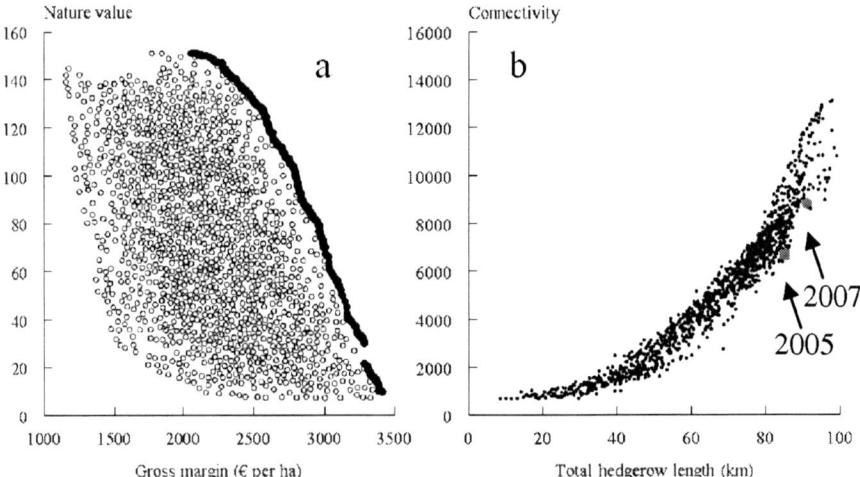

Fig. 5. (a) Solution space (○) and Pareto-optimal frontier (●) for the relation
between gross margin and nature value at landscape scale in application 1, and (b)
the relation between total hedgerow length and connectivity in application 2, generated by the Landscape IMAGES model for a sub-region of the Northern Friesian
Woodlands (The Netherlands). Original landscape (■) and landscape replanted by
NGO (●) are indicated by the grey symbols at the end of the arrows in (b)

terms of land-use on fields and the placement of hedges on field borders. In Fig. 5b this is illustrated for total hedgerow length in the solution space of the second application, which was found to be strongly but not fully correlated with landscape connectivity, a measure of ecological quality at the landscape scale. By identifying in the solution space the 2005 landscape and the restructured 2007 landscape it became clear that the decision rules employed by the NGO impacted positively on connectivity. The results made clear that further improvements would have been feasible without increasing the length (and therefore costs) of hedgerows in the landscape (Fig. 5b). The results also describe the contribution to the various objectives of one additional unit of hedgerow length, thus providing input for negotiations of the NGO with donors about biggest "bang for a buck".

Figure 6 demonstrates the different contributions to landscape quality in terms of porosity, and cultural history in terms of the L/T ratio of two landscapes similar in economic and ecological performance, expressed in hedgerow length and connectivity, respectively. The perception of the two landscapes

(a)

(b)

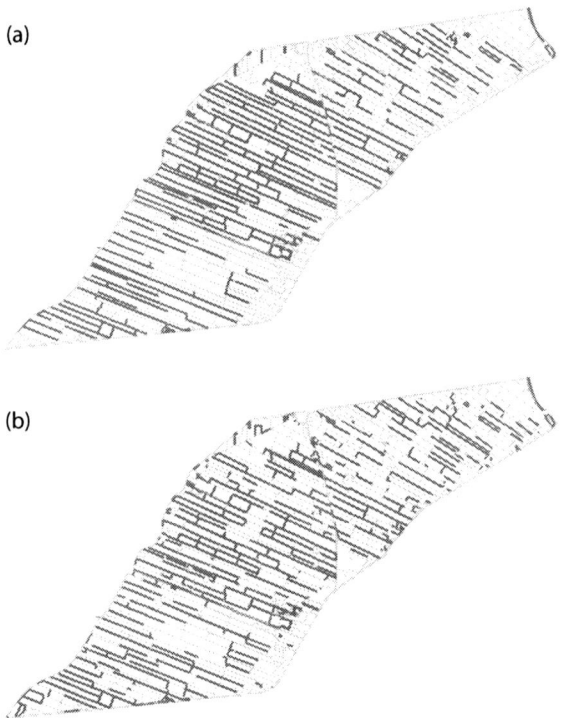

Fig. 6. (a) Original hedgerow configuration in the 873 ha case study region in the Northern Friesian Woodlands, and (b) a generated landscape with similar hedgerow length (85.6 km in a and 85.7 km in b) and connectivity (6.7 and 6.9), but strongly contrasting ratio between longitudinal and transversal hedges (8.79 in a and 5.75 in b) and porosity (45 and 15)

for visitors will be very different, highlighting the need to include multiple objectives in approaches for negotiation support.

4.4 Discussion

By exploring trade-offs among objectives, the Landscape IMAGES modeling instrument aims to reveal the "maneuvering space" of decision makers on land use issues, thus contributing to solutions that do justice to interests of broad groups of stakeholders. We have previously used linear programming based approaches [27], but found generating nearly-optimal solutions [16] more cumbersome than using Pareto-based differential evolution. An additional benefit of the latter method is its tolerance to different specifications of the optimisation problem. Programming methods fall short when faced with large combinatorial problems, as is the case in the applications in this section. Evolutionary computation is a useful compromise for the type of complex decision problems presented here where interest is more in trends and variation in the solutions than in precise optimality.

Problems of crowding have been overcome by selection of solutions in less crowded parts of the solution space based on a crowding metric [3]. This exerts a pressure tangential to the trade-off surface, thus promoting spread over the solution space [12]. In addition, the exploration of the search-spaces of order 10^{115} or larger (in the second application) was improved by enriching the initial population with extreme solutions [8]. Although the approach seems to work (cf. Fig. 5), improvement of the spread after enriching may compromise progress in the direction normal to the trade-off surface [8]. Therefore, more efficient approaches are desirable.

Another area for development is selection of subsets of solutions in the multidimensional solution space that matches viewpoints of stakeholders. Here, preference ranking on the basis of the priority assigned to objectives [1] may provide new directions, which may also be used to connect land use as described in this section to the demand for economic, ecological and social functions by society. Thus, the flexibility of multiobjective evolutionary computation offers opportunities for connecting different spatial scales as well as different scientific disciplines to create new perspectives for sustainable land use. Future applications will rely on increased algorithmic efficiency, particularly in view of sparse solutions spaces at high numbers of objectives, and techniques to select and present relevant solutions in the discussion and negotiation process.

5 Conclusions

These case studies of agricultural systems have identified a number of key points for practitioners. The first is that evolutionary computation obviously "works" across a wide range of problem types. The same certainly cannot be said for competing optimisation methods.

Secondly, we have demonstrated that the exact form of the evolutionary algorithm is not critical, with different versions and implementations all performing acceptably. Similarly, within reasonable bounds, the choice of operational parameters also is not critical. Provided that the population is large enough to contain sufficient genetic material, and that the synergistic processes of recombination and mutation are incorporated, the evolutionary algorithm should converge efficiently.

We have shown that an evolutionary algorithm can be successfully employed to search the feasible solution space for complex systems that involve multiple criteria. The concept of a Pareto-optimal frontier has proved useful for exploring trade-offs among important conflicting criteria (such as economic and environmental consequences) associated with the system design problem. The quantification of attributes of cropping systems that represent Pareto-optimal combinations of economic and environmental objectives provides highly interesting material for informed discussions on decision making with the system managers.

Evolutionary computation has revolutionized the way we conceptualize land use problems, which are typically multiobjective and multiscale and involve stakeholders with objectives that are only partially known. For these types of problems, evolutionary computation provides the opportunity to optimise trade-offs between multiple objectives simultaneously using Pareto-ranking, a feature inherent to the fact that it uses the evolutionary concept of improvement of population performance. It further allows scientists to focus on domain-specific model formulation, rather than having to invest in simplifications imposed by restrictive linear or dynamic programming approaches. And finally, the wealth of optimal and suboptimal solutions offers a rich input to discussion and negotiation processes.

References

1. Anderson SR, Kadirkamanathan V, Chipperfield A, Shafiri V, Swithenbank J (2005) Multi-objective optimization of operational variables in a waste inciration plant. Comput. Chem. Eng. 29:1121–1130
2. Carberry P, Hochman Z, McCown R, Dalgliesh N, Foale M, Poulton P, Hargreaves J, Hargreaves D, Cawthray S, Hillcoat N, Robertson M (2002) The FARMSCAPE approach to decision support: farmers, advisers, researchers monitoring, simulation, communication and performance evaluation. Agricultural Syst. 74:141–177
3. Deb K, Agrawal S, Pratap Meyarivan (2002) A fast elitist non-dominated sorting genetic algorithm for multi-objective optimization: NSGA-II. IEEE Trans. Evol. Comput. 6:182–197
4. deVoil P, Rossing WAH, Hammer GL (2003) Exploring profit – sustainability trade-offs in cropping systems using evolutionary algorithms. Proceedings of International Congress on Modelling and Simulation, Townsville, Australia

5. Gaydon D, Lisson S, Xevi E (2006) Application of APSIM 'multi-paddock' to estimate whole-of-farm water-use efficiency, system water balance and crop production for a rice-based operation in the Coleambally Irrigation District, NSW. Proceedings of 13th Agronomy Conference 2006, Perth, Western Australia
6. Goldberg DE (1989) Genetic Algorithms in Search, Optimization and Machine Learning. Addison-Wesley, Reading, MA
7. Grefenstette JJ (1990) Genesis – GENEtic Search Implementation System, Version 5 (freeware). http://www.genetic-programming.com/c2003 genesisgrefenstette.txt
8. Groot JCJ, Rossing WAH, Jellema A, Van Ittersum MK (2006) Landscape design and agricultural land-use allocation using pareto-based multi-objective differential evolution. In: A. Voinov, AJ Jakeman and A.E. Rizzoli (eds.). Proceedings of the iEMSs Third Biennial Meeting, Summit on Environmental Modelling and Software. International Environmental Modelling and Software Society, Burlington, USA, July 2006
9. Groot JCJ, Rossing WAH, Jellema A, Stobbelaar DJ, Renting H, van Ittersum MK (2007) Exploring multi-scale trade-offs between nature conservation agricultural profits and landscape quality – a methodology to support discussions on land-use perspectives. Agriculture Ecosyst. Environ. 120:58–69
10. Hammer GL, Chapman SC, Snell P (1999). Crop simulation modelling to improve selection efficiency in plant breeding programs. Proceedings of the Ninth Assembly Wheat Breeding Society of Australia, Toowoomba. pp. 79–85
11. Holmes WE (1995) BREEDCOW and DYNAMA – Herd Budgeting Software Package. Queensland Department of Primary Industries, Townsville
12. Khor EF, Tan KC, Lee TH, Goh CK (2005) A study on distribution preservation mechanism in evolutionary multi-objective optimization. Artif. Intell. Rev. 23:31–56
13. Litzkow M, Livny M, Mutka, M (1988) Condor – a hunter of idle workstations. Proceedings of the Eighth International Conference of Distributed Computing Systems, June, 1988, pp. 104–111
14. McCown RL, Hammer GL, Hargreaves GNL, Holzworth DP, Freebairn DM (1996). APSIM: a novel software system for model development, model testing and simulation in agricultural systems research. Agricultural Syst. 50:255–271
15. McPhee MJ, Oltjen JW, Famula TR, Sainz RD (2006) Meta-analysis of factors affecting carcass characteristics of feedlot steers. J. Anim. Sci. 84:3143–3154
16. Makowski D, Hendrix EMT, van Ittersum MK, Rossing WAH (2001) Generation and presentation of nearly optimal solutions for mixed integer linear programming, applied to a case in farming systems design. Eur. J. Oper. Res. 132:182–195
17. Mayer DG (2002) Evolutionary algorithms and agricultural systems. Kluwer, Boston
18. Mayer DG, Schoorl D, Butler DG, Kelly AM (1991) Efficiency and fractal behaviour of optimisation methods on multiple-optima surfaces. Agricultural Syst. 36:315–328
19. Mayer DG, Belward JA, Burrage K (1996) Use of advanced techniques to optimize a multi-dimensional dairy model. Agricultural Syst. 50:239–253
20. Mayer DG, Belward JA, Burrage K (1998) Optimizing simulation models of agricultural systems. Ann. Oper. Res. 82:219–231

21. Mayer DG, Belward JA, Burrage K (1999) Performance of genetic algorithms and simulated annealing in the economic optimization of a herd dynamics model. Environ. Int. 25:899–905
22. Mayer DG, Belward JA, Widell H, Burrage K (1999) Survival of the fittest – genetic algorithms vs evolution strategies in the optimization of systems models. Agricultural Syst. 60:113–122
23. Mayer DG, Belward JA, Burrage K (2001) Robust parameter settings of evolutionary algorithms for the optimisation of agricultural systems models. Agricultural Syst. 69:199–213
24. Mayer DG, Kinghorn BP, Archer AA (2005) Differential evolution – an easy and efficient evolutionary algorithm for model optimisation. Agricultural Syst. 83:315–328
25. Nelson RA, Holzworth DP, Hammer GL, and Hayman PT (2002). Infusing the use of seasonal climate forecasting into crop management practice in North East Australia using discussion support software. Agricultural Syst. 74:393–414
26. Price KV, Storn RM, Lampinen JA (2005) Differential Evolution. A Practical Approach to Global Optimization. Springer-Verlag, Berlin
27. Rossing WAH, Jansma JE, De Ruijter FJ, Schans J (1997) Operationalizing sustainability: Exploring options for environmentally friendly flower bulb production systems. Eur. J. Plant Pathol. 103:217–234
28. Sainz RD, Hasting E (2000) Simulation of the development of adipose tissue in beef cattle. Modelling Nutrient Utilization in Farm Animals, pp. 175–182. CABI Publishing, New York
29. Storn R, Price K (1995) Differential evolution – a simple and efficient adaptive scheme for global optimization over continuous spaces. International Computer Science Institute, Berkeley USA, pp. 12. Technical Report TR-95-012
30. Urban D, Keitt T (2001) Landscape connectivity: a graph-theoretic perspective. Ecology 82:1205–1218
31. Van Ittersum MK, Rabbinge R (1997) Concepts in production ecology for analysis and quantification of agricultural input–output combinations. Field Crops Res. 52:197–208
32. Wang E, Robertson MJ, Hammer GL, Carberry PS, Holzworth D, Meinke H, Chapman SC, Hargreaves JNG, Huth NI, McLean G (2002) Development of a generic crop model template in the cropping system model APSIM. Eur. J. Agron. 18:121–140
33. Widell, H (1998) Genial, A friendly evolution algorithm package for function optimization. http://hjem.get2net.dk/widell/genial.htm

Evolutionary Electronics: Automatic Synthesis of Analog Circuits by GAs

Esteban Tlelo-Cuautle and Miguel A. Duarte-Villaseñor

INAOE. Luis Enrique Erro No. 1. Tonantzintla, Puebla. 72840 Mexico,
e.tlelo@ieee.org

Summary. This paper introduces guidelines for the automatic synthesis of analog circuits by performing evolutionary operations. It is shown that the synthesis of unity-gain cells (UGCs) can be done from nullor-based descriptions. In this manner, UGCs such as: the voltage follower (VF), current follower (CF), voltage mirror (VM), and current mirror (CM), are described using a new genetic representation consisting of ordered genes. Furthermore, a genetic algorithm (GA) is introduced in order to search for the best UGC by performing crossover and mutation operations, and by selecting UGCs by elitism. On the other hand, since GAs operate on the principle of survival of the fittest, the proposed GA has the capability to generate new design solutions, i.e. new UGCs. Additionally, the synthesized UGCs are evolved to design more complex circuits, namely: current conveyors (CCs), inverting CCs (ICCs), and current-feedback amplifiers (CFOAs). Finally, some analog circuit evolution approaches are described to synthesize practical applications such as: active filters, single-resistance controlled oscillators (SRCOs), and chaotic oscillators, which are implemented using UGCs, CCs, and CFOAs.

1 Introduction

The proper aspects of the life, as the self-organization, adaptation and evolution, are giving place to a revolution in various engineering areas. Artificial life (Alife) is the name given to this new discipline which studies natural life to recreate biological phenomena by using computers and other artificial systems. The main goal is to apply biological principles to hardware and software technology, medicine, nanotechnology, and multiple engineering projects such as evolutionary electronics [1–3], where Alife is giving place to an authentic heuristic revolution on the automatic design of electronic circuits [4–12].

Since the world is fundamentally analog in nature, all the electronic systems that interface with the external world require analog circuits, such as: cellular telephones, magnetic disk drives and speech recognition systems [4]. Even though analog design automation (ADA) began four decades ago it is

E. Tlelo-Cuautle and M.A. Duarte-Villaseñor: *Evolutionary Electronics: Automatic Synthesis of Analog Circuits by GAs*, Studies in Computational Intelligence (SCI) **92**, 165–187 (2008)
www.springerlink.com   © Springer-Verlag Berlin Heidelberg 2008

still in its infancy compared to the digital domain [5]. That way, new techniques are required to improve the ADA process in order to shorten the time to market [6], to enhance the quality and optimality of integrated circuits (ICs) designs, and to reduce production costs through increased manufacturability.

Analog design is much amenable for evolutionary techniques, where contrasting with digital design, there is no solid set of design rules or procedures to automate circuit synthesis [3,7]. For instance, in [1,3,7–12], have been introduced synthesis methodologies based on genetic algorithms (GAs). Since GAs operate on the principle of "survival of the fittest" [1], they have the capability to generate new design solutions from a population of existing solutions, and discarding the solutions which have an inferior performance or fitness. GAs begin with an initial collection of random solutions called initial population. Each individual in the population is called chromosome and represents a possible solution to the problem. A chromosome is a chain of symbols called genes, which generally are represented by binary strings. The chromosome evolves through iterations called generations. In each generation the chromosomes are evaluated using an aptitude measure. The next population is formed by descendents created by combining two chromosomes of the current generation using the crossover and the mutation operators. Henceforth, this chapter introduces guidelines to automate the synthesis of practical analog ICs by applying GAs, and by using the nullor element [13] to codify the abstract behavior of analog circuits.

2 Analog Design Automation

In electronics, complete systems that occupied one or more boards are now integrated on a few chips or even on one chip [4,5]. This has put significant pressure on the development of synthesis methodologies to enhance ADA. Furthermore, this section describes the usefulness of intelligent systems for the development of new synthesis tools to design practical analog ICs.

2.1 Circuit Synthesis

The goal of analog circuit synthesis is devoted to make the transition from research to practice. This process seeks to transform abstract descriptions into a working circuit. On this direction, a complex design can be carried-out by partitioning or modularizing the design into blocks or cells that can be shared and reassembled in a manageable and repeteadable form [6–12]. To do that, an ADA developer should to perform the ability to view the design into three levels: system-level, circuit-level and layout synthesis [6]. Elsewhere, analog synthesis can be performed in a top-down approach by beginning with module descriptions, and further by performing a refinement process until obtaining a practical analog IC.

In performing analog circuit synthesis, the selection of a correct topology is one of the main bottlenecks [1–17]. Furthermore, analog design is much more amenable for evolutionary techniques because the search space is smoother [2]. The reason is that the potential for analog signal processing is much greater than what is exploited today. Even a single transistor has many functions that can be exploited to design complex ICs [8–10]. Also, the basic combinations of transistors offer a rich repertoire of linear and nonlinear operators available for analog signal processing applications [9, 12, 14].

2.2 Intelligent Systems

Given the challenges of analog design problems and the shortages of analog design engineers, there are strong and renewed economic pressures to develop new ADA techniques [6, 12]. Many of the ADA techniques have been based on intelligent systems, such as: simulated annealing, fuzzy logic, neural networks, and evolutionary computation [2–8]. The last one is well suited for circuit synthesis because it presents well defined but difficult design problems within the very large space in which analog design operates, e.g. offset, impedances, biasing, sizing, noise restrictions, and several others [4]. In this manner, since topology selection and synthesis are fundamentally optimization-oriented problems, this chapter introduces a new genetic representation [15, 16] to synthesize and to evolve analog cells from abstract descriptions using nullors [13].

The proposed GA begins with the generation of circuits consisting of one to four transistors to synthesize unity-gain cells (UGCs): voltage followers (VFs) [14–16], voltage mirrors (VMs) [18], current followers (CFs) [17], and current mirrors (CMs) [19]. It is shown the evolution of nullor-based circuit descriptions until obtaining known and novel ICs useful for analog signal processing applications. Furthermore, the behavior of the UGCs is modeled by using pathological elements [13–20], and its codification is based on binary strings, but the chromosomes consist of ordered genes. The decodification process consist to synthesize the chromosomes by transistors designed using standard CMOS IC technology [21]. During the synthesis process, SPICE[1] is used to evaluate the fitness, and further the GA performs crossover and mutation operations to generate new circuit topologies which are selected by elitism. The synthesized UGCs are evolved to design amplifiers [22], namely: current conveyors (CCs) [23–25], inverting CCs (ICCs) [20, 26], and current-feedback amplifiers (CFOAs) [27]. At the end of this chapter, some analog IC evolution approaches are described to synthesize active filters [4, 27], single-resistance controlled oscillators (SRCOs) [28, 29], and chaotic oscillators [30, 31].

---

[1] Simulator available in http://bwrc.eecs.berkeley.edu/Classes/IcBook/SPICE/.

3 Synthesis of Unity-Gain Cells by GAs

The nullator (O), the norator (P) and the nullor, are pathological elements which have been quite useful for the development of novel analysis, synthesis and design methodologies [13]. As demonstrated in [22], these elements can be used to represent a variety of different active devices, such as: BJT, MOSFET, Operational amplifiers, CC, VF, CF, etc. Additionally, the VM and CM descriptions have been introduced in [20, 26], to model the behavior or mirroring elements without the use of resistances.

3.1 Synthesis of the Voltage Follower

One or several Os can be used to model the abstract behavior of the VF, as shown in Fig. 1. However, for synthesis purposes each O must be joined with a P, so that there exist three combinations to generate O–P pairs as shown in Fig. 2. Each O–P pair is codified by two bits to create the gene of small-signal (genSS) described in Table 1. Although 00 and 11 are referred to Fig. 2a, 00 means that during the synthesis of the O–P pair, the source (S) of the MOSFET is associated to node i, while 11 to node j, as shown by Fig. 2d.

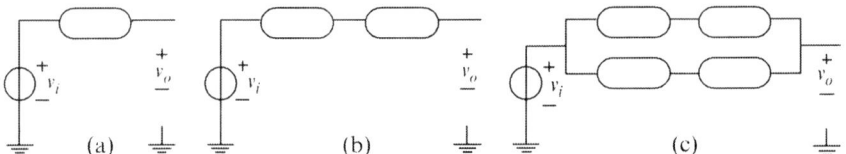

Fig. 1. Modeling the ideal behavior of the VF using nullators

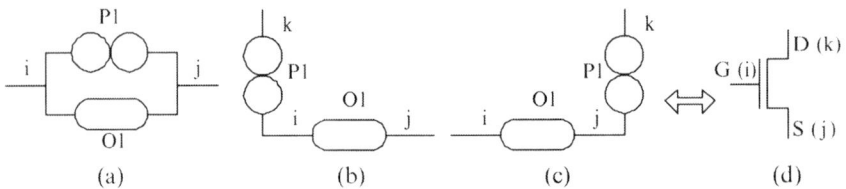

Fig. 2. Forming O–P pairs: (**a**) Adding P between node i and j, (**b**) at node i, (**c**) at node j, and (**d**) synthesis of an O–P pair by a MOSFET

Table 1. Codification of genSS from Fig. 2

bit0	bit1	Union of O–P
0	0	Node i (Fig. 2a)
0	1	Node i (Fig. 2b)
1	0	Node j (Fig. 2c)
1	1	Node j (Fig. 2a)

The synthesis of an O–P pair is codified by one bit to create the gene of synthesis of the MOSFET (genSMos). Thus when genSMos = 0 the O–P pair is synthesized by a N-MOSFET and when genSMos = 1 by a P-MOSFET.

Each MOSFET is biased with the addition of voltage and current sources. This is codified by three bits to create the gene of bias (genBias). Current biases are added to Drain (D) and Source (S) terminals, as shown by Fig. 3.

The synthesis of current biases is codified by a gene of CMs (genCM) whose length depends on the number of available CMs. As a result, the VF is codified by the chromosome consisting of four ordered genes given in (1).

For instance, if all currents in Fig. 3 are synthesized by simple CMs, the synthesized VFs are shown in Fig. 4. Furthermore, the search space of the GA by beginning with Fig. 1c becomes: $K^{8 \times 4 \times 12 \times 1} = 33{,}554{,}432$ individuals, where $K = 2 =$ binary alphabet.

$$Chromosome_{VF} = \underbrace{genSS}_{OPpairs} * \underbrace{genSMos}_{MOSFET} * \underbrace{genBias}_{IdealBias} * \underbrace{genCM}_{KindCM} . \qquad (1)$$

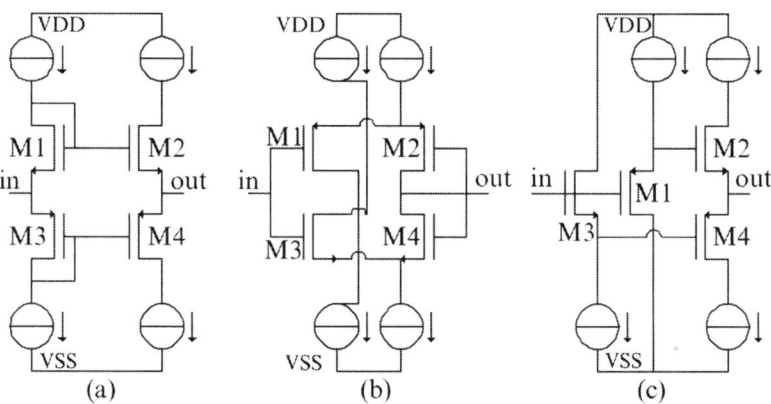

Fig. 3. Addition of current biases to three practical VFs

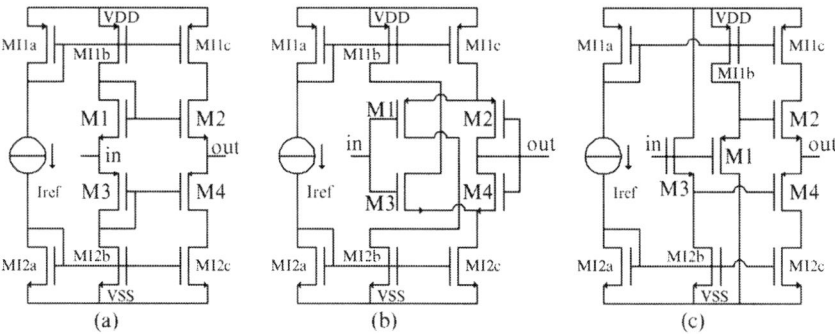

Fig. 4. Synthesis of three practical CMOS compatible VFs from Fig. 1c

The proposed GA consists to implement search strategies in (1) to synthesize practical VFs. Therefore, the GA is summarized as follows [15, 16]:

- A start-population (POP_n) consisting of chromosomes (VF topologies) described by (1), is randomly created.
- Each VF topology is evaluated by applying knowledge rules from [14–16], and using genSS and genBias to verify that neither the input- nor the output-port of the VF be connected to either VDD or VSS.
- Fitness 1: Each selected topology is sized and biased with ideal current sources by using standard CMOS technology, as shown in [21]. Here, genSS, genSMos and genBias are decoded to execute SPICE simulations to select VFs accomplishing (2), where k has a default value of 0.7.

$$v_{out}(s) > k \times v_{in}(s). \tag{2}$$

- Fitness 2: The ideal currents in each selected VF are synthesized by CMOS CMs using genCM. The VFs are simulated again using SPICE to evaluate (2), but now k has a default value of 0.9 to select practical VFs.
- In POP_n (parents), some chromosomes are randomly chosen to recombine two by two, and to mutate some of them to generate a new population (childs) [2], to begin a new cycle until obtaining VFs accomplishing (2). To preserve the average-number of individuals, only 2–4 new chromosomes are created by applying equal probability, and also only 1–4 new chromosomes are mutated randomly by changing one bit [15].

By executing this GA, the behavior on the number of individuals in each generation is shown in Fig. 5a. From the nature of the GAs, our proposed GA generates new and novel VFs from a population of existing solutions codified by (1), i.e. chromosomes that describe a VF topology. The chromosomes evolve through iterations using the aptitude measure given by (2), as shown in Fig. 5b, where fitness 1 is the SPICE simulation using ideal current biases, and fitness 2 is the SPICE simulation when the VF is full-synthesized with

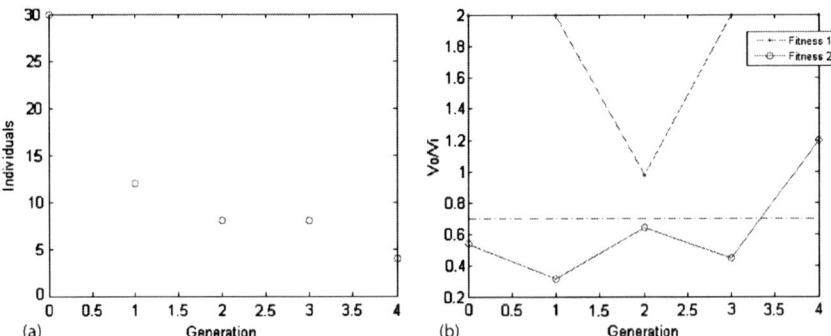

Fig. 5. (a) Individuals in each generation and (b) evaluation of fitness 1 and 2

MOSFETs. As one sees, the first fitness selects VFs accomplishing (2), but they do not accomplish the second fitness when genCM is synthesized by MOSFETs (CMs). In Fig. 5a can be appreciated the reduction on the number of individuals when crossover and mutation operations are applied. This means that the GA has a good convergence behavior, as already shown in [15]. For instance, in [14] are shown six synthesized CMOS compatible VFs, three of them are shown herein in Fig. 4. The one in Fig. 4a is codified by (3) and has been used in [17,21] to design CCs, and in [27,31] to design CFOAs. Also, the one in Fig. 4c has been used in [29] to design SRCOs.

$$Chromosome_{Fig.\,4a} = \underbrace{00,10,00,10}_{OPpairs}\ \underbrace{0,0,1,1}_{MOSFET}\ \underbrace{110,110,011,011}_{IdealBias}\ \underbrace{0}_{KindCM}\ .\quad (3)$$

3.2 Synthesis of the Current Follower

The GA for the synthesis of VFs can be extended to synthesize CFs by beginning with P descriptions, as shown in Fig. 6. Now, each P must be joined with an O, to generate O–P pairs which are codified by two bits to create the gene of small-signal (genSS), described by Table 1.

The synthesis of an O–P pair is also codified by one bit to create the gene of synthesis of the MOSFET (genSMos).

Besides, now each MOSFET is biased with the addition of voltage and current sources. The addition of currents is codified by the gene of Ibias (genIBias), as done for the VF with genBias described by Table 2. On the other hand, voltage biases are added to the gate (G) of the MOSFETs. This process is codified by genVBias with length equal to the number of Ps.

The synthesis of current biases is also codified by genCM. As a result, the CF is codified by the chromosome consisting of five ordered genes and given

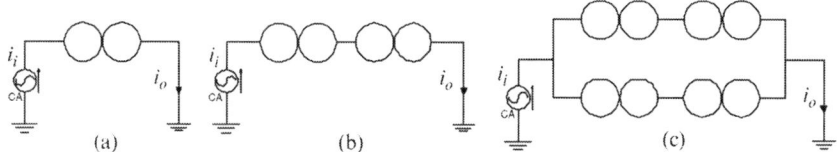

(a)   (b)   (c)

Fig. 6. Modeling the ideal behavior of the CF using norators

Table 2. Codification of genBias

Voltage	Node m	Node n	Biasing MOSFET$_i$
1	0	0	No addition of bias
1	0	1	Addition of current between VDD and node n
0	1	0	Addition of current between node m and VSS
0	1	1	Addition of two currents from VSS to VDD

in (4). Furthermore, the search space of the GA by beginning with Fig. 6c becomes: $K^{8 \times 4 \times 12 \times 4 \times 1} = 536{,}870{,}912$ individuals.

$$Chromosome_{CF} = \underbrace{genSS}_{OPpairs} * \underbrace{genSMos}_{MOSFET} * \underbrace{genIBias}_{IdealIBias} * \underbrace{genVBias}_{IdealVBias} * \underbrace{genCM}_{KindCM}. \qquad (4)$$

As for the VF, the length of (4) increments with the number of available CMs (genCM). Furthermore, the proposed GA implement search strategies in (4) to synthesize practical CFs. The GA begins with a start-population (POP_n) consisting of chromosomes randomly generated. Each CF topology is evaluated by using genSS and genBias. Each selected topology is sized and biased with ideal sources by using genSS, genSMos, genIBias and genVBias to make a SPICE simulation to select those CFs accomplishing (5), where k has a default value of 0.7. The ideal sources of each selected CF are synthesized by CMOS CMs using genCM, so that the CFs are simulated again using SPICE to evaluate (5) with $k = 0.9$. In POP_n (parents), some chromosomes are randomly chosen to recombine two by two, and to mutate some of them to begin a new cycle.

$$i_{out}(s) > k \times i_{in}(s). \qquad (5)$$

By executing this GA, in [17] is presented a practical CMOS compatible CF which is shown in Fig. 7b. The CF in Fig. 7a is the well-known common-gate configuration which was synthesized from Fig. 6a.

3.3 Synthesis of the Voltage Mirror

As demonstrated in [26], despite the ability of Os and Ps to represent analog devices, they fail to represent the positive-type second-generation CC (CCII+), and ICC. That way, in order to eliminate the use of resistors which are combined with Os and Ps to derive the nullor representation of the CCII+ and ICC, the VM and the CM were introduced in [18–20] to describe the behavior of active devices with voltage or current reversing properties.

There are two generic topologies to describe the VM, as shown in Fig. 8 [26,32,33]. Basically, it consists of a VF with two complementary MOSFETs.

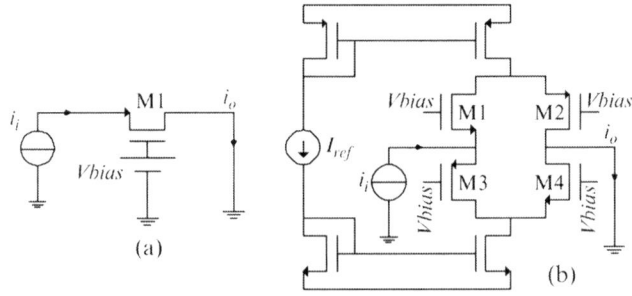

Fig. 7. Synthesis of (a) a CF from Fig. 6a, and (b) from Fig. 6c

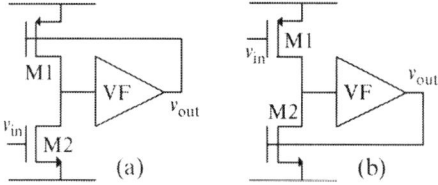

Fig. 8. Synthesis of a VM

In this manner, the genetic representation of the VM includes the chromosome of the VF plus one gene of inversion (genInv) of one bit to codify the connection of the VF to the N- or P-MOSFET. As a result, the VM is codified by the chromosome consisting of five ordered genes given in (6). From Fig. 1c, the search space becomes: $K^{8 \times 4 \times 12 \times 1 \times 1} = 67{,}108{,}864$ individuals.

$$Chromosome_{VM} = \underbrace{genSS}_{OPpairs} * \underbrace{genSMos}_{MOSFET} * \underbrace{genBias}_{IdealBias} * \underbrace{genCM}_{KindCM} * \underbrace{genInv}_{MOSFET} . \quad (6)$$

As for the VF and CF, the length of (6) increments with the number of available CMs (genCM). To synthesize practical VMs the proposed GA begins with a start-population (POP_n) consisting of chromosomes randomly generated. Each VM topology is evaluated by using genSS, genBias and genInv. Each selected topology is sized and biased with ideal sources by using genSS, genSMos, genBias and genInv to make a SPICE simulation to select those VMs accomplishing (7), where k has a default value of 0.7. The ideal sources of each selected VM are synthesized by CMOS CMs using genCM, so that the VMs are simulated again using SPICE to evaluate (7) with $k = 0.9$. In POP_n (parents), some chromosomes are randomly chosen to recombine two by two, and to mutate some of them to begin a new cycle. For instance, by executing this GA, in Fig. 9a is presented a practical CMOS compatible VM which embodies the VF shown in Fig. 4c to synthesize the VM in Fig. 8a. This VM could be used in [32–34].

$$v_{out}(s) < -k \times v_{in}(s). \quad (7)$$

3.4 Synthesis of the Current Mirror

The CM by itself forms a basic building block in IC design [4]. As shown in the previous subsections, it is required to bias analog circuits, e.g. VF, CF and VM. The genetic representation of the CM begins with P descriptions, as shown in Fig. 10a. Again, each O–P pair is codified by two bits to create genSS, described in Table 1. The synthesis of an O–P pair is codified by one bit only to create genSMos, as shown in Fig. 10b. Thus genSMos = 1 synthesizes all O–P pairs by N-MOSFETs, else by P-MOSFETs.

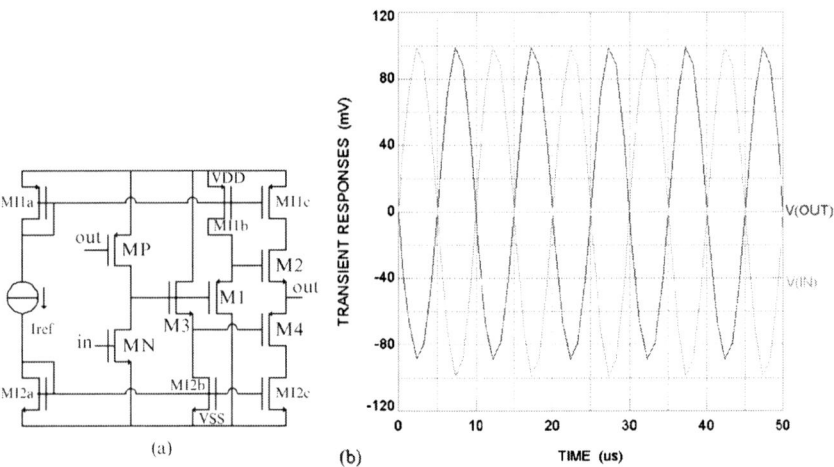

Fig. 9. (a) Synthesized VM and (b) its inverting time response

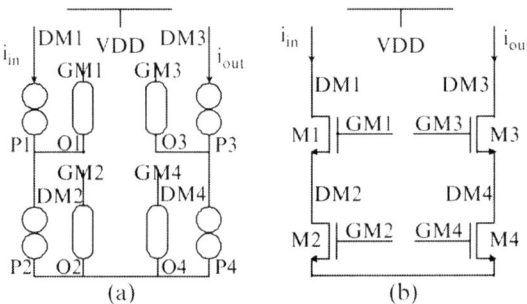

Fig. 10. (a) O–P description of a CM and (b) its transistor equivalent

Two current sources are added to bias the input and output-port of the CM. However, the addition of voltage biases is a complicated task because there are four gates (GMi), and they can be connected to five voltage nodes: V_{DM1}, V_{DM2}, V_{DM3}, V_{DM4}, and V_{DD}. This process is codified by genVBias with length $= 3$ bits for each P in genSS. As a result, the CM is codified by the chromosome consisting of three ordered genes given in (8). By beginning with Fig. 10a the search space becomes: $K^{8\times1\times12} = 2,097,152$ individuals.

$$Chromosome_{CM} = \underbrace{genSS}_{OPpairs} * \underbrace{genSMos}_{MOSFET} * \underbrace{genVBias}_{IdealVBias}. \tag{8}$$

The length of (8) only depends on the number of Ps in genSS. Furthermore, the proposed GA begins with a start-population (POP_n) consisting of chromosomes randomly generated. Each CM topology is evaluated by using only genSS. Each selected topology is sized and biased with ideal sources by using genSS, genSMos, and genVBias to make a SPICE simulation to select CMs

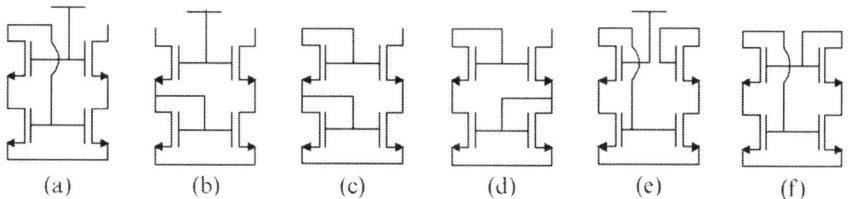

Fig. 11. Six CMs synthesized from Fig. 10a

accomplishing (9), where k has a default value of 0.9. In POP_n (parents), some chromosomes are randomly chosen to recombine two by two, and to mutate some of them to begin a new cycle.

$$i_{out}(s) < -k \times i_{in}(s). \tag{9}$$

By beginning from Fig. 10a, the GA applies two knowledge-rules:

- GM2 and GM4 should be connected together for mirroring behavior.
- Neither V_{GM1} nor V_{GM3}, can be biased to either V_{DM2} or V_{DM4}, because they provide lower voltage bias levels.

In Fig. 11 are shown six synthesized CMs. The CMs (a) and (b) present the lower input and output voltage requirements, (c) and (d) present the higher input voltage requirements, and (e) and (f) the higher output ones. The CM in Fig. 11a is the low-voltage cascode CM, Fig. 11c is the simple cascode CM, and Fig. 11d is the Wilson CM. The remaining topologies are new CMs.

4 Synthesis of Amplifiers by Evolving Unity-Gain Cells

The VF, CF, VM and CM form the basic set of UGCs which can be evolved to design more complex circuits. For instance, the superimposing [9] of a VF with a CF generates the negative-type second generation CC (CCII−) [17]. The connection of a VF with CMs [21] generates the CCII+; and the cascade connection of the CCII+ with a VF synthesizes the CFOA [27].

4.1 Synthesis of Current Conveyors

The CC was introduced by Sedra and Smith in 1968 [23], as a new circuit building block useful to implement a wide variety of analog signal processing applications [4, 11, 17, 24]. In particular, the second generation CC (CCII), provides better performances to implement those analog applications already realized with conventional operational amplifiers (opamps). In this manner, since 1997 [25], the design efforts has been focused to introduce a perfect CMOS CCII to accomplish (10), where K takes two values: K = 1 to describe the ideal behavior of the CCII−, and K = −1 to describe the ideal behavior of the CCII+. The generic CCII is shown in Fig. 12a. As one sees, the main

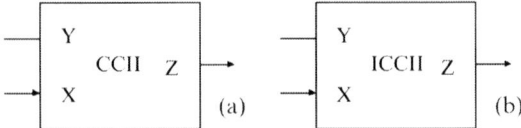

Fig. 12. Description of a: **(a)** CCII and **(b)** inverting CCII (ICCII)

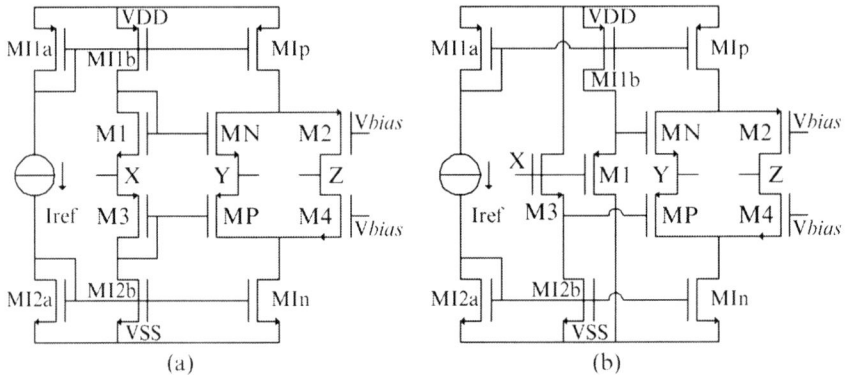

Fig. 13. Two CCII−s synthesized from Fig. 4 and Fig. 7b

difference between the CCII− and CCII+ is the direction of the current in terminal Z. Besides, (10) shows that $i_y = 0$ and $v_x = v_y$ can be synthesized by the VF, and $i_z = K \times i_x$ can be synthesized by looking to (5) and (9), where the CCII− is generated by using CFs, while the CCII+ by using CMs.

$$\begin{bmatrix} i_y \\ v_x \\ i_z \end{bmatrix} = \begin{bmatrix} 0 & 0 & 0 \\ 1 & 0 & 0 \\ 0 & K & 0 \end{bmatrix} \begin{bmatrix} v_y \\ i_x \\ v_z \end{bmatrix}. \tag{10}$$

As one can infer, the simplicity of the architecture of the CCII relies on that it can be synthesized by combining a VF with either a CF or CMs. Furthermore, the search space of the GA is very huge since one should to combine the chromosome of the VF given in (1), with either the chromosome of the CF given in (4), or the chromosome of the CM given in (8). For instance, the superimposing [9] of the VFs in Fig. 4, with the CF in Fig. 7b, generates the two CCII−s shown in Fig. 13. That is, the MOSFETs M2 and M4 in the VFs are superimposed with M1 and M3 in the CF. However, only the VFs in Fig. 4a and Fig. 4c can be superimposed with the CF. In both CCII−s, the superimposed MOSFETs are labeled by MIp, MN, MP and MIn. The CCII− in Fig. 13a has been synthesized in [17], while the one shown in Fig. 13b is a novel and new CMOS compatible CCII−.

On the other hand, the synthesis of the CCII+ can be performed by evolving the chromosome of the VF given in (1). In the first step the GA synthesizes VFs using genSS, genSMos and genBias, as shown in Fig. 3. The second step

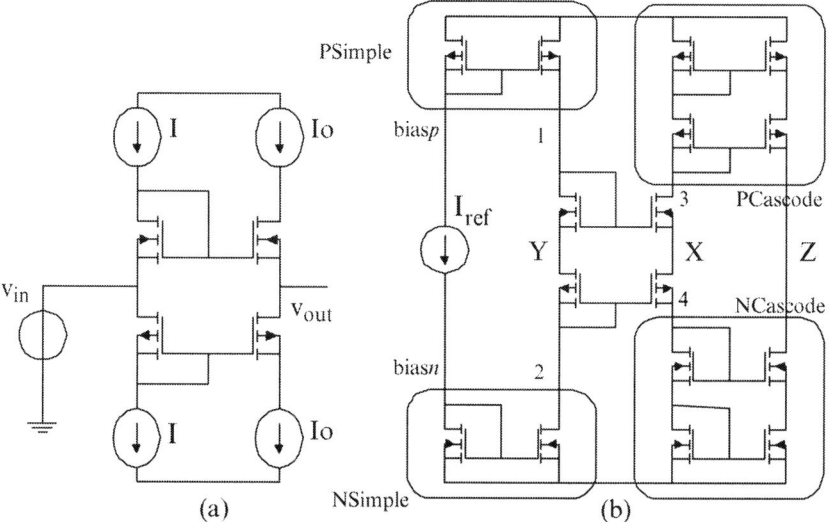

Fig. 14. Evolution of a VF to a CCII+

Table 3. Netlist of the CCII+ from Fig. 14b

Element	Node	Node	Node	Node	Model	Value
M1	1	1	Y	VSS	MODN	
M2	3	1	X	VSS	MODN	
M3	2	2	Y	VDD	MODP	
M4	4	2	X	VDD	MODP	
VDD	VDD	0				1.5
VSS	VSS	0				−1.5
Iref	biasp	biasn				20 u
X1	biasp	1	VDD		PSimple	
X2	biasn	2	VSS		NSimple	
X3	3	Z	VDD		PCascode	
X4	4	Z	VSS		NCascode	

evolves genCM by synthesizing current biases. For example: Fig. 14a shows a VF with ideal biases, and Fig. 14b shows the synthesis of I by simple CMs (PSimple and Nsimple), and the synthesis of Io by simple cascode CMs (PCascode and NCascode). In [31], this process is called synthesis of the CCII+ by sandwiching a VF between two CMs. However, the search space of the GA is very huge since the chromosome of the CCII+ combines the chromosomes of the VF and of CM. Table 3 shows the decodification of Fig. 14b. The SPICE netlist is generated as already shown in [14], but now each ideal current source is declared as a subcircuit, e.g. PSimple, Nsimple, PCascode and NCascode. X.. calls the .SUBCKT SPICE description of each CM. For example, the

module of the CM NCascode in Fig. 14b, is declared as follows:
.SUBCKT NCascode Input Output VBias
M1 Input Input 1 VBias MODN W=12u L=1u
M2 1 1 VBias VBias MODN W=12u L=1u
M3 Output Input 2 VBias MODN W=12u L=1u
M4 2 1 VBias VBias MODN W=12u L=1u
.ENDS

4.2 Synthesis of Inverting Current Conveyors

There have been significant advances in the last four decades in linear analog
ICs applications since the introduction of the opamp [4]. The applications
of the opamps have been enhanced by using current-mode amplifiers [22],
which offers a wider bandwidth than the classical opamp, and additionally,
the CC and ICC do not present the trade-off between the gain-bandwidth
product of the opamp [4,22]. In this manner, both CCs and ICCs have received
considerable attention from analog circuit designers.

A negative-type inverting CCII (ICCII−) was introduced in [26], and its
applications along with that of the inverting CCII+ (ICCII+) are given in
[32–34]. The generic ICCII is shown in Fig. 12b, and its ideal behavior is
modeled by (11), where it can be appreciated that contrary to CCIIs, now
$v_x = -v_y$ which is accomplished by the VM. Also, K takes two values: K = 1
to describe the ICCII−, and K = −1 to the ICCII+. The ICCIIs can be
synthesized by combining a VM with either a CF or CMs. Therefore, the
search space of the GA is very huge because it combines the chromosome of
the VM given in (6), with either that of the CF given in (4), or the CM given
in (8). The superimposing [9] of the VM in Fig. 9a, with the CF in Fig. 7b,
generates the ICCII− shown in Fig. 15a, while by using simple CMs, generates
the ICCII+ shown in Fig. 15b. Both ICCIIs are novel and news.

$$\begin{bmatrix} i_y \\ v_x \\ i_z \end{bmatrix} = \begin{bmatrix} 0 & 0 & 0 \\ -1 & 0 & 0 \\ 0 & K & 0 \end{bmatrix} \begin{bmatrix} v_y \\ i_x \\ v_z \end{bmatrix}. \tag{11}$$

4.3 Synthesis of Current-Feedback Operational Amplifiers

As already shown in [4, 22, 27, 31], the CFOA is synthesized by cascading a
CCII+ with a VF, as shown in Fig. 16a. In this manner, the GA combines
the chromosomes of the CCII+ with the VF. But, since the CCII+ combines
the VF and CM, then it results in a very huge search space because the VF
embedded into the CCII+ can be different of that shown in Fig. 16a. The
CFOA accomplishes (12), where $v_x = v_y$ and $v_w = v_z$ can be synthesized
by VFs, and $i_z = -i_x$ by CMs. For instance, the combination of the VF
synthesized in Fig. 4c with the simple CMs shown in Fig. 14b (PSimple and

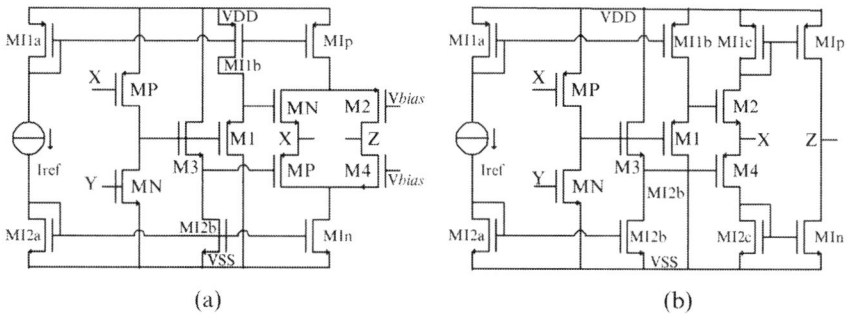

Fig. 15. Synthesis of the (**a**) ICCII− and (**b**) ICCII+

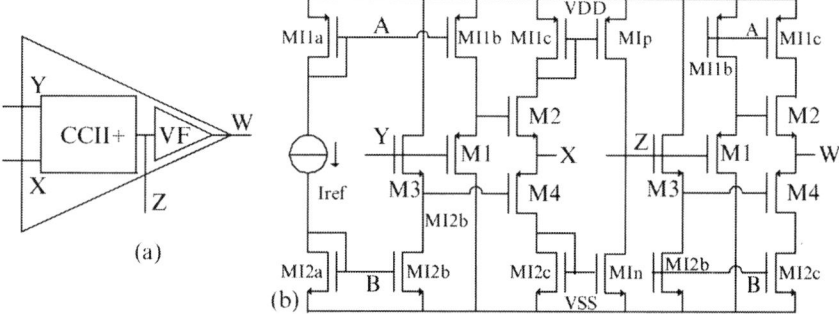

Fig. 16. (**a**) Description of the CFOA and (**b**) a practical CMOS compatible one

NSimple), generates the new and novel CMOS compatible CFOA shown in Fig. 16b.

$$
\begin{bmatrix} i_y \\ v_x \\ i_z \\ v_w \end{bmatrix} = \begin{bmatrix} 0 & 0 & 0 & 0 \\ 1 & 0 & 0 & 0 \\ 0 & -1 & 0 & 0 \\ 0 & 0 & 1 & 0 \end{bmatrix} \begin{bmatrix} v_y \\ i_x \\ v_z \\ i_w \end{bmatrix}.
\tag{12}
$$

It is worthy to mention that in practice, the physical implementation of the CCIIs, ICCIIs and CFOA present voltage and current tracking errors, because the UGCs do not provide unity-gain, and present finite bandwidth. Elsewhere, the gain, bandwidth and other important electrical parameters can be optimized by applying symbolic analysis [35, 36], in a postprocessing step.

5 Practical Synthesis of Analog ICs by Applying GAs

As shown in [12], genetic programming is an extension of the GA, it starts from high-level statements of the requirements of a problem and attempts to automatically create a computer program that solves the problem. Genetic programming has automatically synthesized analog circuits [3, 7, 9, 10, 12]. However, automatic synthesis of analog circuits from high-level specifications is recognized yet as a challenging problem, because the analog IC design process is characterized by a combination of experience and intuition and requires a thorough knowledge of the process characteristics and the detailed specifications of the actual product [2, 12].

Design in the analog domain requires creativity because of the large number of free parameters and the sometimes obscure interactions between them [4, 5]. Besides, this section introduces guidelines to synthesize practical circuits by applying GAs. The examples begin with high-level descriptions of the analog circuit, and a refinement process is executed to synthesize analog behavioral models [6, 29, 30] of current-mode amplifiers [22]. Further, the amplifiers (CCII−, CCII+, ICCII−, ICCII+, CFOA) are synthesized by UGCs (VF, CF, VM, CM), which are codified by the new genetic representation introduced in the previous sections.

5.1 Synthesis of Active Filters

As described in [6], modeling is preliminary work or construction that serves as a plan from which a final product can be made. In this manner, the synthesis of active filters can be done by beginning with high-level descriptions by deriving simple equations [37], and further a refinement process is executed until obtaining the transistor level implementation.

Lets consider the Chevyshev low-pass filter shown in Fig. 17. Its design for a frequency of 10 MHz generates: Rs = 17 k, C1 = 1.999 pF, L2 = 295.211 μH, C3 = 2.809 pF, L4 = 295.211 μH, C5 = 1.999 pF, and RL = 100. By using Verilog [6, 29], the high-level description of a floating inductance is modeled by: $i_L = \frac{1}{L} \int (v_i - v_j) dt$. This equation is synthesized by the circuit in Fig. 18a, which consist of two CCII+s, one dual-output CCII (DOCCII), each CC with parasitic resistance Rx, and one capacitor (Cint), so that the value of L is $L = 2R_x^2 C_{int}$. The GA-based synthesis method executes the following steps:

- The active filter is simulated in Verilog by using simple equations.
- The equation of the inductance is synthesized by using CCIIs.
- Each CCII is synthesized by combining the connection of UGCs.
- The active filter is simulated in SPICE.

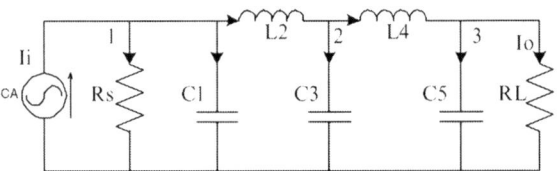

Fig. 17. Current-mode low-pass filter

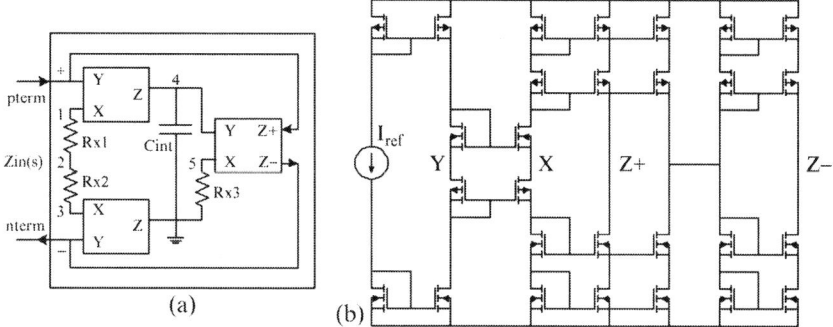

Fig. 18. (a) Simulated inductance using CCIIs and (b) dual-output CCII

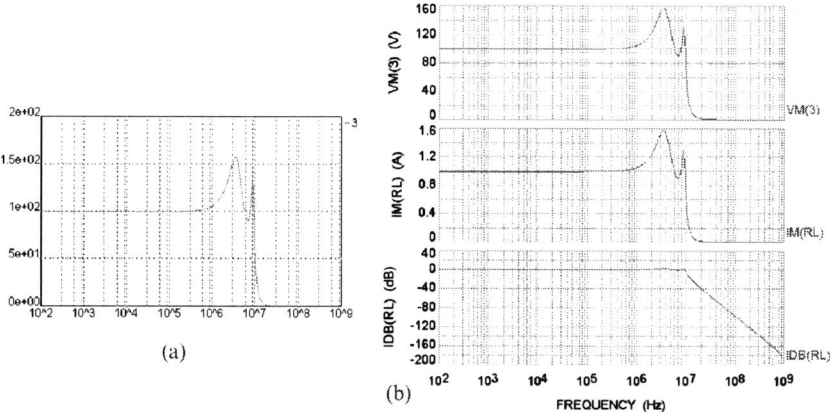

Fig. 19. (a) Verilog simulation and (b) SPICE simulation

In Fig. 18b is shown the synthesized DOCCII by applying the GA described above, and the CCII+s are synthesized as shown in Fig. 14b. The Verilog simulation is shown in Fig. 19a, and the SPICE simulation (at the transistor level) is shown in Fig. 19b.

5.2 Synthesis of Sinusoidal Oscillators

The development of personal wireless communication systems imposes the trend to integrate the radio-frequency front-ends, analogue-to-digital converters and based-band digital signal processors in a single chip. Furthermore, oscillators are critical components of electronic systems since they provide time or frequency reference for the whole chip. In this manner, in [29] is shown the usefulness of Verilog to describe the SRCO at a high-level of abstraction, and SPICE to design the high-level modules down to the transistor level.

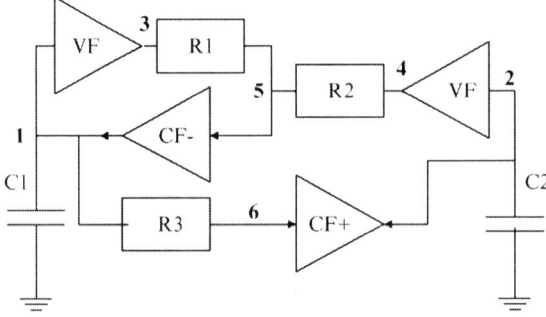

Fig. 20. SRCO designed in [29]

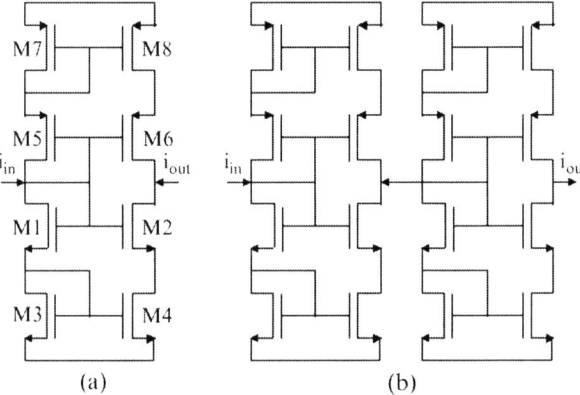

Fig. 21. Synthesis of the (a) CF+ and (b) CF− by using the CM in Fig. 11c

The synthesis method can be adapted to apply GAs to execute the following steps:

- The SRCO is simulated in Verilog using UGCs equations.
- Each UGC is synthesized by transistors as described above.
- The SRCO is simulated in SPICE.

Lets consider the SRCO shown in Fig. 20. It requires two VFs, two CFs and five linear elements (3R, 2C). The GA synthesizes the CFs as shown in Fig. 21, and the VFs as shown in Fig. 4c. By setting R1 = R2 = R3 = 47 k, C1 = C2 = 10 p, and with the initial conditions across C1 and C2 as v(1) = 0.1 and v(2) = 0.1, the resulting Verilog and SPICE simulations are shown in Fig. 22a and Fig. 22b, respectively.

5.3 Synthesis of Chaotic Oscillators

In electronics, chaotic systems find applications in control, secure communications, dry turbulence, visual sensing, neural networks, nonlinear waves and

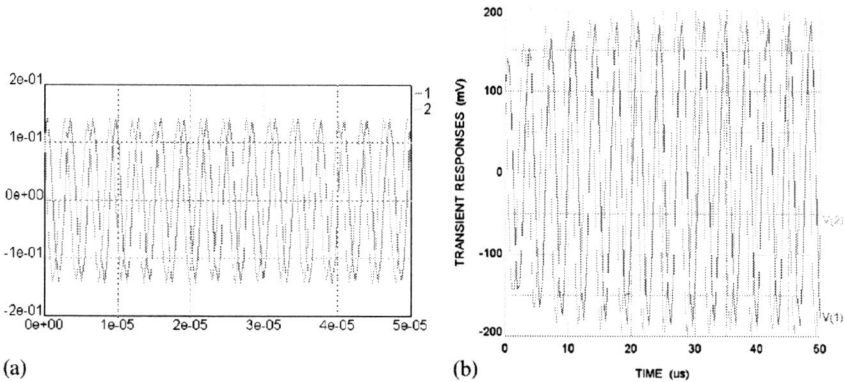

Fig. 22. (a) Verilog simulation and (b) SPICE simulation

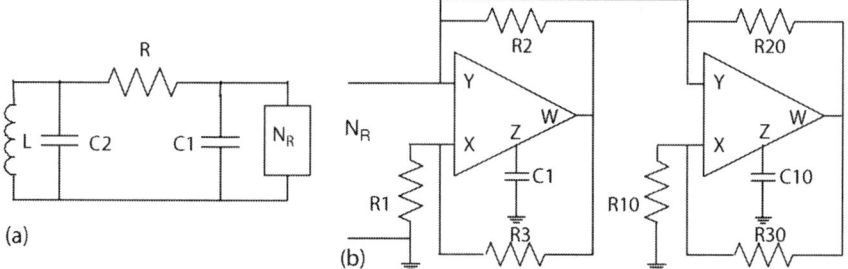

Fig. 23. (a) Chua's circuit and (b) synthesis of Chua's diode (N_R)

music. Besides, Chua's circuit is the only chaotic system which can be easily built, simulated, and tractable mathematically [30, 31]. It consists of the five circuit elements shown in Fig. 23a. Furthermore, in [30] is described the synthesis of Chua's circuit [31], by beginning with high-level simulations, and ending with the design of each individual block by transistors.

Basically, Chua's diode (N_R) is synthesized by the CFOA-based circuit shown in Fig. 23b. The CFOAs are synthesized by applying the GA described above. So that, the GA executes the following steps:

- Chua's circuit is simulated by using high-level descriptions for Chua's diode.
- Chua's diode is synthesized by using CFOAs.
- The CFOA is synthesized by combining the connection of UGCs.
- Chua's circuit is simulated in SPICE.

By setting: C1 = 450 pF, C2 = 1.5 nF, L = 1 mH, the slopes of Chua's diode to g1 = 1/1358, g2 = 1/2464, g3 = 1/1600, with the breaking points BP1 = 0.114 V, BP2 = 0.4 V, as shown in Fig. 24a, VC1(0) = 0.01 V, VC2(0) = 0 V and IL(0) = 0 A, Fig. 24b shows the SPICE simulation for the generation of the double-scroll attractor with R = 1620. In [31] is shown

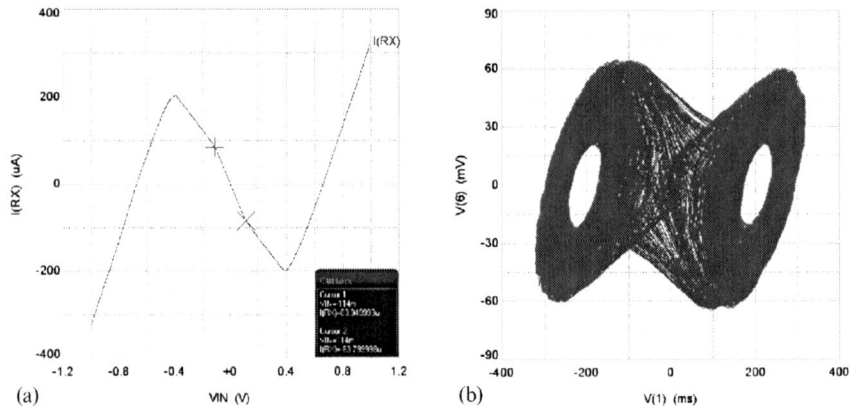

Fig. 24. (a) Piecewise-linear approximation of Chua's diode and (b) double-scroll attractor simulated at the transistor level by SPICE

the sequence of chaotic behaviors by varying R, while in [30] is shown the control of the slopes in N_R.

6 Conclusion

In this chapter it was introduced a new genetic representation using a binary-codification of UGCs (VFs, CFs, VMs and CMs). It was demonstrated that the GA to synthesize VFs, can be adapted to synthesize the others UGCs. Furthermore, the VFs were evolved to synthesize more complex circuits: the superimposing of the VF with the CF generates the CCII−, the combination of a VF with CMs generates the CCII+, the superimposing of the VM with the CF generates the ICCII−, the combination of a VM with CMs generates the ICCII+, and the combination of the CCII+ with a VF generates the CFOA. Further, these current-mode amplifiers were used to synthesize practical analog circuits, such as: active filters, SRCOs and chaotic oscillators. As a result, this chapter introduced guidelines oriented to enhance the ADA process by applying evolutionary techniques.

On the other hand, new open research problems arises: modern synthesis methodologies consist in using hardware description languages, so that an ADA developer will deal with high-level tools, such as Verilog, to model and simulate a system at the most abstract level. ADA synthesis-tools need standards to establish the levels of refinement to model and simulate practical ICs. Finally, the optimization of the synthesized circuits is a hard open research problem to accomplish multiple objectives in the three domains of circuit simulation, namely: DC, AC and time, where the main optimization problems are focused to provide higher performances in RF and high-frequency applications by minimizing noise and distortion.

Acknowledgement

This work is supported by CONACyT/MEXICO with project 48396-Y.

References

1. Grimbleby JB (2000) Automatic analogue circuit synthesis using genetic algorithms. IEE Proceedings Circuits, Devices and Systems 147(6):319–323
2. Salem ZR, Pacheco MA, Vellasco M (2002) Evolutionary Electronics: Automatic Design of Electronic Circuits and Systems by GAs. CRC, New York
3. Mattiussi C (2005) Evolutionary synthesis of analog networks. Ph.D. Dissertation THSE No 3199, Universit degli Studi di Trieste, Lausanne, EPFL, Italie
4. Toumazou Ch, Moschytz G, Gilbert B (2002) Trade-offs in Analog Circuit Design. Kluwer, London
5. Rutenbar RA, Gielen GE, Antao BA (2002) Computer-Aided Design of Analog Integrated Circuits and Systems. IEEE, Piscataway NJ
6. Kundert KS, Zinke O (2004) The Designers Guide to Verilog AMS. Kluwer, Boston
7. Tathagato RD, Chakrabarti PP, Partha R (2005) A synthesis system for analog circuits based on evolutionary search and topological reuse. IEEE Transactions on Evolutionary Computation 9(2):211–224
8. Stefanovic D, Kayal M, Pastre M (2005) PAD: A new interactive knowledge-based analog design approach. Analog Integrated Circuits and Signal Processing 42:291–299
9. Shibata H, Fujii N (2001) Analog circuit synthesis based on reuse of topological features of prototype circuits. IEICE Transactions on Fundamentals of Electronics, Communications and Computer Sciences E84-A(11):2778–2784
10. Alpaydin G, Balkir S, Dundar G (2003) An evolutionary approach to automatic synthesis of high-performance analog integrated circuits. IEEE Transactions on Evolutionary computation 7(3):240–252
11. Aggarwal V (2004) Novel canonic current mode DDCC based SRCO synthesized using a genetic algorithm. Analog Integrated Circuits and Signal Processing 40:83–85
12. Koza JR, Jones LW, Keane MA, Streeter MJ, Al-Sakran AH (2004) Toward automated design of industrial-strength analog circuits by means of genetic programming. In, Genetic Programming Theory and Practice II. Kluwer, Boston, Chap. 8, pp. 121–142
13. Kumar P, Senani R (2002) Bibliography on nullors and their applications in circuit analysis, synthesis and design. Analog Integrated Circuits and Signal Processing 33:65–76
14. Tlelo-Cuautle E, Torres-Munõz D, Torres-Papaqui L (2005) On the computational synthesis of CMOS voltage followers. IEICE Transactions on Fundamentals of Electronics, Communications and Computer Sciences E88-A(12):3479–3484
15. Tlelo-Cuautle E, Duarte-Villaseñor MA, Reyes-Garcia CA, Reyes-Salgado G (2007) Automatic synthesis of electronic circuits using genetic algorithms. Computación y Sistemas, ISSN 1405-5546

16. Tlelo-Cuautle E, Duarte-Villaseñor MA, Reyes-Garcia CA, Sánchez-López C, Reyes-Salgado G, Fakhfakh M, Loulou M (2007) Designing VFs by applying genetic algorithms from nullator-based descriptions. In, 18th European Conference on Circuit Theory and Design, Sevilla, Spain, August 26–30
17. Torres-Papaqui L, Torres-Munõz D, Tlelo-Cuautle E (2006) Synthesis of VFs and CFs by manipulation of generic cells. Analog Integrated Circuits and Signal Processing 46(2):99–102
18. Awad IA, Soliman AM (2002) On the voltage mirrors and the current mirrors. Analog Integrated Circuits and Signal Processing 2:79–81
19. Wang HY, Chang SH, Jeang YL, Huang ChY (2006) Rearrangement of mirror elements. Analog Integrated Circuits and Signal Processing 49:87–90
20. Wang HY, Lee ChT, Huang ChY (2005) Characteristic investigation of new pathological elements. Analog Integrated Circuits and Signal Processing 44:95–102
21. Torres-Muñoz D, Tlelo-Cuautle E (2005) Automatic biasing and sizing of CMOS analog integrated circuits. In, IEEE MWSCAS, 915–918
22. Schmid H (2000) Approximating the universal active element. IEEE Transactions on Circuits and Systems II 47(11):1160–1169
23. Sedra A, Smith KC (1968) The current conveyor: A new circuit building block. Proceedings of the IEEE 56:1368–1369
24. Sedra A, Smith KC (1970) A second generation current conveyor and its application. IEEE Transactions on Circuit Theory 17:132–134
25. Arbel A (1997) Towards a perfect CMOS CCII. Analog Integrated Circuits and Signal Processing 12:119–132
26. Awad A, Soliman AM (1999) Inverting second generation current conveyors: the missing building blocks, CMOS realizations and applications. International Journal of Electronics 86(4):413–432
27. Peña-Perez A, Tlelo-Cuautle E, Diaz-Mendez JA, Sánchez-López C (2007) Design of a CMOS compatible CFOA and its application in analog filtering. IEEE Latin América 5(2):72–76
28. Gupta SS, Senani R (2006) New single-resistance-controlled oscillator configurations using unity-gain cells. Analog Integrated Circuits and Signal Processing 6:111–119
29. Tlelo-Cuautle E, Duarte-Villaseñor MA, Garcia-Ortega JM, Sánchez-López C (2007) Designing SRCOs by combining SPICE and Verilog-A. International Journal of Electronics 94(4):373–379
30. Tlelo-Cuautle E, Muñoz-Pacheco JM (2007) Numerical simulation of Chua's circuit oriented to circuit synthesis. International Journal of Nonlinear Sciences and Numerical Simulation 8(2):249–256
31. Tlelo-Cuautle E, Gaona-Hernández A, Garcia-Delgado J (2006) Implementation of a chaotic oscillator by designing Chua's diode with CMOS CFOAs. Analog Integrated Circuits and Signal Processing 48(2):159–162
32. Chen HP, Lin MT, Yang WS (2006) Novel first-order non-inverting and inverting output of all-pass filter at the same configuration using ICCII. IEICE Transactions on Electronics E89-C(6):865–867
33. Minaei Sh, Yuce E, Cicekoglu O (2006) ICCII-based voltage-mode filter with single input and six outputs employing grounded capacitors. Circuits Systems and Signal Processing 25(4):559–566
34. Toker A, Zeki A (2007) Tunable active network synthesis using ICCIIs. International Journal of Electronics 94(4):335–351

35. Tlelo-Cuautle E, Sánchez-López C, Sandoval-Ibarra F (2005) Computing symbolic expressions in analog circuits using nullors. Computación y Sistemas 9(2):119–132
36. Fakhfakh M, Loulou M, Masmoudi N, Sánchez-López C, Tlelo-Cuautle E (2007) Symbolic noise figure analysis of CCII+s. In, Fourth International Multi-Conference on Systems, Signals and Devices, Hammamet, Tunisia
37. Garcia-Ortega JM, Tlelo-Cuautle E, Sánchez-López C (2007) Design of current-mode Gm-C filters from the transformation of opamp-RC filters. Journal of Applied Sciences 7(9):1321–1326

Intuitive Visualization and Interactive Analysis of Pareto Sets Applied on Production Engineering Systems

H. Müller, D. Biermann, P. Kersting, T. Michelitsch, C. Begau, C. Heuel, R. Joliet, J. Kolanski, M. Kröller, C. Moritz, D. Niggemann, M. Stöber, T. Stönner, J. Varwig, and D. Zhai

Technische Universität Dortmund, Germany

Summary. Applying multi-objective optimization algorithms to practical optimization problems, a high number of multi-dimensional data has to be handled: data of the genotype and phenotype as well as additional information of the optimization problem. This chapter gives an overview of several methods for visualization and analysis which are combined with regard to the characteristics of solution sets generated by evolutionary algorithms in order to get an intuitive instrument for decision making and gaining insight into both - the problem and the algorithm. They are discussed by means of two current production engineering problems providing a high economic potential: the optimization of the five-axis milling process and the design of cooling duct layouts.

1 Introduction

In most practical optimization problems, several conflicting objectives have to be considered. Typically, the costs for the realization of the solution and one or more objectives concerning the quality are regarded. In many cases the objectives are hard to compare and therefore an a priori assessment of weighting factors is difficult. If so, the application of multi-objective optimization methods, which provide a set of e.g. Pareto optimal solutions, is advisable [3]. Due to the high complexity of many practical problems with conflicting objectives, an increasing use of stochastical optimization techniques like multi-objective evolutionary algorithms (MOEA) is observed.

Choosing the appropriate solution from a set generated by a MOEA is up to the user. A well-founded analysis of the solutions is therefore required to guarantee a factual choice. On the one hand, even assuming well-designed objective functions, many solutions of the set can be inconvenient for practical realization. On the other hand, the typically used evaluation models are often

H. Müller et al.: *Intuitive Visualization and Interactive Analysis of Pareto Sets Applied on Production Engineering Systems*, Studies in Computational Intelligence (SCI) **92**, 189–214 (2008)
www.springerlink.com   © Springer-Verlag Berlin Heidelberg 2008

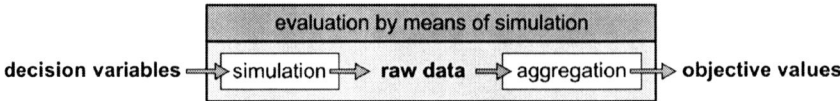

Fig. 1. If simulations are used to evaluate solutions, three spaces can be considered for the analysis of evolutionary generated solution sets: the decision, the objective, and the raw data space

simplified and imprecise. The latter gives reason not to take only the Pareto optimal solutions with regard to the objective functions into account, but also other solutions of the population. A large set of solutions should be evaluated by the user. To facilitate this, support by an analysis software providing a concise visualization and intuitive navigation is highly desirable.

The design of fitness functions, the selection of an adequate MOEA, and its adaptation to the optimization problem is often a very complex task. A far-reaching knowledge of the problem, the optimization method, and their interaction is therefore required. This is the second important use case for a solution set analysis software.

In many real world optimization problems, simulation environments are utilized for the evaluations of the solutions. The simulations provide extensive data sets (called *raw values*). These raw values are aggregated to the objective function values (compare Fig. 1) and normally disregarded afterwards. Considering single solutions, generally only the decision and the objective space is taken into account. This chapter deals with an approach of using the generated raw data and its enormous benefit of taking these data – additionally to the information of the decision and objective space – into account. Methods for visualization and analysis will be itemized and combined with regard to the characteristics of solution sets generated by evolutionary algorithms in order to get an intuitive instrument for decision making and gaining insight into both – the problem and the algorithm. They are discussed by means of two current production engineering problems providing a high economic potential: the optimization of the five-axis milling process and the design of cooling duct layouts. Like many other applied problems, these challenging multi-objective optimization problems are subject to several constraints, have to cope with incomplete models, use simulations and surrogate functions for evaluation, and their fitness function design depends particularly on the run-time of the simulation. Due to the fact that the solutions are close to the restriction areas in many cases, both optimization applications are faced with the difficulty of getting well-designed solutions.

These two optimization problems have exemplarily been chosen in this chapter since they show primary similarities, but are completely different with regard to their problem encoding and their solution space characteristics. Due to certain problem-specific properties, these problems turn out to be quite difficult in their own way. Whereas the highly dimensional decision space of the milling optimization application is rather hard to handle, the optimization of cooling duct layouts, which has a manageable decision space, is strongly multi-modal and has a discontinuous objective function.

In this chapter both, the methods for the visualization and interactive analysis of solution sets as well as the evolutionary optimization approaches for the two actual problems, will be presented. The successful application of the visualization and analysis techniques to the algorithm design will be shown. The benefit of using clustering approaches for the decision making process, in particular dealing with simulation raw data, is discussed exemplarily on the basis of the real world applications.

2 Data Visualization and Analysis

Although both mentioned real world optimization problems are completely different with regard to their encoding, the characteristics of their solution space, and their applied simulation methods, significant similarities are identified. Since all three spaces – the decision, objective, and raw data space – may be described with numerical values, accepted plotting methods can be utilized. General purpose techniques like for example parallel coordinates, scatterplots, and projection techniques [10,33] are proven methods to illustrate evolutionary generated high dimensional data. Typically, solution sets are only presented in the objective space [12,19,31]. The application of well-established visualization methods in all three spaces and the interaction between each other is presented in this chapter.

The geometric background of production engineering eases the mapping of the decision space variables to the three-dimensional space. Furthermore, the raw data space can often be illustrated as free-form surface depicted with different values, e.g. temperature distribution and resulting gradients or collision marks. Although the dimension of the decision space is high, the mapping to the three-dimensional space not only allows a concise presentation of single solutions but even of solution sets and subsets.

Both, the end user choosing a single solution from a given set and the scientist wishing to gain insight into the characteristics of the problem and the behavior of an applied optimization algorithm, demand for a far-reaching analysis of the solutions. The simultaneous presentation of all three spaces and the interactive navigation through the solution set is therefore realized.

For the (semi-)automatic analysis of large solution sets, a combination with clustering algorithms [21, 32, 34] is reasonable. The new aspect is not only to cluster within the objective space, but within any of the three spaces independently or combined. This allows for an automatic distinction of certain types of solutions as already mentioned in the example before. Up to now it was necessary to manually interpret the entire solutions one by one, in such a case. Especially presenting the clustering results of one space in another one is very helpful for the analysis.

In order to present the designer of multi-objective evolutionary algorithms and the user of optimization applications an intuitive instrument for decision making and gaining insight into the optimization problem and the algorithm, different techniques are integrated in a software tool called PAVEl (*Pareto Set Analysis, Visualization, and Evaluation*), which has been developed by a projectgroup of eleven students of the University of Dortmund. This general purpose visualization, classification, and analysis tool offers the import of any data in form of simple formatted ASCII files. In addition, it is easily adaptable to problem specific representations via a plug-in interface.

The basic principles of the applied visualization and clustering approaches are briefly explained in Sects. 2.1 and 2.2. The application of this tool as well as the description of specific methods is discussed on the basis of the two introduced real world applications in Sects. 3 and 4. Examples of the successful use during the algorithm design as well as during the decision making process are given.

2.1 Visualization

In order to visualize sets of values of the objective, decision, or raw data space, different techniques can be used. The easiest way to display the numerical values of each set is to use a simple table. Determining the quality of a numerical value in a given context is much more difficult compared to graphical representations. To intuitively identify interesting solutions and discover relations between sets, it is advisable to colorize the table cells by mapping numerical values of any dimension to a color scale.

A proven technique for illustrating high-dimensional data are parallel plots [33]. This kind of plot maps the dimensions of the given data to vertical axes. Each d-dimensional solution in the set is displayed as a horizontal line set (see Fig. 2a). The intersection of this line with an axis corresponds to the value of the solution in the dimension represented by this axis.

To be able to intuitively understand the given data, several ways of fitting the visualization to individual needs are necessary. For parallel plots the most important means to this are exchanging or hiding axes or adjusting their scaling. Using these options can alter the view on the data dramatically and thereby offers new insights into properties and characteristics of the displayed solution set.

Another proven technique are scatterplots (see Fig. 2b). A scatterplot consists of a two- or three-dimensional coordinate system with the domains of the selected dimensions mapped to its axes. Individual solutions are represented by points in this coordinate system and clouds formed by these points indicate correlations in the data set. High density in these clouds can make it impossible to identify single points unless the viewport may be zoomed in on regions of interest. The main disadvantage of scatterplots is their restriction on the number of presentable dimensions. Using spatial coordinates allows two or

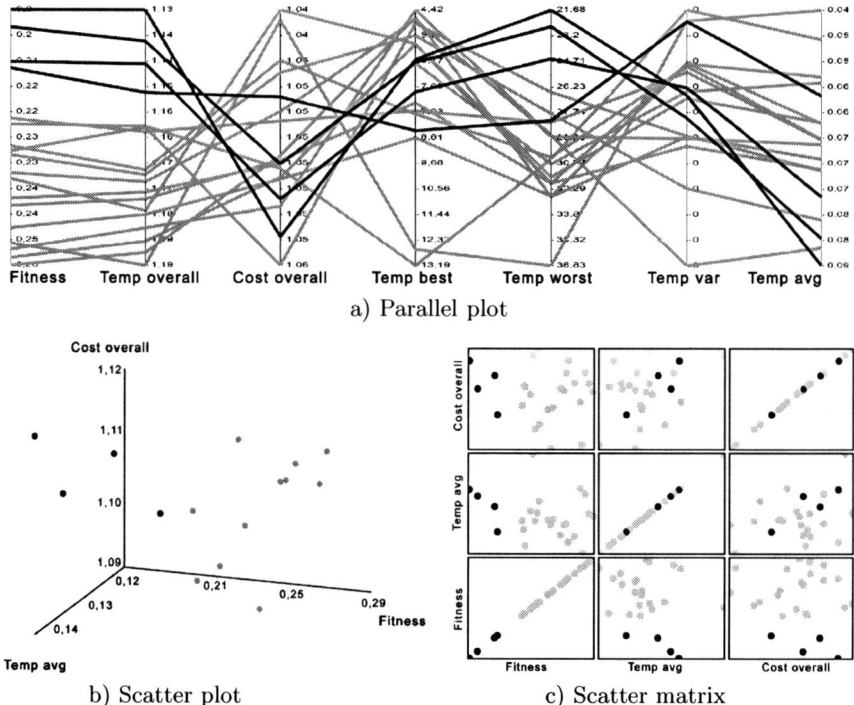

a) Parallel plot

b) Scatter plot   c) Scatter matrix

Fig. 2. Examples of three different visualization techniques displaying the objective space of a solution set. The parallel plot is able to display all seven dimensions of the space, the scatter plot can only display three. The scatter matrix is shown here comparing the same dimensions as the scatter plot for the sake of clarity, theoretically its number of presentable dimensions is unlimited. A subset of four solutions currently chosen by the user is colored black

three, colorizing the points an additional fourth dimension to be displayed. If the number of dimensions is higher than four, scatter matrices can be used.

A scatter matrix is a variant of two-dimensional scatterplots in which all possible pairs of dimensions are displayed as two-dimensional plots arranged in a matrix (see Fig. 2c). This makes it very easy to spot correlations or patterns in the data. However, their quadratic size makes scatter matrices infeasible for solutions with many dimensions.

2.2 Clustering

Clustering is vital to the analysis of large data sets. On the one hand, the sets have to be reduced to speed up the visualization and processing of solutions to allow real-time interaction. On the other hand, solution sets may contain many similar solutions. To make these more manageable for the user, solutions that can be considered similar with regard to a certain distance function need to be aggregated. In clustering, the most common approach is to use Euclidean distances between solutions in a certain space. Due to the fact that these measure is commonly used for evaluating object proximity in two- or three-dimensional space, this distance is very intuitive. However, it is preferable to take advantage of a space's known semantics by utilizing non-Euclidean distance functions on solution values [32]. That way, it is possible to define similarity of solutions in a manner that is closer to the actual problem, a manner that could not be expressed by treating the solutions' values as abstract spatial coordinates.

Jain, Murty, and Flynn [7] as well as Xu and Wunsch [32] give a detailed review of clustering techniques. Generally, these techniques are divided into hierarchical and partitional approaches. Hierarchical methods create a nested series of partitions, whereas partitional approaches produce only one [7]. Due to its easy structure and its linear time complexity, the simplest and most commonly used partitional clustering algorithm is k-means. The basic idea is to start with a random initial partition and to assign each pattern to its closest cluster center. Afterwards, the cluster centers are recalculated. This procedure is repeated until a convergence criterion is met [7]. In contrast, the basic idea of hierarchical agglomerative clustering methods (*HACM* for short) is to initially create a cluster for every pattern and recursively select and merge pairs of clusters afterwards.

For the purpose of analyzing data sets optimized by MOEA, clustering algorithms, which are able to cope with large data sets and high dimensionality, are necessary. Most of the common clustering algorithms are not suited to handle this [32]. Classical HACM variants for example are inappropriate due to their quadratic complexity. The main drawback of k-means is its fixed number of clusters, which have to be determined in advance. Due to this, several approaches have been developed and are known in literature, which were specifically designed for large data sets. Those are based on randomization, condensation, densities, or grids to achieve high performance. However, many of them are not well-suited to high-dimensional data. The general problem is that computed distances become meaningless for high dimensions [7]. Therefore, clustering in subspaces and with aggregated axes is very important. Alternatively, a mapping of high dimensional data to lower dimensions by use of nonlinear functions is advisable [32]. Afterwards, common clustering techniques can be applied.

Cluster representatives have special characteristics that are not present in plain numerical solutions. Additional ways of interacting and displaying clusters are necessary to allow the user to utilize these special characteristics.

3 Optimizing the Five-Axis Milling Process

Milling is a material-working process in which the desired workpiece form is produced by eliminating small pieces (chips) from the raw material using a rotating cutting tool. It is used in many manufacturing industries. Several

a)   b)

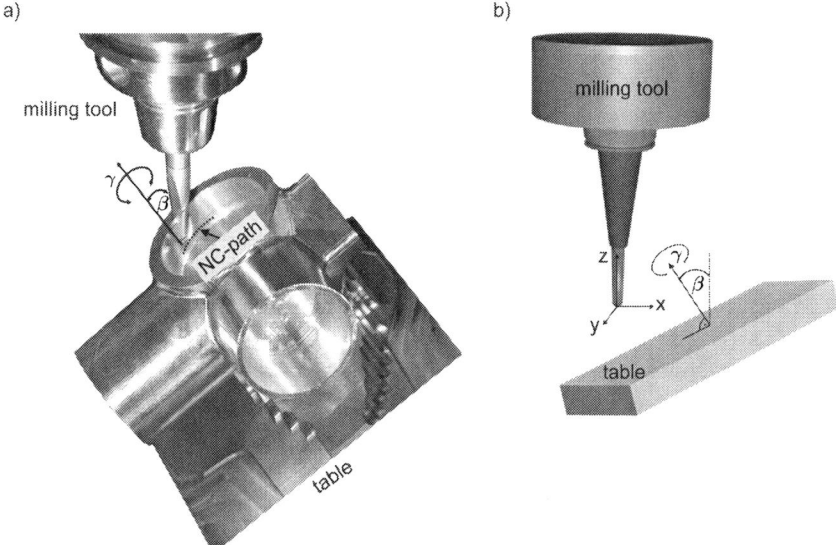

Fig. 3. Five-axis concept of a selected milling machine demonstrating the definition of the tilting and rotating angles (**a**). The milling tool can be moved along the three coordinate axes in this machine design. In contrast to other concepts where the milling tool can also be rotated, the two additional degrees of freedom are provided by rotating the machine table (**b**)

objects of the everyday life are machined by milling tools or by instruments which are themselves produced by this machining process. Prostheses (for example dental or hip joint endoprostheses), turbine blades, or impellors as well as molds and dies are examples of five-axis machined free-formed surfaces. Automotive bodies and many aircraft components are formed by the deep drawing process [18]. The forming die, which is used in order to form the sheet metal blank, is also produced in a milling process.

In the five-axis process the three translational degrees of freedom of conventional three-axis milling are extended by the tilting and the rotating angle of the machine table (Fig. 3) or the tool spindle [29]. Using these two additional degrees of freedom, it is possible to influence the engagement conditions of the tool edges. Furthermore, the milling tools can be used in areas which are difficult to access like for example undercuts, steep walls, or deep cavities. Of course, undercuts can also be produced by three-axis machines, but for this the workpiece has to be clamped in different positions. This is more complicated, less precise, and time-consuming. When machining deep cavities in a three-axis way, large and thereby unstable cutting tools have to be used. In contrast, the five-axis process offers the possibility of using shorter and thereby stiffer milling tools, which show less tendency of deflection and lead to an improved surface quality [30].

Today, milling machines are computer numerical control (CNC)-driven and all the information required for controlling the machine (for example information on the milling tool properties and the tool locations) is contained in the numerical control (NC)-programs which are generated by a computer aided design (CAD)/computer aided manufacturing (CAM)-system. These NC-paths are usually based on the target geometry of the desired workpiece. Although the five-axis milling process offers different advantages and provides the opportunity of increasing the process efficiency, the additional degrees of freedom lead to a more complex process [30]. Particularly the generation of tool paths, which are free of collision and adapted to the specific machine as well as to the process characteristics, is significantly more complicated, especially if complex free-formed surfaces have to be machined.

3.1 Optimization of NC-Paths for Five-Axis Milling

Due to the specified application areas and the existing problems, the optimization of NC-paths for five-axis milling is a relevant topic of interest and several research publications are known in literature aiming at different optimization objectives. In addition to several deterministic approaches [1,4,9,23], several works on optimizing and analyzing the milling process using evolutionary approaches exist. Vigouroux et al. [27] for example introduce an approach for the optimization of machining parameters under uncertainties using evolutionary algorithms. Tandon et al. [24] optimize multiple machining parameters simultaneously for the milling process using particle swarm optimization [11]. Valvo, Martuscelli, and Piacentini [26] also use the particle swarm optimization in the context of determining optimal cutting parameters of the milling process. Kovacic et al. [13] developed an evolutionary approach for cutting force prediction and Brezocnik et al. [2] for surface roughness prediction based on genetic programming. In contrast to all these research works, the optimization approach at hand deals with the optimization of the milling process by modifying given NC-paths with respect to the angle values taking these process parameters into account. The combination with a simulation environment for the simultaneous five-axis milling process allows a direct validation of the optimization solutions.

3.2 Milling Simulation

A simulation environment containing different simulation types is necessary during the optimization approach. The main purpose of a simulation system in this case is the imitation of the real process behavior in order to gain insight into the system and to get an improved process understanding. Within the integrated optimization approach, well-suited parameters are selected and verified simulatively. Afterwards, these simulation results are translated to the actual process. In case of the milling approach, a simulation of the material removal process is required, but it does not need to be as detailed as possible.

Especially the efficient and precise treatment of undercuts plays a decisive role [35].

The optimization approach at hand is integrated in a simulation environment for the simultaneous five-axis milling process, which depends on a syntactic model of engagement conditions (i.e. uncut chip forms). In order to allow for an exact calculation of the chip thickness and cutting forces, these chips are constructed by constructive solid geometry (CSG, [6]). Interpreting NC-paths for the considered workpiece and using further information about the characteristics of the used milling machine, the raw material, and the process parameters, the simulation is able to model the real milling process accurately [23]. Furthermore, the simulation of the kinematic behavior of the milling tool and a geometric simulation for an efficient and fast collision detection are required.

3.3 Optimization Problem Design

The purpose of the optimization problem at hand is the generation of optimized NC-paths for simultaneous five-axis milling with respect to the milling tool engagement [30]. The variables in this case are the tilting and rotating angle of the milling tool at each of the thousands of tool positions of the NC-path (Fig. 3). These result in thousands of degrees of freedom in the optimization problem at hand, which makes this problem quite difficult. Therefore, it is not advisable to optimize the tool orientation simultaneously for all NC-path points, but to optimize only a section of the NC-path at once. So one individual is composed of values for β and γ for consecutive milling tool positions.

The optimization of the introduced application turns out to be quite complex. In order to receive good results, an adaptation of established algorithms, which are known in literature, to the specific problem is necessary. For this purpose a visualization and analysis tool is very valuable to the algorithm designer.

3.4 Fitness Function Design

In order to evaluate individuals, very often different criteria can be identified. In the case of the milling optimization approach, the minimization of the movement changes, of the processing time, and of process forces or the improvement of the tool engagement are appropriate objectives. Sometimes criteria depend on each other and can then be combined to one factor. (An improvement of the tool engagement can for example result in lower process forces.) However, to some extent the dependency of the selected criteria is a priori not obvious. For this purpose it is beneficial to use visualization techniques in order to identify these dependencies. Parallel plots (see Sect. 2.1) are for example appropriate for the illustration of multi-dimensional data sets. In Fig. 4, a parallel plot is exemplarily shown, which visualizes the evaluation

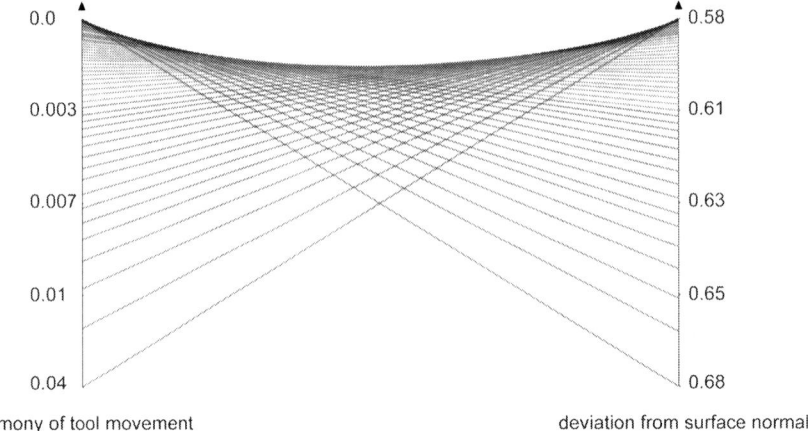

Fig. 4. Parallel plot of the evaluation of two main objectives of the milling optimization approach. The intersection structure indicates the conflicting relation of the two applied criteria

of two main objectives of the milling optimization approach. An intersection structure can be identified, which indicates the conflicting relation of the two applied criteria.

These two criteria are the optimization of the movement of the milling tool and the minimization of the deviation from the detected tool orientation to the surface normals [30]. The normals describe the contour of the specified workpiece. This criterion is utilized in order to adjust the tool orientation to the particular workpiece and to improve the tool engagement. It is important to take the tool movement into consideration because abrupt changes of the moving direction of the milling tool could result in surface defects [30].

The fitness values at hand are normalized and restricted to the interval [0.0, 1.0]. The better the evaluation of the criterion, the smaller is the fitness value. Several restrictions have to be regarded, which makes the generation of feasible solutions quite difficult. Therefore, an efficient collision detection approach plays a decisive role in order to prevent collisions between the milling tool and the workpiece. Furthermore, the optimization is always carried out considering production limitations [30].

3.5 Multi-National Multi-Objective Evolutionary Algorithm

For optimizing the five-axis process, the principal tasks are an efficient design of the fitness function evaluation as well as an effective treatment of restriction areas. Therefore, a hybrid multi-objective evolutionary algorithm has been developed and applied to the optimization problem at hand. The algorithm is based on the state-of-the-art evolutionary algorithm SMS-EMOA (*S-metric Selection Evolutionary Multi-Objective Algorithm*) introduced by

Emmerich, Beume, and Naujoks [5]. This Pareto-based technique at first performs a non-dominated sorting of the individuals and secondly sorts according to the hypervolume measure contribution (S-metric) if necessary. This well-applied measure takes convergence as well as distribution properties of the Pareto front approximation into account [36, 37].

This established evolutionary algorithm is combined with deterministic approaches, which use the information of the specific optimization application. A basic problem while optimizing the tilting and rotating angles of the milling tool is the difficulty of dealing with restriction areas. This problem can be overcome in this case since all restriction areas can be precalculated [28] using a fast collision detection approach, which generates a collision profile for each milling tool position. In Fig. 5b an example of a collision profile is visualized. This profile specifies which of the angle combinations lead to collisions between the workpiece surface and the milling tool and which are free of collisions. Using these profiles accelerates the evaluation process because no time consuming simulations are required for collision detection – the collision information can be easily obtained from the collision profiles.

While optimizing the NC-path of a milling tool, it has been a great benefit to visualize the angles of the individuals to the β-γ-area (Fig. 5c). Earlier plots

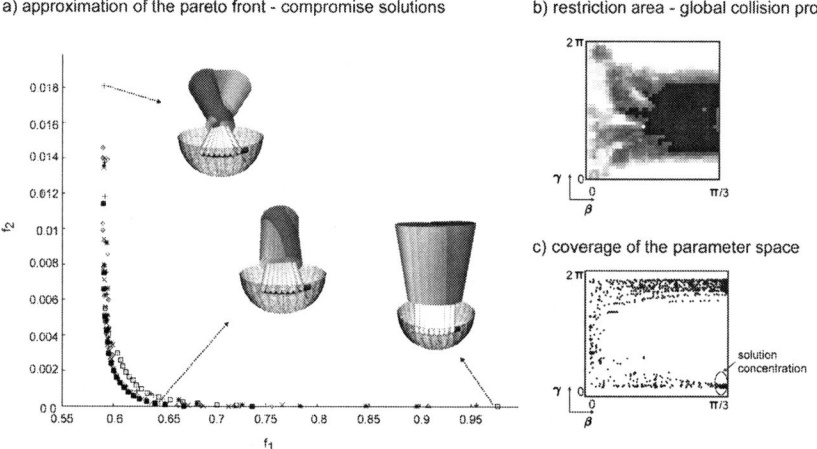

Fig. 5. Results of evolutionary runs optimizing a section of a NC-path for a hemisphere-shaped cavity. In (**a**) an approximation of the Pareto front after several runs is shown. f_1 marks the objective of normal-orientation, f_2 denotes the harmonization objective. Exemplarily, three selected solutions of the optimization process are visualized. Consecutive tool locations are shown simultaneously. In (**b**) the restriction area corresponding to the optimized NC-path part is visualized. The white area shows collision-free tool positions. The darker the color, the more NC-path steps lead to collisions. (**c**) Mapping of the solutions to the β-γ-area. All tilting and rotating angles are shown. It can be seen that the algorithm generates individuals which cover the whole feasible (collision-free) area

illustrated the effect that the solutions were concentrated on a special area of the angle domain. This phenomenon results from the difficulty of getting out of collision areas. If the algorithm starts with valid solutions and gets into restriction areas due to mutation and recombination, it normally returns to the same collision-free area where it came from. Another restriction-free area corresponding to another structure of the NC-path design was not found.

In order to improve this and to increase the diversity of the results, a multi-national approach was applied. This technique is based on the multi-national evolutionary algorithms presented by Ursem [25]. Their basic idea is to improve the results for a problem by dividing the population into smaller sub-populations each searching a particular segment of the search space [25]. The division into sub-populations is motivated by the topology of the fitness landscape.

In contrast to Ursem's research, the problem-specific restriction, which is known in the specific case of the optimization problem at hand, is taken into account instead of the topology of the particular fitness landscape. The principal idea is to start one evolutionary run in each of the collision-free areas of the collision profiles. As illustrated in Fig. 5c, this multi-national approach leads to a comprehensive covering of the restriction-free area. It can be shown that this approach increases the diversity of the NC-path design and can also improve the corresponding Pareto fronts (Fig. 5).

In order to select one result of the Pareto approximated solutions, it is important to visualize the individuals. In the optimization problem at hand, an individual consists of a defined number of angle values, namely the β_i and γ_i of the considered NC-path positions. β_i identifies the particular tilting angle of position p_i and γ_i denotes the particular rotating angle. Choosing one solution from a number of individuals is very difficult. Due to the fact that one individual consists of so many (angle) values, a visualization using parallel plots is advisable. In Fig. 6 there is exemplarily one plot for the tilting angles and one for the rotating angles of all individuals of an archive of an evolutionary run after 500,000 generations. The corresponding β- and γ-progressions can be indicated by the choice of color.

A horizontal trend (compare line (1) in Fig. 6a) indicates that the angle values do not change during the milling process. This movement is very harmonical. This statement could mislead the user to choose those solutions with horizontal β- and γ-trends. But this is only advisable in case of milling straight lines. Otherwise, if always the same angle values are used, the advantages of the five-axis process (see Sect. 3) are not used. The almost monotone changing of the tilting angles shown in Fig. 6 reflects the circular contour of the workpiece surface.

Applying the visualization and analysis tool, a navigation through the different spaces is possible. Choosing a line in the parallel plot leads to an illustration of the individual's phenotype (see Fig. 6 illustration (1) and (2)). This visualization is very intuitive. However, it is easier to notice the details in the parallel plot. This depiction of the individuals' values (Fig. 6) is not

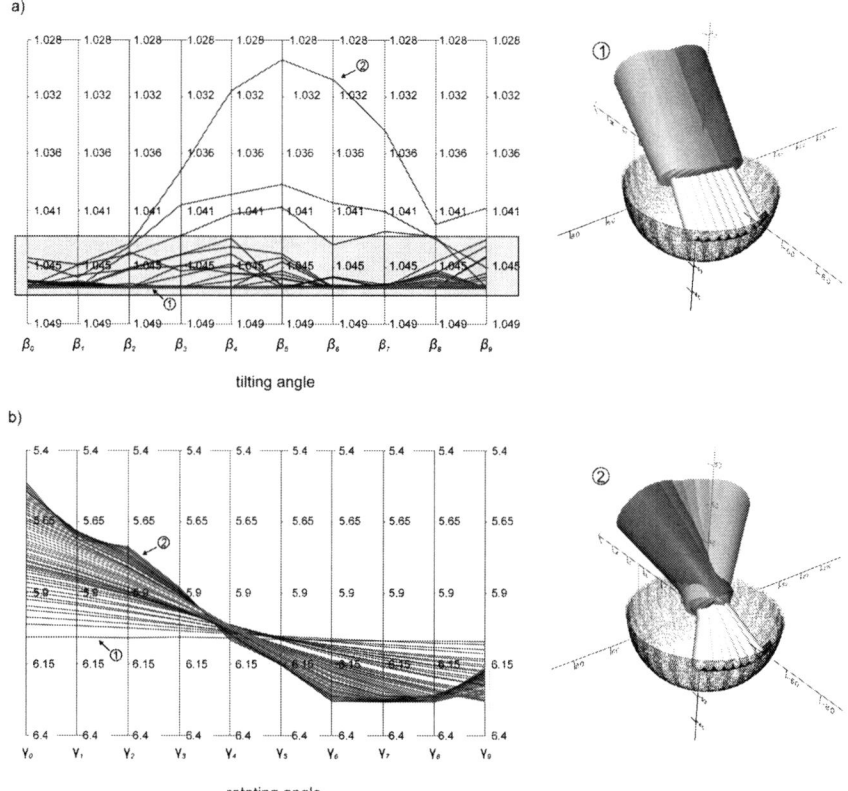

Fig. 6. Visualization of the solutions. Due to the high number of variables of one individual – namely the angle values β_i and γ_i – a visualization using parallel plots is advisable. The tilting (**a**) and the rotating (**b**) angle values are shown separately. A horizontal trend (compare individual (**1**)) indicates a harmonical movement. The almost monotone changing of the tilting angles (**2**) in figure (**b**) reflects the circular contour of the workpiece surface

only very valuable in the decision making process, but also for the algorithm designer. For example, the changes of the tilting angles, which are marked with a gray background in Fig. 6a, are only in a range of smaller than a half degree. In this case it is advisable to modify the optimization algorithm in such a way that those small changes are smoothed and the variables are set to the same values. This modification should always be carried out under the constraint of being free of collisions and technical well suited.

For the designer of problem-specific optimization algorithms, it is also important to analyze the time-dependant behavior of the designed algorithm. For this purpose scatter plots are suitable. In the case of two or three criteria, the two-/three-dimensional plots can be used. If there are more than three criteria, scatter matrices can be applied (see Sect. 2.1). A time-dependant

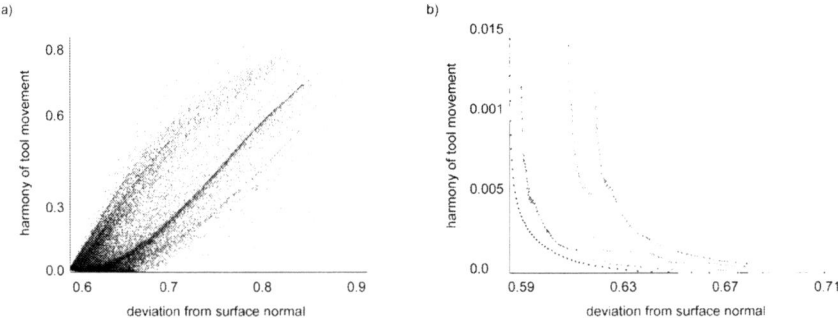

Fig. 7. Scatter plots. (**a**) Time-dependant plot of an evolutionary run of the milling optimization problem. (**b**) Classification of the results after different generations using gray scale values

plot of an evolutionary run of the milling optimization is shown in Fig. 7a. The algorithm designer can easily recognize a convergence (or stagnation) of the algorithm. Furthermore, the proposed visualization and analysis tool offers the possibility of plotting the results of different generations in different colors (Fig. 7b).

In Fig. 5, results of several evolutionary runs optimizing a fragment of a hemisphere are visualized. An approximation of the Pareto front after several runs is shown in Fig. 5a. In this case an interactive navigation handling is very helpful. Picking one point of the Pareto front approximation results in the visualization of all the information about the selected solution. In Fig. 5a, additionally to the Pareto front, three solutions of the optimization process are exemplarily visualized. The individual shown on the bottom is optimal with regard to the second criteria, which describes the movement harmony of the milling tool. In this case the tilting angle does not change, which leads to an improved surface. The solution visualized on the top is normal-oriented whereas the individual shown in the middle offers both. All three solutions are machineable and the engineer has to select one of these solutions. The visualization of the phenotype additionally to the Pareto front gives a realistic and intuitive impression of the solutions.

4 Design of Cooling Ducts for Molding Tools

Whereas the optimization of a production process has been discussed in the previous section, the problem dealt with in this section is related to a construction process.

4.1 Technical Background

Injection molding and die casting are very common manufacturing processes for the mass production of plastic and metal parts, which allow for complex

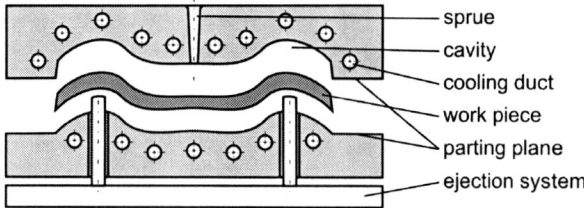

Fig. 8. A simple molding tool with cooling ducts orthogonal to the viewing plane

designs and even fragile shapes. Assuming a well adjusted process, the produced parts provide a high precision and surface quality – often no subsequent finishing is required.

In both processes – injection molding and die casting – hot molten material is injected into a mold (see Fig. 8), solidifies, and is ejected afterwards. The cycle time and thus the production rate of a molding tool is primarily limited by the time required by the part to solidify. In general, this time span is called cooling time. In order to achieve low cooling times, molds are provided with active cooling systems [14]. Cooling ducts within a mold are realized by sets of intersecting deep hole bores, which are either plugged or provided with inlet or outlet connectors. The connectors are attached to an external cooling device providing a cooling fluid that circulates through the ducts.

The layout of the cooling ducts has a decisive effect on the achievable cooling times. Due to the high cost pressure in industries of mass production, the optimization of cooling ducts is of high interest. In addition to the cooling times, the cooling system does also affect the product quality and the process reliability by counteracting thermally induced stresses resulting in a distortion of the product's shape, an increased tool wear, or even a breakage of the tool. The main objectives when designing cooling ducts are therefore to achieve an efficient heat removal and simultaneously to reduce thermally induced stresses. Although the thermal conditions within a mold result from complex dependencies between the parts of the tool in the molding process, the layout of the cooling duct is commonly determined only by the designer's experience and intuition. Whereas computational fluid dynamics simulations are more and more used to analyze the filling process, it is still rather uncommon to perform thermal finite element simulations with regard to the cooling ducts. Even if so, the main purpose is to verify that the final layout does not show severe flaws – a systematic optimization of cooling ducts utilizing simulations is not yet realized.

Another important design objective is to reduce the manufacturing costs of the cooling ducts. In practice the designer aims to reduce the amount of bores used and their lengths. Additionally, regularly shaped cooling duct systems are preferred since they are easier to design and also cheaper to manufacture in general. Whereas product shapes tend to get more and more complex, the cooling ducts covering the cavity do, too. The significance of bores placed diagonally in the mold is therefore increasing. For these bores, limitations of rotational axes of the particular drilling machines used should be taken into account. Already slight modifications of location or direction of a bore may lead to dramatically lower manufacturing costs because single production steps like preliminary milling and pilot drilling or reclamping of the tool are not required under certain geometric conditions. Due to these discontinuities of the cost function, examining the effects resulting from the particular machine's characteristics is very challenging, but may yield significant savings of manufacturing costs.

In the design process of a molding tool, the cooling ducts are usually one of the last parts added. Therefore next to the cavity, several other functional parts – as the sprue, the ejection system, and leader pins as well as slides and cam pins used for parts with undercuts – are to be regarded as geometric obstacles from the cooling duct designer's point of view. Due to mechanical and thermal reasons, different clearance distances are to be considered for these parts. From the designer's point of view, assuring the clearance distances is a must and therefore a design constraint. Additionally, there are several other geometry-dependent constraints to be regarded: the cooling system must be located within the mold, the cooling ducts may not show short circuits, in- and outlet connectors are restricted to certain faces of the mold, and the space for mounting the connectors and plugs must be sufficient. As a result of these constraints and the limited help offered by current CAD-systems for examining them, the design of cooling duct systems is a time-consuming and error-prone process.

4.2 Geometric Model of the Environment and Solutions

In order to provide a simulation based optimization system for the design of cooling ducts, the first step is to define a model for the given static geometric environment primarily representing the currently examined half of the mold. The mold itself is often essentially box-shaped and aligned to the axes of coordinate system and can therefore easily be defined by the two extremal corner points. For more intricately shaped molds, the cavity surface and possible obstacles within the mold a more general representation is required. While most CAD-systems support volumetric models of free form shapes internally, common non-proprietary file formats are surface-based. The optimization system at hand makes use of the STL-format (surface tessellation language format) which is a very simple but common file format describing free form surfaces by an unstructured set of triangles. Using approximations of the surfaces by using triangle sets allows utilizing highly efficient algorithms to determine geometric relations (e.g. the distance) between the surface and the cooling ducts for the simulation and evaluation. However, two essential disadvantages have to be considered: triangle sets exported by common CAD-systems often show small gaps and due to the unstructured surface

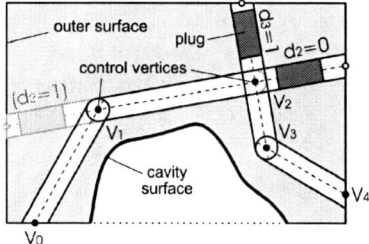

Fig. 9. Compact representation of a cooling system by a set of control vertices V_i and bore directions d_i

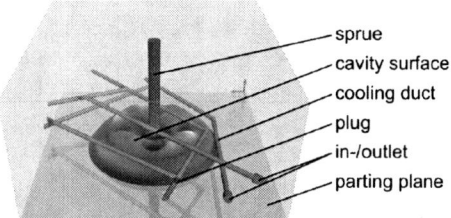

Fig. 10. A mold for bottle-bottoms with an evolutionary designed cooling system. The simulated temperature derivation of the cavity surface is gray-scale coded

representation, it is difficult to distinguish the inner and outer side of the surface.

Since the layout of the cooling ducts shall be subject to an evolutionary optimization, the needs for the cooling system model are different from the mold. A compact representation is desirable here. At an early stage the cooling systems was modeled as a set of independent bores, in order to ease the observance of production-related constraints for certain machines. Due to severe problems with not or not properly intersecting bores occurring during the optimization process a new model for the cooling system has been designed. As Fig. 9 shows, a cooling circuit with m bores is represented by $m + 1$ control vertices V_i with $i \in \{0, \ldots, m\}$ defining the in- and outlet and the intersection points of adjacent bores. Additionally, $m - 1$ Boolean variables d_i with $i \in \{1, \ldots, m - 1\}$ define whether the direction of the bore used to manufacture a cooling duct is identical to the flow direction of the cooling fluid or not. Thus, this new model is very compact and consistent and can easily be utilized by a general purpose EA.

In addition to the shape of the mold and the structure of the cooling ducts, further geometric data is required. To cite some examples, the diameter of the bores and the length of the plugs and connectors have to be specified as well as the size of the safety margins of the various faces of the mold. Along with the model of the mold and the cooling system, it is possible to carry out geometry-based simulations and to visualize the solution. Figure 10 depicts a complete three-dimensional model of a mold for the bottom of bottles with a cooling system designed by the optimization system at hand. The brightness variation of the cavity surface indicates the temperatures simulated by the approach discussed in the following.

4.3 Simulation and Evaluation

The demands on simulation systems utilized by evolutionary algorithms show some special features. Since several thousand evaluations are performed during an evolutionary optimization run, low simulation run times are crucial. Another important aspect is the need for the ability to also evaluate degenerated and infeasible solutions. However, the fact that the accuracy requirements are rather low and that a multiplicity of rather similar solutions is to be evaluated may offer several opportunities to significantly increase the performance of the simulation approaches like utilizing surrogate functions, by precalculating reused values, or by providing promising initial values for search heuristics.

For the optimization problem at hand, analyses of the thermal properties of cooling systems and of the expected production costs as well as analyses regarding the geometric constraints are to be performed. The expected production costs of a cooling system depend on several factors, which all can easily and quickly be determined: the amount and length of the bores, the angles between the bores and the corresponding outer surfaces as well as the relative locations of the bores.

The situation is different with the thermal properties. Due to the complex thermal effects occurring even for simple molds, a more sophisticated simulation approach is required. Conventional general purpose systems for computational fluid dynamics simulations allow for detailed analyses of the filling process on one side and of the flow of the cooling liquid on the other side, but have high demands concerning the computation time. It is similar with finite element simulation systems used to analyze the heat conduction process within the mold. Depending on the accuracy and the complexity of a mold, such simulations last several minutes to hours on a standard personal computer. Therefore, a special surrogate method to estimate the thermal influence of cooling bores on the cavity surface has been developed. The radiation approach is motivated by several commonalities between heat conduction and heat radiation processes. For every cooling segment (the part of the bore passing the cooling liquid) of the cooling system geometrically defined by two adjacent control vertices (V_i and V_{i+1}), the so-called cooling effect $e_{i,j}$ on a point C_j of the cavity surface is determined by

$$e_{i,j} = \int_0^1 \frac{1}{(C_j - (V_i - t \cdot (V_{i+1} - V_i)))^2} dt. \tag{1}$$

This formula takes the location and orientation of the cooling segment relative to the surface point as well as the length of the segment into account.

To approximate the temperature derivation, the cooling effects of all cooling segments are determined and superposed for a large set of points on the cavity surface. Depending on the mold's shape, assuming a linear superposition of the cooling effects of different cooling segments is sufficient. Otherwise, more realistic non-linear approaches should be utilized. Large cooling effect values indicate low cavity surface temperatures, which are represented by brighter gray-scale values in the figures (e.g. Fig. 10). Comparisons with finite element simulations show rather good qualitative correlations, which are in the majority of cases sufficient to distinguish good from less adequate cooling systems as required by an evolutionary algorithm. Typical speed-up factors of about 100,000 can be realized by utilizing the radiation approach instead of finite element simulations, resulting in run-times of less than $1\,\mu s$ for rather simple shapes. Due to its simplicity the radiation approach allows the evaluation of infeasible solutions for example with bores outside the mold or intersecting the cavity.

For the feasibility-analysis of a cooling system, several geometric properties like the distance to the cavity surface and obstacles as well as the distance between non-adjacent bores have to be regarded. Since the triangle meshes of the free form surfaces usually consist of several thousand triangles, brute force approaches would lead to high computation times. Optimization techniques used in computer graphics have been adopted to this problem, delivering a significantly improved performance, resulting again in run-times of less than $1\,\mu s$ for rather simple shapes. Experiments in exploiting the correlation of related solutions provided by EAs delivered promising results. By utilizing common bounding objects for similar bores efficient quick tests regarding collisions or possible exceeding of safety margins can be performed.

After the simulation has been performed, the resulting values are aggregated to characteristic objective values, which are to be minimized during the optimization process. In addition to the production costs, two characteristic values derived from the simulated cavity surface temperature distribution are regarded: the cooling efficiency with the average and the maximum temperature as indicators and the cooling homogeneity derived from the temperature standard derivation. Depending on the amount of objectives independently regarded by the optimization algorithms, these objective values can be aggregated further. At the end a penalty value indicating the degree of constraint violation is added to all objective values.

4.4 Evolutionary Optimization

For the design of cooling duct systems, both single- and multi-objective evolutionary algorithms have been applied [15]. For the single-objective optimization an evolution strategy (ES) with self-adaptive mutation mechanism proposed by Schwefel [20] has been applied. At first, it showed poor performance due to the fact that good solutions of the problem at hand are often close to the constraint bounds. Since infeasible solutions are more likely to

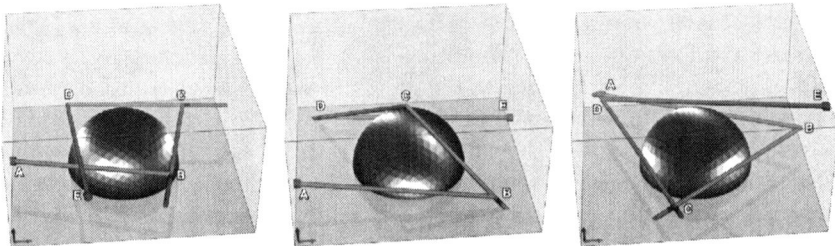

Fig. 11. A multi-modal problem: with just four bores several reasonable cooling duct systems can be designed [16]

occur at high mutation rates in this case, they are avoided by the algorithm, which leads to a stagnation of the evolutionary process. The introduction of a lower bound for the mutation rate decreasing over the generations resulted in a performance boost of the algorithm for the current problem. This approach has been applied to design a cooling systems consisting of up to eight cooling circuits and 50 bores for a rather complex hot stamping tool [22].

In order to get insight into the optimization problem's characteristics, 200 evolutionary runs of a (1,100)-ES over 500 generations have been executed for a simple hemispherically shaped cavity. An analysis with the proposed visualization tool indicates that nearly all runs show reasonable results and many do cover the cavity surface extremely well. Figure 11 depicts that very different reasonable structures can be generated with only four bores. In addition, a huge amount of variants can be derived from these solutions by rotation or symmetric transformations. For cooling systems with more bores, the amount of reasonable solutions will drastically increase. Since cooling systems in practice mostly contain 3–50 bores, the multi-modal character of the problem is very challenging. In order to cope with this, the evolutionary algorithm has been extended to a multi-national algorithm allowing solutions with initially disadvantageous yet promising structure to develop better.

The utilization of a single-objective EA presumes a user with good specific knowledge regarding the optimization problem and the algorithm, since the design of the fitness function aggregating the different objective values has a crucial effect on the solutions provided by the algorithm. To avoid this, a multi-objective evolutionary algorithm on basis of the SMS-EMOA [17] has been implemented. Initially, the run-time required to determine the S-metric values and to perform the non-dominated sorting of the individuals exceeded the time required by the simulation and evaluation by far. In order to significantly improve the speed, two modifications were carried out. First, by regarding only two objective functions, a fast non-dominated sorting algorithm by Jensen [8] with a run-time complexity of $O(n \log n)$ has been utilized. Second, the $(\mu+1)$ selection operator of the SMS-EMOA has been extended to a $(\mu+\lambda)$ and (μ, λ) selection. This allows to evaluate more (λ) new individuals before another pass of the non-dominated sorting algorithm is required. Additionally, the average ratio of S-metric evaluations to newly generated individuals decreases.

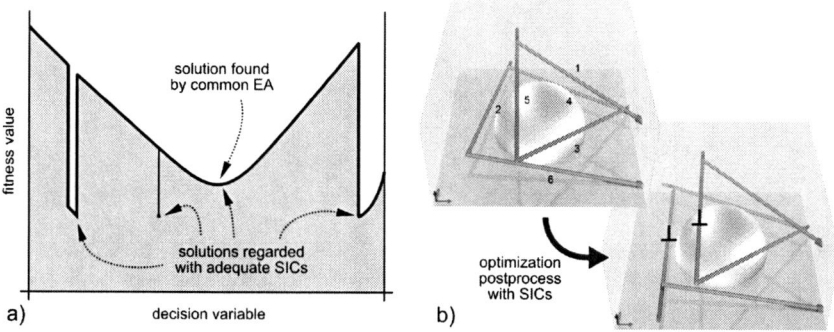

Fig. 12. By utilizing problem-specific self-imposed constraints (SICs), a general purpose EA can be extended to cope with discontinuous objective functions [15]

4.5 Optimization with Self-Imposed Constraints

The objective function regarding the manufacturing costs is highly discontinuous. If certain constraints are regarded, single manufacturing steps can be skipped or more efficient tools can be used. By orienting a bore orthogonal to the outer surface, no pilot drilling preceding the deep hole drilling is required. By positioning a set of bores within a common plane, no reclamping of the mold is required for the production of these bores. There are several further conditions allowing similar cost savings.

Since general purpose EA can hardly deal with discontinuous objective functions, the concept of self-imposed constraints (SICs) [16] has been developed. Problem-specific knowledge of possibly advantageous locations in the decision space is modeled in form of constraint templates. For the optimization a nested EA is utilized: in the outer loop a certain set of SICs is derived from the constraint templates – evolutionary or by specific heuristics. In the inner loop a standard EA is executed, which is modified in such a way, that after every variation of the solutions the current set of SICs is applied.

The concept of self-imposed constraints has been applied to the problem at hand. Significantly more cost-efficient variants of solutions found by common EA could be generated by applying a SIC-EA in a postprocess (Fig. 12b).

4.6 Interactive Data Analysis

Due to the use of the imprecise surrogate function for the thermal simulation and possibly incomplete geometric constraint models, not only the solutions rated Pareto optimal should be regarded, but also other solutions with slightly inferior evaluation and in particular those which are of different structure. Thus, for the final selection and verification of the solutions, the end user is typically confronted with hundreds to thousands of solution candidates. In addition to a meaningful visualization of single solutions, a software support

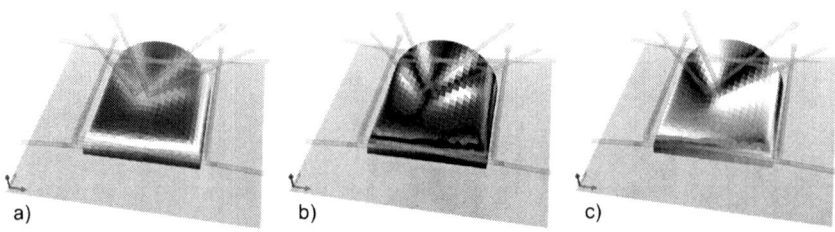

Fig. 13. Two solutions A and B are compared by combining the raw values representing the simulated cavity surface temperature. The corresponding cooling duct systems are displayed semitransparent. (**a**) Maximum max (A+B): Dark areas are not covered by any of the solutions. (**b**) Absolute difference |A−B|: The brightness indicates high temperature differences. (**c**) Difference A−B: The brightness indicates better cooling of solution A than of B

for the navigation through the solution set and classification of the solutions is desirable.

As illustrated in Fig. 11, single solutions of the current problem can be visualized as three-dimensional model. In order to provide a good impression of the location of the bores in the three-dimensional space, an exaggerated perspective, an irregular illumination and a soft depth fog as well as the separation plane reflecting the cooling system are utilized. An optional stereographic visualization of the solutions delivers an even better impression of the model.

For the comparison of two or more solutions, the visualization has been extended as shown in Fig. 13. The semitransparent visualization of the cooling systems in conjunction with an adequate combination of the simulation raw data of the solutions can help to identify cavity surface areas not covered properly by any solution or to highlight the differences of similar solutions.

In order to deal with extensive solution sets, the clustering algorithms of PAVEl have been utilized. As an exemplary data set, the final solutions of the 200 evolutionary runs mentioned before have been used and have been partitioned into 50 clusters with the HACM. Figure 14a shows two typical solutions of a single cluster in case of the decision space has been regarded by the clustering algorithm. Taking a user-defined function – the angle between two adjacent bores – into account, solutions of similar structure are clustered even if the orientation is different (Fig. 14b). Examining the simulation raw data allows to identify different structures causing similar thermal effects (Fig. 15). As Fig. 15 shows, detailed considerations are required if data of different types are regarded by the clustering algorithm simultaneously.

5 Summary and Conclusion

In this chapter, the successful application of evolutionary algorithms to two current production engineering problems was shown. Characteristical for both problems is – as for many other applications – their subject to multiple

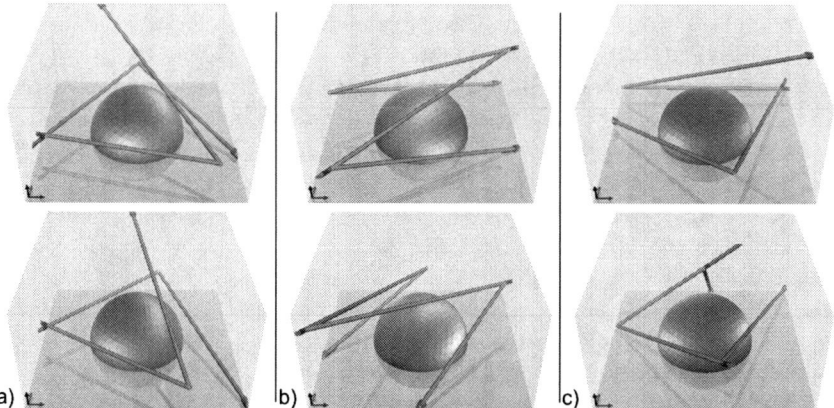

Fig. 14. Depending on the criteria the cluster algorithm groups different solutions. (a) Decision variable values: similar in structure and position; (b) angle between adjacent bores: similar structure, arbitrary position and (c) simulation raw data: arbitrary structures causing similar temperature distribution

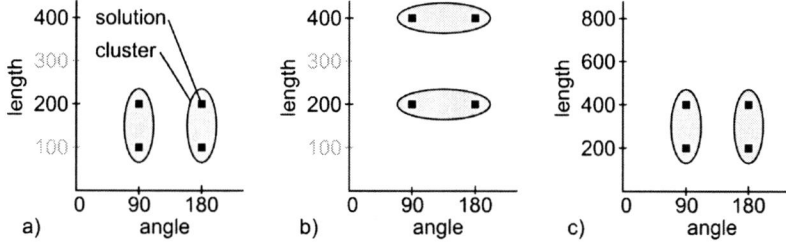

Fig. 15. (a) The length of a bore and the angle between two bores are regarded by an algorithm clustering four solutions. (b) If the model is scaled by the factor two, the length increases but the angle does not, resulting in different clusters. (c) Instead of the original values, the geometric distances of solutions according to the visualization can be taken into account, so the user can influence the clustering results by changing the view scale (e.g. by applying a scale factor of two)

objectives and restrictions as well as the use of simulations and of surrogate functions for evaluation. Due to the high complexity of the introduced real world problems, which results especially from the high-dimensional decision space in the case of the milling optimization application and from the strongly multi-modal problem of optimizing cooling duct layouts, adaptations of established algorithms to the problem-specific requirements have been realized.

Due to the fact that several thousand evaluations have to be performed during an evolutionary optimization run, there are high demands on the run-time of the used simulation systems. For both optimization applications, approaches which significantly increase the simulation performance have been shown. Hybrid algorithms are used as well as surrogate functions and methods for the precalculation of reused values.

In order to gain insight into the characteristics of the optimization problem, it is advisable to visualize all the information which results from the evolutionary process. It is insufficient to consider only a few solutions. The simultaneous presentation of all three spaces – the decision, objective, as well as the raw data space – and the interactive navigation through the solution sets can help to detect process-specific behavior. In the case of optimizing the NC-paths for five-axis milling, the visualization of the solution mapping to the β–γ-area has shown that the solutions are concentrated only on a restricted area of the whole solution space. The development of a multi-national approach solves this problem.

Due to the fact that typically used evaluation models are often simplified and imprecise, not only the solutions, which are evaluated best, should be considered. To deal with this, bringing in the expert's knowledge is advisable even for single-objective problems. As the example applications show, optimization algorithms for real world problems may provide extensive solution sets from which the most suitable has to be selected. For this purpose an appropriate data visualization and preparation has proven to be helpful.

References

1. Anotaipaiboon W, Makhanov SS (2005) Tool path generation for five-axis NC machining using adaptive space-filling curves. In: International Journal of Production Research, 43(8):1643–1665
2. Brezocnik M, Kovacic M, Ficko M (2004) Prediction of surface roughness with genetic programming. In: Journal of Materials Processing Technologies, 157–158:28–36
3. Coello Coello CA, Van Veldhuizen DA, Lamond GB (2002) Evolutionary Algorithms for Solving Multi-Objective Problems. Kluwer, Drodretch
4. Dragomatz D, Mann SA (1997) Classified bibliography of literature on NC tool path generation. In: Computer-Aided Design, 29(3):239–247
5. Emmerich M, Beume N, Naujoks B (2005) An EMO algorithm using the hyper-volume measure as selection criterion. In: Evolutionary Multi-Criterion Optimization: Third International Conference, EMO 2005:62–76, Springer
6. Foley JD, Van Dam A, Feiner SK, Hughes JF (1995) Computer Graphics, Principles and Practice. Addison-Wesley, Reading, MA
7. Jain AK, Murty MN, Flynn PJ (1999) Data clustering: A review. In: ACM Computing Surveys, 31(3):264–323
8. Jensen MT (2003) Reducing the run-time complexity of multiobjective EAs: The NSGA-II and other algorithms. In: IEEE Transactions on Evolutionary Computation, 7(5):503–515
9. Jun CS, Cha K, Lee YS (2003) Optimizing tool orientations for 5-axis machining by configuration-space search method. In: Computer-Aided Design, 35:549–566
10. Keim DA (2000) Designing Pixel-Oriented Visualization Techniques: Theory and Applications. IEEE Transactions on Visualization and Computer Graphics, 6(1)

11. Kennedy J, Eberhart R (1995) Particle Swarm Optimization. In: IEEE International Conference on Neural Networks, IV:1942–1948, IEEE Service Center
12. Kerren A, Egger T (2005) EAVis: A Visualization Tool for Evolutionary Algorithms. Procedings of the 2005 IEEE Symposium on Visual Languages and Human-Centric Computation, pp 299–301
13. Kovacic M, Balic J, Brezocnik M (2004) Evolutionary approach for cutting force prediction in milling. In: Journal of Materials Processing Technology, 155–156 (part 2):1647–1652
14. Menges G, Michaeli W, Mohren P (2001) How to Make Injection Molds. Hanser
15. Michelitsch T (2007) Fertigungsgerechtes Optimieren von Temperierbohrungssystemen. Innovative Prozesse im Werkzeug- und Formenbau, pp 131–149
16. Michelitsch T, Mehnen J (2006) Optimization of production engineering problems with discontinuous cost-functions. Proceedings of the 5th CIRP ICME, pp 275–280
17. Naujoks B, Beume N, Emmerich M (2005) Multi-objective optimisation using S-metric Selection: Application to three-dimensional Solution Spaces. In: Proceedings of the Congress on Evolutionary Computation, 2:1282–1289, IEEE Press
18. Paulsen P (2001) Dictionary of Production Engineering, Metal Forming 1. Springer, Berlin Heidelberg New York
19. Pohlheim H (1999) Visualization of Evolutionary Algorithms – Set of Standard Techniques and Multidimensional Visualization. Proceedings of the GECCO, pp 533–540
20. Schwefel HP (1981) Numerical Optimization of Computer Models. Wiley
21. Silverman BW (1986) Density estimation for statistics and data analysis. Chapman and Hall, UK
22. Steinbeiss H, Hyunwoo S, Michelitsch T, Hoffmann H (2007) Method for optimizing the cooling design of hot stamping tools. Production Engineering, DOI 10.1007/s11740-007-0010-3, Springer, Berlin Heidelberg New York
23. Surmann T, Kalveram M, Weinert K (2005) Simulation of cutting tool vibrations for the milling of free formed surfaces. In: Proceedings of the 8th CIRP International Workshop on Modeling of Machining Operations, pp 175–182
24. Tandon V, El-Mounayri H, Kishawy H (2002) NC end milling optimization using evolutionary computation. In: International Journal of Machine Tools and Manufacture, 42:595–605
25. Ursem R (1999) Multinational Evolutionary Algorithms. In: Proceedings of the Congress on Evolutionary Computation, 3:1633–1640
26. Valvo EL, Martuscelli B, Piacentini M (2004) NC End Milling Optimization within CAD/CAM System Using Particle Swarm Optimization. In: Proceedings of 4th CIRP ICME
27. Vigouroux JL, Deshayes L, Foufou S, Welsh LA (2007) An approach for optimization of machining parameters under uncertainties using intervals and evolutionary algorithms. International Conference On Smart Machining Systems
28. Weinert K, Kersting P (2007) Effiziente Kollisionsberechnung optimiert das 5-Achs-Fräsen. In: MM Maschinenmarkt, 4:28–33
29. Weinert K, Zabel A (2001) Modeling, Simulation, and Visualization of Simultaneous Five-Axis Milling with a Hexapod Machine Tool. In: Simulation in Industry. 13th European Simulation Symposium, pp 344–348, SCS

30. Weinert K, Zabel A, Müller H, Kersting P (2006) Optimizing NC Tool Paths for Five-Axis Milling using Evolutionary Algorithms on Wavelets. In: Proceedings of the Genetic and Evolutionary Computation Conference, pp 1809–1816
31. Wu AS, De Jong KA, Burke DS, Grefenstette JJ, Ramsey CL (1999) Visual analysis of evolutionary algorithms. In: Proceedings of the CEC, 2:1419–1425
32. Xu R, Wunsch DII (2005) Survey of clustering algorithms. In: IEEE Transactions on Neural Networks, 16(3):645–678
33. Yang J, Ward MO, Rundensteiner EA (2003) Interactive hierarchical displays: a general framework for visualization and exploration of large multivariate data sets. In: Computers and Graphics, 27(2):265–283
34. Yoshikawa T, Yamashiro D, Furuhashi T (2007) Visualization of multi-objective pareto solutions – devolopment of mining technique for solutions. Late Breaking Papers of the Fourth International Conference on Evolutionary Multi-Criterion Optimization, pp 1–6
35. Zabel A, Müller H, Stautner M, Kersting P (2006) Improvement of machine tool movement for simultaneous five-axes milling. In: Proceedings of 5th CIRP ICME
36. Zitzler E, Thiele L (1998) Multiobjective optimization using evolutionary algorithms – a comparative case study. Eiben A. E. (ed), Parallel Problem Solving from Nature V, pp 292–301, Springer, Berlin Heidelberg New York
37. Zitzler E, Thiele L, Laumanns M, Fonseca CM, Grunert da Fonseca V (2003) performance assessment of multiobjective optimizers: an analysis and review. In: IEEE Transactions on Evolutionary Computation, 7(2):117–132

Privacy Protection with Genetic Algorithms

Agusti Solanas

CRISES Research Group.
UNESCO Chair in Data Privacy
Dept. Computer Engineering and Maths.
Rovira i Virgili University. Tarragona. Catalonia (Spain)
agusti.solanas@urv.cat

Summary. In this chapter we describe some of the most important privacy problems that the Information Society has to face in a variety of fields, namely ecommerce, health care, etc. We next propose a solution based on data aggregation, which is known to be NP-hard when applied to multivariate data. Due to this computational complexity, it is necessary to use heuristics to solve the problem. We propose the use of a Genetic Algorithm (GA) conveniently tuned up for tackling the problem properly.

1 Introduction

> *"Seduced by the effortless gathering of data, we discount the costs of turning data into information, information into knowledge and knowledge into wisdom."*

> **B. Harris, 1987**

When Harris wrote the above sentence in 1987 he could not imagine how easy it would be in the twenty-first century to gather huge amounts of data whilst, at the same time, transforming them into information, and next into knowledge. The last step, to turn knowledge into wisdom, is still not reached by computers and is even difficult to be achieved by most humans.

We want to obtain the information we need as fast as possible, from everywhere, and at any time. To that end, our storage and communication methods have evolved e.g. from the classical paper-based mail carried by postmen to the e-mail that fleetingly crosses the world through optical fibre. This achievement is the result of years of research and efforts. Research on information and communications technologies (ICTs) is a keystone of the Europe's 7th Research Framework Programme [31] that will run until 2013. The European Union acknowledges the importance of ICTs for the growth and competitiveness of Europe, and plans to invest over € 9 billion in research on ICTs. This investment will lead to the development of more efficient techniques for gathering, storing and sharing data by using a variety of devices, namely computers,

A. Solanas: *Privacy Protection with Genetic Algorithms*, Studies in Computational Intelligence (SCI) **92**, 215–237 (2008)
www.springerlink.com

mobile phones, smart cards, PDAs and even the interactive digital terrestrial television (DTT).

The great advances achieved on ICTs in the last few years, and the new ones that we envisage, pave the way for the storage and analysis of a huge amount of information. This information, which is growing with each passing day, is gathered by a plethora of institutions, namely statistical agencies, private companies, health-care institutions and universities among others.

The balance between the right of the society to information and the right of the individuals to privacy is on the line. A number of examples of this contradictory situation can be found in our daily life.

- *Commercial information.* Supermarkets may gather information from the fidelity cards of their customers in order to improve the services they offer to them or to increase their annual benefits. On the other hand, if this information is misused, the privacy of the costumers could be in jeopardy because e.g. their consumer habits could be inferred from the data and they can be flooded with undesired advertisements.
- *E-commerce.* Electronic commerce results in the automated collection of large amounts of consumer data. This wealth of information is very useful to companies, which are often interested in sharing it with their subsidiaries or partners. Such consumer information transfer should not result in public profiling of individuals and is subject to strict regulation.
- *Statistical Agencies.* Most countries have legislation which compels national statistical agencies to guarantee statistical confidentiality when they release data collected from citizens or companies; see [25] for regulations in the European Union, [26] for regulations in Canada, and [32] for regulations in the U.S.
- *Health information.* This is one of the most sensitive areas regarding privacy. For example, in the U.S., the Privacy Rule of the Health Insurance Portability and Accountability Act (HIPAA, [13]) requires the strict regulation of protected health information for use in medical research. In most western countries, the situation is similar, which has caused e-health to become a hot topic in privacy research (see e.g. [1]).
- *Location-Based Services.* Cell phones have become ubiquitous and services related to the current position of the user are growing fast. If the queries that a user makes to a location-based server are not securely managed, it could be possible to infer the consumer habits of the user, e.g. a third party could know that a given user likes Chinese food if she usually makes queries from a Chinese restaurant.
- *RFID technology.* The massive deployment of the Radio Frequency IDentification (RFID) technology is a reality. In the near future almost every object will contain an RFID tag capable of transmitting information under the request of an RFID reader. On the one hand, this technology will increase the efficiency of supply chains and will eventually replace bar codes. On the other hand, the existence of RFID tags in almost every

object that we have, could be a privacy problem e.g. an eavesdropper in the street will be able to scan the RFID tags of someone close to him and he will be able to determine the amount of money that she has, what kind of clothes she wears and even whether she has suffered from a hip operation and she has a prosthesis.

These problems directly affect each of us in almost every aspect of our daily life. If the proper actions are not taken, our personal data could be in jeopardy due to the data storage and the advances in data mining techniques.

Releasing statistical information for public use can jeopardise the privacy of individual respondents on which the data were collected. A possible solution leveraged from the field of Statistical Disclosure Control (SDC) is the aggregation of the original data before its publication. By aggregating the data, it becomes statistically unprovable to link the released data with a given respondent and, at the same time, data utility is preserved. Unfortunately, this solution is not straightforward because the elements composing the data must be properly clustered prior to the aggregation process. The clustering of data with constraints on the size of the clusters is known to be an NP-Hard problem when the data is multivariate. Thus, it is necessary to make use of heuristics in order to properly aggregate multivariate data.

1.1 Contribution and Plan of the Chapter

In this chapter we tackle the privacy problems that the Information Society has to face. In order to address them, we propose the use of micro-aggregation as the method for distorting data and guaranteeing respondents privacy. Unfortunately, optimal micro-aggregation is known to be NP-Hard and heuristics have to be used.

We suggest the use of a GA for solving the micro-aggregation problem. Moreover, we show how to tune up a classical GA, i.e. how to codify the chromosomes, how to choose the most effective parameters for the mutation and crossover rates, etc.

The rest of the chapter is organised as follows. Section 2 gives an overview of the micro-aggregation problem and introduces some important concepts and renown micro-aggregation methods. In Sect. 3, we show how to solve the micro-aggregation problem by using a GA. Section 4 presents an improvement of the method in Sect. 3 that permits the aggregation of large files. Section 5 provides some experimental results and, finally, Sect. 6 finishes with some conclusions and future challenges.

2 Basics of Micro-Aggregation

Statistical Disclosure Control (SDC), also known as Statistical Disclosure Limitation (SDL), seeks to transform data in such a way that they can be publicly released whilst preserving data utility and statistical confidentiality, where the latter means avoiding disclosure of information that can be linked to specific individual or corporate respondent entities.

The *SDC problem* is to modify data in such a way that sufficient protection is provided whilst minimising information loss, i.e. the loss of the accuracy sought by users.

Micro-aggregation is an SDC technique consisting in the aggregation of individual data. It can be considered as an SDC subdiscipline devoted to the protection of individual data, also called micro-data. It is only recently that data collectors (statistical agencies and the like) have been persuaded to publish micro-data. Therefore, micro-data protection is the youngest SDC subdiscipline and it is experiencing continuous evolution in the last years. Micro-aggregation can be seen as a clustering problem with constraints on the size of the clusters. It is somehow related to other clustering problems (e.g. dimension reduction or minimum squares design of clusters). However, the main difference of the micro-aggregation problem is that it does not consider the number of clusters to generate or the number of dimensions to reduce, but only the minimum number of elements that are grouped in the same cluster.

When we micro-aggregate data we have to keep two goals in mind:

- *Preserving data utility.* To do so, we should introduce as less noise as possible into the data, i.e. we should aggregate similar elements instead of different ones. In the example given in Fig. 1 for a security parameter $k = 3$, groups of three elements are built and aggregated. Note that elements in the same aggregation group are similar.
- *Protecting the privacy of the respondents.* Data have to be sufficiently modified to make re-identification difficult, i.e. by increasing the number

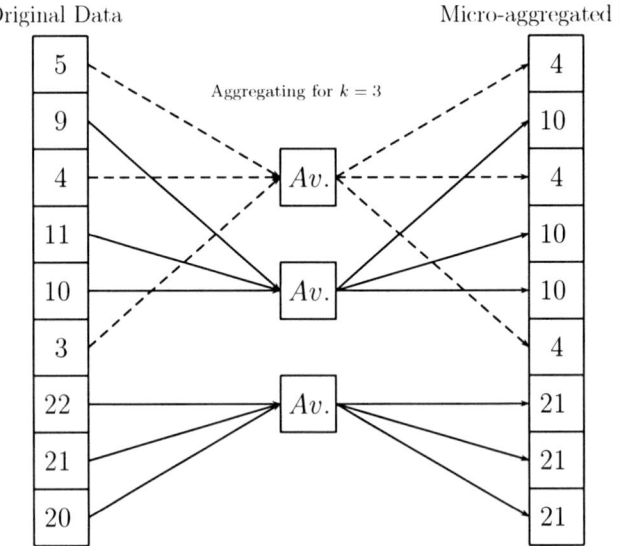

Fig. 1. k-Aggregation example with $k = 3$

of aggregated elements, we increase data privacy. In the example given in Fig. 1, after aggregating the chosen elements, it is impossible to distinguish them, so that the probability of linking any respondent is inversely proportional to the number of aggregated elements.

In order to determine whether two elements are similar, a similarity function such as the Euclidean Distance can be used. By means of this similarity function, we can define a homogeneity measure **HOM** for a given univariate data set $\mathbf{D_{uni}}$ such as the one in Expression 1.

$$\mathbf{HOM_{D_{uni}}} = \sum_{i=1}^{n} \mathbf{x_i} - \bar{\mathbf{x}}, \tag{1}$$

where n is the number of elements in $\mathbf{D_{uni}}$, $\mathbf{x_i}$ is the i-th element in $\mathbf{D_{uni}}$ and $\bar{\mathbf{x}}$ is the average element, i.e. $\bar{\mathbf{x}} = \frac{\sum_{i=0}^{n} \mathbf{x_i}}{n}$. This expression can be extended to multivariate data as follows:

$$\mathbf{HOM_{D_{multi}}} = \sum_{i=1}^{n} \sum_{j=1}^{d} \mathbf{x_{ij}} - \bar{\mathbf{x}_j}, \tag{2}$$

where n is the number of elements in $\mathbf{D_{multi}}$, d is the number of dimensions of each element, $\mathbf{x_{ij}}$ is the j-th component of the i-th element in $\mathbf{D_{multi}}$ and $\bar{\mathbf{x}_j}$ is the j-th component of the average element.

Expression 2 can be extended to multiple subsets and it is known as the Sum of Squared Errors (SSE) cf. Expression 3.

$$\mathbf{SSE} = \sum_{i=1}^{s} \sum_{j=1}^{n_i} (\mathbf{x_{ij}} - \bar{\mathbf{x}_i})'(\mathbf{x_{ij}} - \bar{\mathbf{x}_i}), \tag{3}$$

where s is the number of subsets, n_i is the number of elements in the i-th subset, $\mathbf{x_{ij}}$ is the j-th element in the i-th subset and $\bar{\mathbf{x}_i}$ is the average element of the i-th subset.

Given a homogeneity measure such as the SSE and a security parameter k, which determines the minimum cardinality of the subsets, the microaggregation or k-micro-aggregation problem can be enunciated as follows:

> *Given a data set* **D** *built up of* **n** *elements in a characteristic space* $\mathbb{R}^d$, *the problem consists in obtaining a k-partition[1] $\mathcal{P}$ of* **D** *so that the homogeneity of $\mathcal{P}$ is maximised. Once $\mathcal{P}$ is obtained, each element of every part of $\mathcal{P}$ is replaced by the average element of the part.*

This problem is known to be NP-hard [23] for multivariate data sets so heuristic methods must be used to solve it.

---

[1] A k-partition of **D** is a partition where its parts have, at least, k elements of **D**.

2.1 Multivariate Micro-Aggregation Methods

There is a plethora of methods to address the multivariate micro-aggregation problem (see e.g. [33] *for a more detailed approach*). In this section we give a brief overview of the most relevant techniques and we point out some of their shortcomings.

Minimum Spanning Tree Partitioning (MSTP)

The MSTP algorithm for micro-aggregation was proposed in [18] as a variable-size multivariate micro-aggregation method. The method is based in the well-known Minimum Spanning Tree (MST) problem. However, the standard MST partitioning algorithm does not solve the micro-aggregation problem since it does not consider the group size constraints.

In order to address this problem, the authors proposed a modification of the MSTP algorithm that takes into account the constraints on the size of the clusters. The algorithm is divided in three main steps:

1. MST construction: Constructs a MST by using the method of Prim and a Euclidean metric. Moreover, the authors augment the algorithm by adding a priority queue based on the edge length and a descendant count, that determines the number of nodes under a given node, in each node of the tree.
2. Edge cutting: The algorithm iteratively visits every MST edge in length order, from longest to shortest, and deletes the removable edges whilst retaining the remaining edges. An edge is said to be removable if the forest that results from the cutting is acceptable, i.e. each tree in the forest has at least k nodes. Note that this condition guarantees the final partition to be a k-partition.
3. Cluster generation: Once the forest has been constructed, a recursive algorithm that visits every node in the forest is applied in order to assign every node (*point*) to a cluster (*part*) (see [18] for further details). At the end of this process a k-partition is obtained.

The main limitation of this algorithm remains in its foundation, i.e. in the minimum spanning tree. Although it could be very useful when the data are distributed in clusters, it fails to properly adapt to the data when the points (elements) are distributed in a scattered way.

Maximum Distance Method (MD)

The MD was proposed in [7] as a multivariate micro-aggregation method. The main advantage of this method is its simplicity. Moreover, it achieves very good results in most data sets.

The MD algorithm builds a k-partition as follows:

1. Let r and s be the two most distant records in the data set, using the Euclidean Distance; form a group with r and the $k-1$ records closest to r; form a group with s and the $k-1$ records closest to s.
2. If there are at least $2k$ records which do not belong to any of the groups formed in Step 1, go to Step 1 taking as the new data set the previous data set minus the groups formed in the previous instance of Step 1.
3. If there are between k and $2k-1$ records which do not belong to any of the groups formed in Step 1, form a new group with those points and exit the algorithm.
4. If there are less than k records which do not belong to any of the groups formed in Step 1, add them to the closest group formed in Step 1.

At this moment, we have a k-partition of the data set. As explained above, given a k-partition, the micro-aggregated data are computed by replacing each record by the centroid of the group to which it belongs.

The main shortcoming of this method is its computational complexity, i.e. $O(n^3)$.

Maximum Distance to Average Vector Method (MDAV)

The main drawback of MD is its computational complexity because of the cost of finding the most distant records at each iteration. The Maximum Distance to Average Vector method (MDAV) improves on MD in terms of computational complexity whilst maintaining the performance in terms of resulting SSE.

MDAV was proposed in [9, 15] as part of a multivariate micro-aggregation method implemented in the μ-Argus package for statistical disclosure control. A slight modification of the same construction was later described in [18] under the name centroid-based fixed size micro-aggregation.

MDAV is the most popular method used for micro-aggregating data sets. Thus, we will explain it in detail and we will compare it with our genetic approach.

MDAV builds a k-partition as follows. First, a square matrix of distances between all records is computed (line 1 in Algorithm 1). Two main approaches can be adopted to perform these distance calculations. The first approach is to compute and store the distances at the beginning of the micro-aggregation process. The second approach consists in computing the distances on the fly when they are needed. The first approach is computationally cheaper but it requires too much memory when the number of records in the data set is large.

After computing the matrix of distances, MDAV iterates (lines 4-13) and builds two groups at each iteration. In order to build these groups, the centroid c – i.e. the average vector – of the remaining records – those which have not yet been assigned to any group – is computed at the beginning of each iteration (line 5). Then the most distant record r from c is taken (line 6) and a group of

---

Algorithm 1: Maximum Distance to Average Vector (MDAV)

    Data: Data set $\mathbf{X}$, Integer k, Integer n.
    Result: k-partition.
1 ComputeDistanceMatrix($\mathbf{X}$);
2 $RR = n$; //Remaining Records
3 $i = 0$;
4 **while** $RR > 2k - 1$ **do**
5   c = ComputeNewCentroid($\mathbf{X}$);
6   r = GetFarthestRecordFromCentroid($\mathbf{X}$,c);
7   s = GetFarthestRecordFromRecord ($\mathbf{X}$,r);
8   g_i = BuildGroupAroundRecord(r,$\mathbf{X}$,k);
9     $i = i + 1$;
10   g_i = BuildGroupAroundRecord(s,$\mathbf{X}$,k);
11    $i = i + 1$;
12    $RR = RR - 2k$;
13 **end**
14 $g_1 \ldots g_s$ = AssignRemainingRec($\mathbf{X}$, $g_1 \ldots g_s$);
15 **return** $g_1 \ldots g_s$

---

k records is built around r (line 8). The group of k records around r is formed by r and the $k - 1$ closest records to r. Next, the most distant record s from r is taken (line 7) and a group of k records is built around s (line 10). The generation of groups continues until the number of remaining records (RR) is less than $2k$. When this condition is met, two cases are possible, namely $RR < k$ or $RR \geq k$. In the first case, the remaining records are assigned to their closest group. In the second case, a new group is built with all the remaining records. Note that all groups have k elements except perhaps the last one.

Finally, given the k-partition obtained by MDAV, a micro-aggregated data set is computed by replacing each record in the original data set by the centroid of the group to which it belongs. The main problem of this algorithm is its lack of flexibility. It only generates subsets of fixed cardinality k and, due to this limitation, it can suffer from a performance reduction when applied to data sets where the points are distributed in clusters of variable cardinality.

Variable-MDAV

MDAV generates groups of fixed size k and, thus, it lacks flexibility for adapting the group size to the distribution of the records in the data set, which may result in poor within-group homogeneity.

Variable-size MDAV (V-MDAV) [28] algorithm intends to overcome this limitation by computing a variable-size k-partition with a computational cost similar to the MDAV cost.

The algorithm for building a k-partition using V-MDAV is as follows:

1. Compute the distances between the records and store them in a distance matrix.
2. Compute the centroid c of the data set.
3. Whilst there are more than $k - 1$ records not yet assigned to a group do:
   a) Let e be the most distant record to c.
   b) Form a group around e which contains the $k - 1$ records closest to e.
   c) Extend the group, which consists of adding to the current group up to $k - 1$ records using these steps:
      i. Find the unassigned record e_{min} which is closest to any of the records of the current group and let d_{in} be the distance between e_{min} and the group;
      ii. Let d_{out} be the shortest distance from e_{min} to the other unassigned records;
      iii. If $d_{in} < \gamma d_{out}$ then assign e_{min} to the current group. γ is a gain factor that has to be tuned according to the data set.
4. If less than k records remain unassigned, no new group can be formed. In that case, assign the remaining records to their closest group.

At the time of writing this chapter, MDAV is the most popular algorithm used to micro-aggregate multivariate data sets. Thus, we will use it as a reference to compare with our genetic approach.

3 Privacy Protection Through Genetic-Algorithm-Based Micro-aggregation

As we have shown in previous sections, the micro-aggregation problem is known to be NP-hard. Looking for optimal solutions in a multivariate data set is a very difficult task. Genetic algorithms are very well suited for searching solutions in complex search spaces. Therefore, a genetic-based technique seems to be a good candidate for solving this privacy problem.

We use the concept of GA described by Holland in [14]. A GA is a method for moving from one population of chromosomes to a new population by using a kind of natural selection together with the genetics-inspired operators of crossover, mutation, and inversion. Each chromosome consists of genes, each gene being an instance of a particular allele [22]. A GA depends on a number of parameters, namely population size, mutation rate, crossover rate and so on. The value of these parameters is typically initialised by following the recommendations of experts [5,10], although statistical techniques can be used to tune these parameters [4].

In a nutshell, a GA is a function optimiser that can be described in terms of a population of chromosomes, a coding sequence, a selection algorithm, a fitness function and some genetic operators. The chromosomes are candidate solutions to the problem which are encoded by means of the coding sequence. The selection algorithm determines which chromosomes will generate offspring. The fitness function evaluates how good a chromosome is and the genetic operators are used to mix the genetic information of the selected chromosomes in order to generate the offspring.

In order to address the micro-aggregation problem, a GA must be properly modified. The modifications introduced in our scheme with respect to a classical GA are intended to boost performance when micro-aggregating multivariate data. We next deal with those modifications in detail.

3.1 The Coding Sequence

The way in which candidate solutions are encoded is a key factor in the success of a GA [22].

Searching a proper coding could be a bewildering task due to the array of possible coding alternatives. However, at the same time, looking for a good coding among a plethora of alphabets is an invigorating challenge.

Codings can be mainly classified according to their alphabet in three main categories:

- Binary codings: They are based on a two-symbol alphabet (e.g. 0 and 1) and have been widely used because they were the codings initially introduced by Holland in his pioneering work [14]. Some recent examples are [19, 24, 30]. Moreover, lots of heuristics based on binary codings have been used as meta-heuristics in order to determine the value of parameters such as the crossover rate and the mutation rate [4, 12, 27].
- N-ary codings: They differ from binary codings in that each gene can take N different alleles[2]. As it will be shown next, this coding is our choice for addressing the multivariate micro-aggregation problem.
- Real-valued codings: These codings assign a real value to each gene. This kind of coding is basically used when the representation of a real number in a binary alphabet is awkward, this is, when the number's precision or range are not previously and clearly known [3, 21, 35].

In [7] a binary coding was used to solve the problem of **univariate** micro-aggregation. The idea was to sort univariate records in ascending order and to represent the i-th record by the i-th symbol in a binary string (chromosome). Thus, a chromosome represents a k-partition as follows. If a cluster starts at the i-th record, its bit in the chromosome is initialised with a "1" symbol; otherwise it is set to "0". This coding makes sense for univariate data sets because they can be sorted. Unfortunately, no generalisation to the multivariate case is possible because there is no way of sorting multivariate records

---

[2] It is clear that all problems that can be encoded using N-ary codings can be also encoded by using a binary alphabet which would encode the N-ary alphabet. However, some problems are more easily and naturally encoded by using an N-ary alphabet [17, 34].

without giving a higher priority to one of the attributes. To overcome this limitation, we propose a new coding that can be applied to data sets containing any number of dimensions.

Micro-aggregation requires that each group in a k-partition contain at least k records. Thus, the maximum number G of groups in a k-partition is

$$G = \left\lfloor \frac{n}{k} \right\rfloor, \tag{4}$$

where n is the total number of records to be micro-aggregated. The maximum number of groups G is the number of different alleles which must be defined in order to be able to encode a valid solution of the problem within each chromosome. Thus, the cardinality of our alphabet is G and each symbol of the alphabet represents one group/part of the k-partition.

We consider that each chromosome has a fixed number of genes equalling the number of records in the data set. This is, the value of the i-th gene in a chromosome defines the group of the k-partition which the i-th record in the data set belongs to.

Example 1. Taking the data in Table 1 and considering $k = 3$ we use (4) to get that $G = 3$. We then define a 3-character alphabet (e.g. A, B and C) that will be used to encode our chromosomes. Since we have 11 records in the data set, each chromosome is built up of 11 genes. A valid chromosome could be $AAABBBCCCAA$ which encodes the following 3-partition: group $A = \{1,2,3,10,11\}$, group $B = \{4,5,6\}$ and group $C = \{7,8,9\}$

Note that using alphabets with more than G symbols is conceivable, but the *principle of minimum alphabets* states that the smallest alphabet that permits a natural expression of the problem should be selected [11].

Table 1. Example: SME data

Company name	Surface (m²)	Employees	Turnover (€)	Net profit (€)
A&A Ltd	790	55	32,12,334	3,13,250
B&B SpA	710	44	22,83,340	2,99,876
C&C Inc	730	32	19,89,233	2,00,213
D&D BV	810	17	9,84,983	1,43,211
E&E SL	950	3	1,94,232	51,233
F&F GmbH	510	25	1,19,332	20,333
G&G AG	400	45	30,12,444	5,01,233
H&H SA	330	50	42,33,312	7,77,882
I&I LLC	510	5	1,59,999	60,388
J&J Co	760	52	53,33,442	10,01,233
K&K Sarl	50	12	6,45,223	3,33,010

"Company name" is an identifier to be suppressed before publishing the data set

3.2 Initialising the Population

The initialisation of the population is not a key issue when all the possible combinations of the symbols in the alphabet are candidate solutions. Generally, the initialisation of the chromosomes in the population is performed at random, i.e. by using a pseudo-random generator mapped over the alphabet of the coding. Pseudo-random chromosome initialisation with n records and G different alphabet symbols results in

$$V'_{(G,n)} = G^n \tag{5}$$

possible chromosomes. Unfortunately, only a small fraction of such chromosomes meet the cardinality constraints to represent valid k-partitions.

Domingo-Ferrer and Mateo-Sanz [7] proved that

Result 1 *In an optimal k-partition, each group has between k and $2k - 1$ records.*

From Result 1 the minimum number of groups in a k-partitions is

$$g = \left\lceil \frac{n}{2k - 1} \right\rceil . \tag{6}$$

The probability of randomly obtaining a k-partition candidate to optimality (that is, meeting the cardinality constraints of Result 1) is the total number of candidate optimal k-partitions divided by the number of possible chromosomes given in (5). Computing the exact number of k-partitions candidate to optimality is not straightforward. However, in [29] an upper bound in the number of candidates is given by following the next steps.

1. First, the number $a_{n,k}$ of *representatives*[3] of optimal k-partitions given n and k is determined by using the next recursive definition for $a_{n,k}$

$$a_{n,p} = 0 \ \text{if} \ n < p, k \le p \le 2k - 1;$$

$$a_{n,p} = 1 \ \text{if} \ p \le n \le 2k - 1, k \le p \le 2k - 1;$$

otherwise, if $n \ge 2k$

$$a_{n,p} = a_{n-p,p} + a_{n-(p+1),p+1} + a_{n-(p+2),p+2} + \dots$$

$$\dots + a_{n-(2k-1),2k-1}, k \le p \le 2k - 1; \tag{7}$$

where $a_{n,p}$ denotes the number of representatives with n elements with the first subset having a cardinality $\ge p$.

Table 2 lists the values of $a_{n,k}$ for several choices of k and n.

---

[3] A representative is a k-partition with its groups enumerated in ascending order of cardinality. Thus, the k-partition with three groups of cardinalities 3, 4 and 5, respectively, represents all k-partitions formed by three groups with the same cardinalities (i.e. $\{3, 4, 5\}$, $\{3, 5, 4\}$, $\{4, 3, 5\}$, $\{4, 5, 3\}$, $\{5, 3, 4\}$ and $\{5, 4, 3\}$).

Table 2. Values of $a_{n,k}$ for several choices of k and n

n	$k = 3$	$k = 4$	$k = 5$
10	2	2	1
20	6	6	5
30	11	14	14
40	18	27	29
50	26	45	56
100	94	272	520
500	2,134	26,477	1,97,549
1,000	8,434	2,05,037	29,53,389

2. The representative with the maximal number G of groups is taken and the number μ of different candidate optimal k-partitions it represents is obtained by using the following multinomial-inspired formula.

$$\mu = \frac{n!}{k_1! k_2! \ldots k_G!} \frac{1}{G!},\tag{8}$$

where k_m is the cardinality of the m-th group in the k-partition. Clearly, μ is an upper bound for the number of candidate optimal k-partition represented by any representative.

3. Finally, multiplying μ by the number of representatives $a_{n,k}$ and dividing by the number G^n of random chromosomes, an upper bound on the probability $P(\text{random candidate optimal})$ of hitting a candidate optimal k-partition by random initialisation is obtained:

$$P(\text{random candidate optimal}) \leq \frac{a_{n,k}\mu}{G^n}.\tag{9}$$

Example 2. Considering a data set with $n = 30$ records/elements and cardinality constraint $k = 3$, the probability of obtaining a candidate to optimal chromosome at random is

$$P(\text{random candidate optimal}) \leq 13,2977212023656 \times 10^{-12} \approx 0$$

From Example 2, it becomes clear that random initialisation is not suitable to obtain candidate optimal k-partitions. The cardinality constraints stated in Result 1 must be embedded in the initialisation procedure. Thus, Algorithm 2 is used to initialise the chromosomes with solutions that are candidate to optimality.

The application of Algorithm 2 guarantees that each group/part has at most $2k - 1$ elements. However, groups with less than k elements can exist. In order to avoid this problem, elements from the most populated groups are randomly moved to the groups with less than k elements. At the end of this process, chromosomes representing k-partitions which are candidate to be optimal are obtained.

Algorithm 2: Chromosome initialisation with constraints

        Data: ***inout*** chromosome, ***in*** k.
        Result: k-partition.
1   gene_number = 0
2   **while** *gene_number < chromosome_size* **do**
3   group_number = ***rand() mod*** G
4   **if** *records_in_group[group_number]*< $2k - 1$ **then**
5   chromosome[gene_number]←group_number
6   records_in_group[group_number] += 1
7   gene_number += 1
8   **end**
9   **end**

3.3 The Fitness Function

The evaluation of a chromosome in the population is the costliest part in the whole evolution process of the GA. We want to obtain a measure of the homogeneity of the groups in the k-partition represented by a given chromosome.

As we have introduced in Sect. 2, the most frequently used homogeneity measure for k-partitions is the within-group sum of squares SSE [7, 16]. The SSE is the sum of square distances from the centroid of each group to every element (i.e. record) in the group. For a k-partition, SSE can be computed as:

$$SSE = \sum_{i=1}^{s} \sum_{j=1}^{n_i} (\mathbf{x}_{ij} - \bar{\mathbf{x}}_i)'(\mathbf{x}_{ij} - \bar{\mathbf{x}}_i), \qquad (10)$$

where s is the number of groups in the k-partition and n_i is the number of records in the i-th group. Expression (10) is applied to the data after standardising them, this is, after subtracting from the values of each attribute the attribute mean and dividing the result by the attribute standard deviation[4].

During the evolution process, some chromosomes can appear which are not candidate optimal k-partitions due to the effect of genetic operators. This is, some groups can appear whose cardinality is no longer in the range $[k, 2k-1]$. In these cases, we do not remove the chromosome from the population; instead, we penalise it in order to nearly preclude its reproduction.

Our goal is to minimise SSE. Thus, the fitness value that we finally assign to a chromosome is

$$Fitness = \frac{1}{SSE + 1}. \qquad (11)$$

---

[4] Standardisation does not affect attribute correlations.

3.4 Selection Scheme and Genetic Operators

Our GA uses the well-known roulette-wheel selection algorithm. This selection algorithm is based on weighting the chromosomes of the population according to their fitness. Thus, the probability of a chromosome for being selected to pass to the next generation is proportional to its fitness. This method tends to converge quickly because it does not preserve the diversity of the population. Even though this entails some risk of missing good solutions, experimental results show that this risk is small and it can be afforded for the sake of speed.

We have used the most common genetic operators, this is, one-point crossover and mutation.

4 A Hybrid Approach

In the previous section, we have introduced how to build a GA to let it properly face the problem of micro-aggregating a multivariate data set. Although the proposed method is correct, it suffers from some performance limitations when the number of elements in the data set is large.

On the one hand, the GA approach is very efficient in terms of SSE, but it cannot deal with large data sets. On the other, distance-based methods like MDAV can deal with very large data sets, but they are not as good as the GA in terms of SSE.

The solution consists in taking the advantages of both methods and mix them to obtain a new hybrid method able to micro-aggregate large data sets, whilst maintaining a good performance in terms of SSE.

The hybrid method, that we call two-step partitioning [20], can be summarised as follows:

1. Let k be a small value (e.g. $k = 3$).
2. Let K be larger than k and divisible by k, small enough to be suitable for our GA (e.g. $K = 21$).
3. Use a fixed-size heuristic to build a k-partition.
4. Taking as input records the average vectors obtained in the previous step, apply the fixed-size heuristic to build *macro-groups* (i.e. sets of average vectors) of size K/k.
5. For each given *macro-group*, replace the average vectors by the k original records to obtain a K-partition.

Figure 2 shows two macro-groups obtained by the two-step partitioning method using MDAV two times with $K = 24$ and $k = 3$.

Finally, apply the GA to each macro-group in the K-partition in order to generate an optimal or near optimal k-partition of the macro-group. The composition of the k-partitions of all macro-groups yields a k-partition for the entire data set. An important issue when using the GA on a macro-group is to include the fixed-size k-partition of the macro-group induced by the

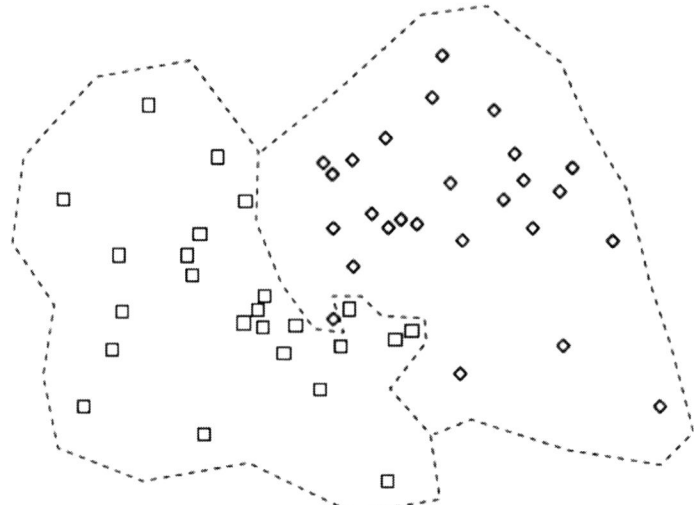

Fig. 2. Two macro-groups obtained by the two-step partitioning method using MDAV two times with $K = 24$ and $k = 3$

k-partition of the entire data set computed in Step 3 above as one of the chromosomes of the initial population fed to the GA.

The following result holds:

Lemma 1. *The SSE of the k-partition obtained using a two-step partition based on a fixed-size heuristic followed by a GA is not greater than the SSE of the k-partition output by the fixed-size heuristic alone.*

Proof. By assumption, the fixed-size k-partition of each macro-group induced by the fixed-size k-partition of the data set obtained at Step 3 is one of the chromosomes of the initial population fed to the GA. Therefore, for each macro-group, the k-partition output by the GA is at least as good as the fixed-size k-partition in terms of the fitness function. This implies that the *SSE* of the k-partition of the entire data set obtained by composition of the macro-group k-partitions output by the GA is not greater than the *SSE* of the k-partition of Step 3.   □

5 Experimental Results

In this section we summarise the most relevant experimental results obtained with the proposed GA-based micro-aggregation algorithm and the hybrid micro-aggregation approach. We compare these results with the ones obtained with the MDAV algorithm and by exhaustive search (ES). Note that ES is only possible with tiny data sets of up to 11 elements.

We classify the experiments in three main groups:

1. Experiments with the example data set: These experiments have been carried out in order to clearly show the main properties of the GA approach in a tiny data set. The obtained results are not statistically relevant, but they provide a first intuition of what the GA approach is able to do.
2. Experiments with small data sets: This set of experiments are intended for a complete analysis of the behaviour of the GA-based micro-aggregation.
3. Experiments with large and real data sets: In these case, the hybrid approach is on the line. Well-known data sets are used to demonstrate that the hybrid approach performs properly.

Each experiment consists of several runs of the GA. At each run of the GA, a different combination of parameters is used i.e. the Mutation rate (M_r) has been varied from 0 to 1, with a step of 0.1, the Crossover rate (C_r) has been varied from 0 to 1, with a step of 0.1 and, the population size has been varied from 10 to 100 chromosomes with a step of 10.

The above implies 1210 different parameter settings. For each setting, the GA was run 10 times, so that altogether each test comprised 12,100 runs of the GA[5].

5.1 Results for the Running Example

With the aim of giving a first intuition of how our GA-based micro-aggregation method performs, we have used it to micro-aggregate the data set in Table 1.

Table 3 gives a comparison between ES, MDAV and GA in terms of running time and *SSE* for the example data set in Table 1. The running time of the GA depends on the number of generations. Figure 3 shows a histogram of the convergence generation of the tests. It can be observed that most of the tests converge in less than 5,000 iterations. Thus, the running time for the GA in Table 3 corresponds to 5,000 iterations.

Even though MDAV is faster than the GA approach, the SSE obtained with the GA is better. In fact, in 91% of the 12,100 runs, the GA reaches the optimal $SSE = 14.82$.

Table 3. Performance of ES, MDAV and GA on the running example

$n = 11, d = 4$	ES	MDAV	GA
Time	717 s	<1 s	1.4 s
SSE	14.82	18.29	14.82

---

[5] Tests were performed on a Pentium 4 PC running at 3 GHz with a Debian Linux operating system. Programs were coded in C++.

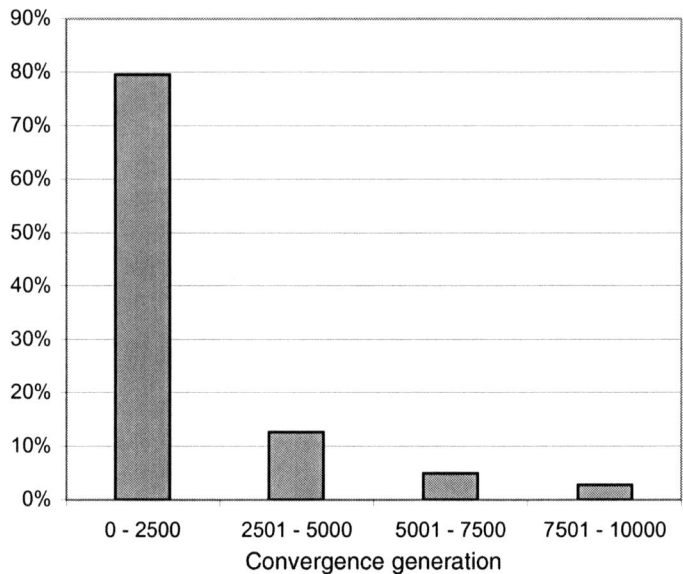

Fig. 3. Histogram of the convergence generation of the tests on the example data set

Table 4. Minimum SSE values for tests with larger data sets

Data set	MDAV	GA_m	M_r	C_r	pop.size
20×2	5.39	4.29	0.1	0.4	60
35×2	5.01	3.13	0.1	0.1	90
50×2	5.92	4.02	0.1	0.1	50
100×2	5.48	15.37	0.1	0.2	50

5.2 Results in Small Data Sets

After observing the promising results achieved by the GA approach over the data set in Table 1, we have carried out a deeper analysis of its performance on a variety of small data sets.

Random data sets $n \times d$ have been constructed for several numbers n of records and several numbers of attributes d. Attribute values were drawn from the $[-10000, 10000]$ interval by simple random sampling. For data sets with $n = 11$, $n = 20$ and $n = 35$ the number of generations (number of iterations) was 10^4. For the data set with $n = 50$, the number of generations was 10^5. Note that for micro-data sets with $n = 11$ records, the optimal k-partition could be found by ES in a reasonable time. For $n > 11$ the *SSE* obtained with GA was only compared to the *SSE* obtained using MDAV.

Table 4 shows the best results obtained with the GA over data sets of fixed dimensionality $d = 2$ and a variable number of elements (i.e. 20, 35, 50 and 100). From these results it becomes clear that the mutation rate should

Table 5. Average of the minimum SSE values for tests with larger data sets

Data set	MDAV	GA_A	M_r	C_r
20×2	5.39	4.61	0.1	0.7
35×2	5.01	3.59	0.1	0.5
50×2	5.92	4.67	0.1	0.5
100×2	5.48	23.78	0.1	0.0

Table 6. SSE for data sets with $n = 11$ and different d values

Data set	SE	MDAV	GA	Av.Conv.E.
11×5	29.36	39.11	29.36	931
11×7	43.28	52.47	43.28	839
11×10	68.66	71.14	68.66	705

be low[6] (e.g. 0.1). Moreover, it is also clear that the GA improves the MDAV heuristic when the number of elements is lower than 100. This result must be emphasised because it represents the main limitation of the proposed method (i.e. it cannot deal with large data sets).

Table 5 shows the averages of the minimum values obtained in the same data sets and confirms the previous conclusions.

In both the 11×2 and 11×3 data sets the optimal SSE was obtained. In order to check whether the optimal SSE is obtained for larger values of d, we have tested our GA over random data sets with $d = 5$, $d = 7$ and $d = 10$. In all cases, the k-partition generated for $M_r \neq 0$ and $C_r \neq 0$ equals the optimal k-partition output by ES. Table 6 shows the results of these tests. In all three data sets, around a 90% of the tests output the optimal SSE value.

In [29] a detailed sensitivity analysis of the parameters of the GA based on the Analysis Of Variance (ANOVA) and the Tukey and Scheffé test is given.

5.3 Results in Real and Large Data Sets

The main limitation of our GA-based micro-aggregation technique is that it cannot deal with large data sets. Thus, the hybrid technique is needed.

In order to test the hybrid technique we have used the next data sets:

- Synthetic data sets :"Scattered" and "Clustered" are data sets that contain $n = 1,000$ records with $d = 2$ attributes. The former is *scattered* in the sense that no natural clusters are apparent. Attribute values were drawn from the $[-10000, 10000]$ range by simple random sampling. The second data set is considered to be *clustered* because its records are grouped in natural clusters. The attribute values were uniformly drawn from $[-10000, 10000]$ as in the previous case, but records were then group-wise shifted by a random amount in order to form clusters of size between

---

[6] The use of higher values of the mutation rate leads to the generation of too much non-optimal chromosomes.

Table 7. Summary of experimental results with the hybrid approach ($k = 3$)

Data set	K	SSE_{MDAV}	Two-step MDAV partitioning		
			# clus.	SSE	% impr.(%)
"Scattered"	12	4.71	83	4.28	9
	18	4.71	56	4.33	8
	27	4.71	38	4.58	3
"Clustered"	12	3.57	84	2.84	20
	18	3.57	56	2.14	40
	27	3.57	41	2.09	41
"Census"	12	799	91	768	4
	18	799	61	767	4
	27	799	41	789	1
"EIA"	12	217	342	189	13
	18	217	228	186	14
	27	217	152	197	9

3 and 5 records each; more details on the generation of this data sets can be found in [6].

- Real data sets: "Census" is a data set that contains 1,080 records with 13 numerical attributes. "EIA" contains 4092 records with 11 numerical attributes. These data sets were proposed as reference micro-data sets during the CASC project [2] and have been widely used [7–9, 18].

Table 7 provides a summary of the obtained results. It can be observed that the hybrid method always improves MDAV. The main improvement takes place when the data set is clustered. This is due to the fact that MDAV is a fixed-size heuristic i.e. it generates groups of k elements. This lack of flexibility results in a poor performance when the elements in the micro-aggregated data set are clustered in groups of variable size. On the contrary, the hybrid approach makes use of the flexibility of the GA and clearly improves the SSE results (cf. [20] for details).

6 Conclusions and Future Challenges

In this chapter we have shown some of the privacy problems that the Information Society has to face. Some of these privacy threads can be tackled by using well-known techniques from the field of statistical disclosure control such as micro-aggregation.

We have introduced the main concepts of micro-aggregation and we have shown that, due to the complexity of the problem, no optimal methods exist and novel heuristics must be developed.

We have pointed out the possibility of using a GA to solve the micro-aggregation problem and we have shown how to modify a classical GA in order to let it cope with the micro-aggregation problem. Moreover, we have reported the main limitation of the GA approach (i.e. it can only deal with small data sets) and we have proposed an improvement that consists in the mixture of our GA and a classical distance-based heuristic like MDAV.

The reported experimental results demonstrate the usefulness of the proposed methods and open the door to an invigorating research line. Lots of questions remain open, next we conclude with some future challenges:

- Look for better codings.
- Test the efficiency of other selection algorithms.
- Evaluate the importance of genetic operators such as multiple-point crossover or inversion.

Disclaimer and Acknowledgements

The author is solely responsible for the views expressed in this paper, which do not necessarily reflect the position of UNESCO nor commit that organisation. This work was partly supported by the Spanish Ministry of Education through projects TSI2007-65406-C03-01 "E-AEGIS" and CONSOLIDER CSD2007-00004 "ARES", and by the Government of Catalonia under grant 2005 SGR 00446.

References

1. C. Boyens, R. Krishnan, and R. Padman. On privacy-preserving access to distributed heterogeneous healthcare information. In IEEE Computer Society, (ed.), *Proceedings of the 37th Hawaii International Conference on System Sciences HICSS-37*, Big Island, HI, 2004.
2. R. Brand, J. Domingo-Ferrer, and J. M. Mateo-Sanz. Reference data sets to test and compare SDC methods for protection of numerical microdata, 2002. European Project IST-2000-25069 CASC, http://neon.vb.cbs.nl/casc.
3. W. Chien and C.-C. Chiu. Using nu-ssga to reduce the searching time in inverse problem of a buried metallic object. *IEEE Transactions on Antennas and Propagation*, 53(10):3128–3134, 2005.
4. A. Czarn, C. MacNish, K. Vijayan, B. Turlach, and R. Gupta. Statistical exploratory analysis of genetic algorithms. *IEEE Transactions on Evolutionary Computation*, 8(4):405–421, 2004.
5. K. A. DeJong. *Analysis of the Behaviour of a Class of Genetic Adaptive Systems.* PhD thesis, Dept. Comput. Commun. Sci., University of Michigan, 1975.
6. J. Domingo-Ferrer, A. Martínez-Ballesté, J. M. Mateo-Sanz, and F. Sebé. Efficient multivariate data-oriented microaggregation. *The VLDB Journal*, 15(4):355–369, 2006.
7. J. Domingo-Ferrer and J. M. Mateo-Sanz. Practical data-oriented microaggregation for statistical disclosure control. *IEEE Transactions on Knowledge and Data Engineering*, 14(1):189–201, 2002.

8. J. Domingo-Ferrer, F. Sebe, and A. Solanas. A polynomial-time approximation to optimal multivariate microaggregation. *Computers & Mathematics with Applications, In Press*, Corrected Proof, Available online 19 July 2007. (http://www.sciencedirect.com/science/article/B6TYJ-4P77G9K-1/2/65a8e45541c49e7862a14870253124ba)

9. J. Domingo-Ferrer and V. Torra. Ordinal, continuous and heterogenerous k-anonymity through microaggregation. *Data Mining and Knowledge Discovery*, 11(2):195–212, 2005.

10. B. Freisleben and M. Härtfelder. Optimization of genetic algorithms by genetic algorithms. In *In Proc. Int. Conf. Artificial Neural Networks and Genetic Algorithms*, pages 392–399, Vienna, 1993.

11. D. E. Goldberg. *Genetic Algorithms in Search, Optimization & Machine Learning*. Addison-Wesley Publishing, 1989.

12. J. Grefenstette. Optimization of control parameters for genetic algorithms. *IEEE Transactions on Systems, Man and Cybernetics*, 16(1):122–128, 1986.

13. HIPAA. Health insurance portability and accountability act, 2004. http://www.hhs.gov/ocr/hipaa/.

14. J. H. Holland. *Adaptation in Natural and Artificial Systems*. University of Michigan Press, 1975. Second edition: MIT, 1992.

15. A. Hundepool, A. Van de Wetering, R. Ramaswamy, L. Franconi, A. Capobianchi, P.-P. DeWolf, J. Domingo-Ferrer, V. Torra, R. Brand, and S. Giessing. *μ-ARGUS version 4.0 Software and User's Manual*. Statistics Netherlands, Voorburg NL, may 2005. http://neon.vb.cbs.nl/casc.

16. Y. Jung, H. Park, D.-Z. Du, and B. L. Drake. A decision criterion for the optimal number of clusters in hierarchical clustering. *Journal of Global Optimization*, 25:91–111, 2005.

17. K. Knodler, J. Poland, A. Mitterer, and A. Zell. Optimizing data measurements at test beds using multi-step genetic algorithms. In N. Mastorakis, (ed.), *Advances In Fuzzy Systems and Evolutionary Computation (Proceedings of WSES EC 2001)*, pages 277–282, 2001.

18. M. Laszlo and S. Mukherjee. Minimum spanning tree partitioning algorithm for microaggregation. *IEEE Transactions on Knowledge and Data Engineering*, 17(7):902–911, 2005.

19. W. C. Liu. Design of a cpw-fed notched planar monopole antenna for multiband operations using a genetic algorithm. *IEE Proceedings on Microwaves, Antennas and Propagation*, 152(4):273–277, 2005.

20. A. Martinez-Balleste, A. Solanas, J. Domingo-Ferrer, and J. M. Mateo-Sanz. A genetic approach to multivariate microaggregation for database privacy. In *23rd Internationl Conference on Data Engineering. Workshop on Privacy Data Management*, pages 180–185. IEEE Computer Society, April 2007.

21. G. M. Megson and I. M. Bland. Synthesis of a systolic array genetic algorithm. In *IPPS/SPDP*, pages 316–320, 1998.

22. M. Mitchell. *An Introduction to Genetic Algorithms*. MIT, 1999. Fifth printing.

23. A. Oganian and J. Domingo-Ferrer. On the complexity of optimal microaggregation for statistical disclosure control. *Statistical Journal of the United Nations Economic Comission for Europe*, 18(4):345–354, 2001.

24. S. J. Ovaska, T. Bose, and O. Vainio. Genetic algorithm-assisted design of adaptive predictive filters for 50/60 hz power systems instrumentation. *IEEE Transactions on Instrumentation and Measurement*, 54(5):2041–2048, 2005.

25. European Parliament. DIRECTIVE 2002/58/EC of the European Parliament and Council of 12 july 2002 concerning the processing of personal data and the protection of privacy in the electronic communications sector (Directive on privacy and electronic communications), 2002. `http://europa.eu.int/eur-lex/pri/en/oj/dat/2002/l_201/l_20120020731en00370047.pdf`.

26. Canadian Privacy. Canadian privacy regulations, 2005. http://www.media-awareness.ca/english/issues/privacy/canadian_legislation_privacy.cfm.

27. J. D. Schaffer, R. A. Caruana, L. J. Eshelman, and R. Das. A study of control parameters affecting online performance of genetic algorithms for function optimization. In *Proceedings of the Third International Conference on Genetic algorithms*, pages 51–60, San Francisco, CA, USA, 1989. Morgan Kaufmann Publishers Inc.

28. A. Solanas and A. Martínez-Ballesté. V-MDAV: Variable group size multivariate microaggregation. In *COMPSTAT'2006*, pages 917–925, Rome, 2006.

29. A. Solanas, A. Martínez-Ballesté, J. M. Mateo-Sanz, and J. Domingo-Ferrer. Multivariate microaggregation based on genetic algorithms. In *3rd IEEE Conference On Intelligent Systems (IEEE IS'2006)*, pages 65–70, Westminster, 2006. IEEE Computer Society Press.

30. A. Solanas, E. Romero, S. Gómez, J. M. Sopena, R. Alquezar, and J. Domingo-Ferrer. Feature selection and oulier detection with genetic algorithms and neural networks. In P. Radeva B. López, J. Meléndez and J. Vitrià, (eds.), *Artificial Intelligence Research and Developement*, pages 41–48, Amsterdam, NL, 2005. IOS Press.

31. European Union. The Seventh Framework Programme (FP7), 2006. http://ec.europa.eu/research/fp7/understanding/index.html.

32. USPrivacy. U.S. privacy regulations, 2005. `http://www.media-awareness.ca/english/issues/privacy/us_legislation_privacy.cfm`.

33. L. Willenborg and T. DeWaal. *Elements of Statistical Disclosure Control*. Springer-Verlag, New York, 2001.

34. A. S. Wu and I. I. Garibay. The proportional genetic algorithm: Gene expression in a genetic algorithm. *Genetic Programming and Evolvable Machines*, 3(2):157–192, 2002.

35. Y. K. Yu, K. H. Wong, and M. M. Y. Chang. Pose estimation for augmented reality applications using genetic algorithm. *IEEE Transactions on Systems, Man and Cybernetics, Part B*, 35(6):1295–1301, 2005.

A Revision of Evolutionary Computation Techniques in Telecommunications and an Application for the Network Global Planning Problem

Pablo Cortés, Luis Onieva, Jesús Muñuzuri, and Jose Guadix

Ingeniería de Organización. Escuela Técnica Superior de Ingenieros, University of Seville, C/Camino de los Descubrimientos s/n 41092, Sevilla, Spain,
pca@esi.us.es

Summary. This chapter consists of two parts. Firstly, a revision of the state-of-the-art is presented for the application of EC techniques to several problems arising in telecommunications. The revision includes wired based networks, wireless networks considering both planning and operation stages. Secondly, an evolutionary computation approach is proposed to solve the very complex telecommunications network global planning problem. The model is based on a hubbing topology that considers two hierarchical levels for low and high speed telecommunication networks. EC results are compared with industry heuristic solutions showing a much better behaviour, outperforming it in all the analysed ranges.

1 Introduction

Evolutionary computation (EC) is a methodology with wide application to the telecommunications industry. This paper is divided into two differentiated parts. The first one includes a quick revision referring the research undertaken in telecommunications using EC methodologies. It includes the applications of EC in the telecommunication planning stage, as well as the applications of EC to deal with real time network operation. And it also includes EC application to wired, wireless, mobile and satellite telecommunication systems.

The second part is devoted to an application of EC to solve the telecommunications network global planning problem. Firstly the problem is described. Secondly a hierarchical optimization approach based on an EC strategy is presented in depth. Thirdly the results so obtained are compared with other heuristic methods that are usually applied in real practice. Finally a conclusions section summarizes the main aspects of the paper.

P. Cortés et al.: *A Revision of Evolutionary Computation Techniques in Telecommunications and an Application for the Network Global Planning Problem*, Studies in Computational Intelligence (SCI) **92**, 239–262 (2008)
www.springerlink.com

2 A Revision of EC Techniques in Telecommunications

The application of EC methodologies has been successfully tested in numerous optimization problems arising in telecommunications. In this section, we present a quick revision that collects some of the most representative applications and that tries to cover the major part of the problems that can be found in the literature and that have been solved using EC methodologies. We have divided the section attending to the typology of the telecommunication network, i.e., wire based networks, wireless and mobile networks, or satellite based networks. And we also consider the optimization stage that is tackled in the problem, i.e., the design or planning stage, and the real time operation stage. Next Table 1 classifies the main problems showing the references that approach the problem by EC techniques.

2.1 Topological Design of Wire Based Telecommunication Networks

The topological design of wire based telecommunication networks includes a great variety of different problems. They are the basic network infrastructure that corresponds to the ditch and conduit placement problem, the link dimensioning and capacity assignment problem that is dealt with multicommodity flows, the design of hierarchical network structures (what are actually deployed in real practice), the consideration of survivability and

Table 1. A classification of EC in telecommunications research

Problem	References
Topological design of wire based telecommunication networks	
Basic network infrastructure (conduit installation)	[1, 8, 13, 39]
Capacity assignment (approaches based on multicommodity flows)	[3]
Global design (approaches based on hierarchical networks)	[10]
Survivable and reliable networks	[18, 19, 34]
Self-Healing ring networks	[4, 20]
Real time operation	
Dynamic routing	[9, 25, 35]
Communication admission control	[30, 32, 36]
Dynamic bandwidth allocation	[14, 28]
Multicasting transmission	[7, 17, 38]
Wavelength division multiplexing	[11, 12, 24]
Wireless and mobile communication	
Infrastructure design	[2, 33]
Bandwidth allocation	[31]
Frequency assignment	[22]
Satellite based communication	
Satellite constellation design	[23]

reliability conditions in broadband networks, and the consideration of specific self-healing ring topologies that provide adequate reliability for local/urban networks.

In the *basic network infrastructure design problem* Abuali et al. [1] have dealt with the probabilistic minimum spanning tree problem that addresses the circumstances that arise when not all nodes are deterministically present but, rather, nodes are present with known probabilities. Also, Elbaum and Sidi [13] showed one of the first applications of GA to the topological design of *local area networks*.

Cortés et al. [8] have developed an evolutionary algorithm to design the infrastructure of Hybrid Fiber Coaxial networks in urban areas based on the Steiner problem. The model and results refer to the deployment of such network inside a city from small to medium size. Lastly, Zhou and Gen [39] have presented a tree-like telecommunication network design problem with a limitation on the number of terminals. The problem is formulated as a leaf-constrained minimum spanning tree. Solutions provided by genetic algorithms were compared with heuristic solutions.

Once the basic infrastructure is designed the telecommunication links have to be dimensioned, and capacity must be assigned to those links. This problem has been historically dealt with *multicommodity flows*. Albiñana et al. [3] have presented a two-stage stochastic mixed 0–1 dynamic multicommodity model and algorithm for determining the enrouting protocol in telecommunications networks under uncertainty. The problem is solved by using a mixture of an enrouting arc generation scheme and a genetic algorithm for obtaining the enrouting protocols over the scenarios.

In real practice, telecommunication network deployment is undertaken by means of *hierarchical network* structures. Cortés et al. [10] have dealt with a case study applied to Andalucía (Spain). The authors describe the design of a tool capable of evaluating the deployment cost of a network that was not to be limited only to interconnect large cities, but also to include smaller towns, with respect to the objective of preventing them from staying behind the progress of the Information Society.

One of the problems in modern telecommunication networks is that the disconnection of a key trunk link due to a failure could provoke the loss of a great amount of data or voice traffic. In order to avoid such situations, *survivable networks* are designed trying to provide additional reliability to the telecommunication network.

So, the global design proposed by Cortés et al. [10] includes grade-2 survivability conditions. In other line, Huang et al. [19] have proposed a genetic algorithm for three connected telecommunication network designs. The genetic algorithm used a solution representation, in which constraints such as diameter and connectivity constraints were encoded and two-point crossover with the operation of swapping duplicated nodes ensured feasible solutions. Also, Hazem et al. [18] have proposed a hybrid genetic algorithm with local search for a survivability study of broadband communication networks. Finally, Wille et al. [34] presented a topological design of survivable IP Networks by using EC techniques.

It is worth to mention the specific topologies of *self-healing networks*. They are ring topologies that provide reliability to urban or local networks. Armony et al. [4] implemented a genetic algorithm to determine how traffic should be routed on self-healing rings with the objective of optimizing the trade-off between the cost of equipment to implement the rings and the cost of exchanging traffic among rings. Previously, Karunanithi and Carpenter [20] had presented a genetic algorithm for traffic load assignment in SONET rings.

2.2 Real Time Operation

Optimization in real time operation of telecommunication networks arises when communications have to be routed across the network from the origin point to the destination. This problem is known as routing and can be undertaken in a static or dynamic way.

The static routing consist of making use of the routing tables that can be constructed as an output from the capacity assignment and multicommodity flow problem of the previous stage. The *dynamic routing* arises when the messages are routed in real time trying to minimize the losses or to minimize the delays experimented by the messages. Medhi and Tipper [25] have considered solution approaches to a multihour combined capacity design and routing problem which arises in the design of dynamically reconfigurable broadband communication networks that uses the virtual path concept. They compared genetic algorithm, Lagrangian relaxation based subgradient optimization method, generalized proximal point algorithm with subgradient optimization, and a hybrid approach where the subgradient based method is combined with a genetic algorithm. More recently, Cortés et al. [9] have presented a genetic algorithm for dynamic routing in ATM networks under Constant Bit Rate traffic conditions. The model objective was to minimise the cell loss subject to strict delay conditions. In other line, Wille et al. [35] showed models and algorithms considering at the same time the dynamics of packet network and the quality of service (QoS) user-layer requirements.

Other problem in real time operation corresponds to *communication admission control*. This admission control is a deterministic and informed decision that is made before a communication is established and is based on whether the required network resources are available to provide suitable QoS for the new communication. In this line, Yener and Rose [36] have presented a paper dealing with the cellular call admission control. They define local policies based on genetic algorithms. For the mobile network cases a threshold-adaptive call admission control based on genetic algorithms for multiclass services has been successfully tested by Sheng et al. [30].

The *dynamic bandwidth allocation* is a problem by which traffic bandwidth in a shared telecommunications medium can be allocated on demand and fairly between different users of that bandwidth. Pitsillides et al. [28] analysed the performance of traditional constrained optimisation algorithm compared to a new constrained optimisation genetic algorithm for solving the bandwidth allocation for virtual paths problem. Comparisons were done with respect to throughput, fairness and time complexity. More recently, El-Madbouly [14] has proposed a genetic algorithm for the design of an optimization algorithm to achieve the bandwidth allocation of ATM networks.

The *multicasting transmission* is a method which enables a single source node to communicate with two or more destination nodes using a single transmission. Zhang and Leung [38] proposed an orthogonal genetic algorithm to deal with the problem of determining the multicast tree of minimal cost that connects the source node to the destination nodes subject to delay constraints on multimedia communication. Later, Guimarães and Rodrigues [17] have presented a genetic algorithm to solve the problem of mapping a large set of subscriptions into a fixed, smaller, set of multicast groups in order to support efficiently the dissemination of events. More recently, Chen and Xu [7] have proposed a multicast routing algorithm based on genetic algorithm to construct degree-delay-constrained least-cost multicast routing Tree.

The problem of *wavelength division multiplexing* (WDM) consists of the simultaneous transmission of more than one information-carrying channel on a single fiber lightside using two or more light sources of different wavelengths. Today is one of the most efficient modes for the transmission of massive communication. Specially, WDM technology has emerged as a promising technology for backbone networks. The set of all-optical communication channels (lightpaths) in the optical layer defines the virtual topology for the upper layer applications. Since the traffic demand of upper layer applications fluctuates from time to time, it is required to reconfigure the underlying virtual topology in the optical layer accordingly. Din [12] has proposed a genetic algorithm to solve the WDM virtual topology configuration transition problem. Previously Din [11] had tackled the optimal multiple multicast problem on wavelength division multiplexing ring networks without wavelength conversion using genetic algorithms showing their robustness. Also, Le et al. [24] have dealt with the dynamic survivable routing problem, both in optical networks without wavelength conversion and in optical networks with sparse wavelength conversion. They make use of a hybrid algorithm based on the combination of mobile agents technique and genetic algorithms.

2.3 Wireless and Mobile Communication and Satellite Communication

The wireless and mobile communication category includes those problems that are specific from wireless and mobile networks. As we have seen some of the problems corresponding to real time operation can be applied to wire based networks and wireless networks. Here we address only those problems corresponding specifically to wireless and mobile networks.

Within the *wireless infrastructure design* Watanabe et al. [33] proposed a new type of parallel genetic algorithm model for multiobjective optimization problems. It was applied to solve the antenna arrangement and placement problem in mobile communications. Later, the same problem has been also approached by Alba and Chicano [2].

Sherif et al. [31] proposed a generic call admission control scheme based on genetic algorithms for *bandwidth allocation*. The scheme is applied to a wireless network using an adaptive resource allocation framework. A multimedia connection is represented in terms of three substreams (video, audio and data) each with a pre-specified range of acceptable QoS requirements. This range of requirements make the application adaptive in the sense that each of its substreams specify to the network a number of acceptable QoS levels instead of just a single one.

Another problem arising in wireless networks is the *frequency assignment*. Kim et al. [22] reveals a problem corresponding to the minimum span frequency assignment in wireless communications networks. They use a memetic algorithm combining a genetic algorithm global search with a computationally efficient local search method.

Finally, we address the problem of *satellite constellations design*. This problem is a complex one because it depends on a lot of parameters with many possible combinations, such as the total number of satellites, the orbital parameters of each individual satellite, the number of orbital planes, the number of satellites in each plane, the spacings between satellites of each plane, the spacings between orbital planes, or the relative phasings between consecutive orbital planes. Lansard et al. [23] have tackled the problem using genetic algorithms and they state that the results so obtained outperform other heuristic methods applied in practice.

3 An EC Approach to Solve the Telecommunications Network Global Planning Problem

Here, we address the global planning problem arising in the telecommunication network planning stage. To do so, we consider a hubbing topology based on two hierarchical levels. The first level corresponds to the low speed network; flows are first transported from the initial switching office to the regional hub to be switched there. The second level corresponds to the high-speed network. Here, the demand flows arriving to the hub are sorted and rerouted to their specific destinations. To deal with this arrangement in the hubs, transmission equipment such as Digital Cross-Connect System has to be installed at each hub. So called hubbing topologies are adequate for such fiber optic backbone networks, see [6, 10, 15, 21, 37].

Such fiber optic backbone network can be represented as two networks overlapped. The first of them represents the physical conduit network and the second one represents the logical network. The physical network is defined by the physical conduit placement over the underlying transport network. The *physical network* includes the urban nodes, the passive nodes (typically intersection nodes in the transport network), and the direct connections among all these nodes that we call arcs. The *logical network* is defined by the transmission links and the paths transporting the origin-destination demand. The logical network identifies as nodes only those representing hubs or terminal nodes. The transmission links will be different from arcs and will represent the hub–hub connections or the terminal-hub connections. Terminal–hub connections are referred to the own dedicated fiber cables, and hub–hub connections are referred to the shared high capacity fiber cables.

Figure 1 depicts such logical and physical networks. The figure includes the existence of three categories of nodes. The lowest category consists of passive nodes that represent crosses in the network infrastructure (e.g., node 1). They are simple points where no information is added or dropped. The intermediate category is the urban node, and they are quite similar to Central Offices or Serving Wire Centers in certain telecommunication literature (e.g., nodes A or B). At last, the third category of nodes is called hub, and is quite similar in their role to carrier's Points of Presence representing the highest level of communication (e.g., nodes h_1, h_2 or h_3). Figure 1 points out the differences between the logical and the physical networks, as well as between the arc and the transmission link concepts. Note that transmission link can be the direct connections between one hub and the terminals of its region (within-region links by own dedicated fiber cables), or the connection among hubs from different regions (among the regions links by shared high

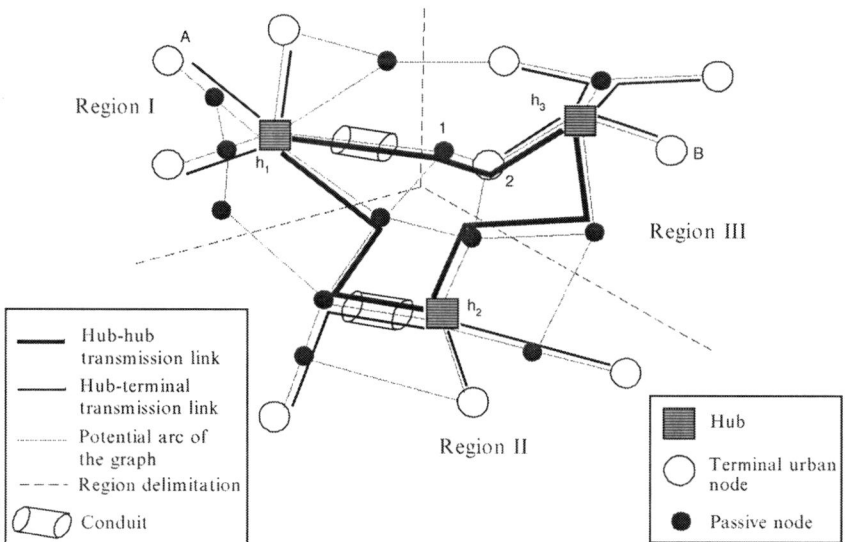

Fig. 1. Logical network vs. physical network topologies

capacity fiber cables). So, the link concept is associated to provide telematic services. On the other hand, arc refers to the connection joining each pair of adjacent nodes, independently of their switching capacities. These nodes could be hubs, terminals or passive nodes. So, the arc concept is associated to the conduit placement. For example the communication between node A and B follows the next process: first the message is sent through the intra-regional link (A, h_1), after that it follows the inter-regional link (h_1, h_3) that includes the arcs $(h_1, 1)$, $(1, 2)$ and $(2, h_3)$. Finally the communication is routed onto the intra-region link (h_3, B).

Typically, the problem we have described intertwines three well-known subproblems: hub location, conduit network design and cable installation problems. Approaches considering this kind of global problem are not common in the bibliography, and most of the authors tend to make simplifications that allow dealing with less ambitious problems. So, Kerner et al. [21] was one of the previous approaches to the hierarchical telecommunication networks problem, but it was previously to the great deployment of the ATM or SDH proposals. Just a bit later, Cardwell et al. [6] focused on the computer networks case. More close to our proposal, Yoon et al. [37] dealt with a relatively similar problem, although their model considered neither survivability conditions nor delay limitation constraints nor passive nodes. Moreover their model did not differentiate between transmission links and simple arcs of the graph, so results from Yoon et al. [37] are not directly comparable with our results. The output of the problem includes the location of switching nodes (hubs), the location of transmission links and their capacities, the paths to transport the origin-destination demand pairs and the costs associated to the network planning and dimensioning. In other line, Gódor and Magyar [15] have proposed a hierarchical network design based on hubbing topologies, however their work do not consider the high level clustering connections or the survivability conditions as ours. Finally, Cortés et al. [10] approached a case study related to a global design problem using genetic algorithms.

To deal with a tractable problem, we make the following hypothesis:

- All the urban nodes are the initial switching offices in the network and can be considered as hub candidate. Each pair of these urban nodes determines a communication origin-destination pair.
- The whole geographical area is partitioned into a certain number of regions. Only one hub is to be set up into each region. Each switching office of a region is connected to its regional hub by an own dedicated fiber cable. Shared high capacity fiber cables connect the regional hubs with each other benefiting from the multiplexing process.
- The total cost function has three main components: the cost of activating one hub on a hub candidate urban node, the infrastructure cost associated to prepare a conduit and the cost of placing a transmission link with a specific length and capacity STM-u level (unit of measure for SDH).

- We consider reliability grade 2 for inter-regional communication (it means the connection of a hub by two links at least) and reliability grade 1 for intra-regional communication (it means the connection of a terminal by one link at least) following [16].

3.1 EC Approach

The huge size of the problem we are considering lead us to consider a step-by-step approach. We follow different subproblems ordered by importance of each subproblem on the total investment required to deploy such network.

Given a hub location, the set of feasible links and paths is stated as in Fig. 1 was done. However, one of the things that the whole topological design problem must state is indeed the hub location. Even more, the hub location drastically affects the network topology because each feasible hub combination has repercussions on the arrangement of potential links with a different costs structure. An in-depth discussion about hub arrangements and its repercussion can be read in [10].

To explore different hub location possibilities we make use of an EC approach, where the hub structure defines the chromosome of the individuals in the population. Figure 2 shows the chromosome encoding of the individuals. The main characteristics of the algorithm are described next. The selection of the parents is random to enrich the population genetic variety. The algorithm uses uniform crossover operators. So, parents having the hub located in the same urban node of the region give place to an offspring that maintains the hub in such location. Otherwise the hub is located in each parents' position with the same a half probability. We also consider a mutation operator that relocates the regional hub from one node to another. The individuals' replacement is based on the fitness ranking using a hypergeometric function allowing more probability of replacement to individuals with worse fitness and less proba-

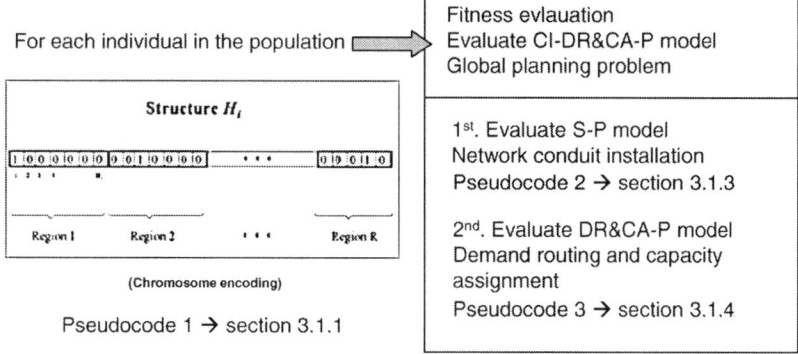

Fig. 2. Individual chromosome encoding and hierarchical fitness evaluation

bility of replacement to individuals with better fitness. So, the individual in ranking position-i, have a replacement probability equal to $q(1\text{-}q)^i$, being q the replacement probability of the worst individual. To evaluate the individuals' fitness that corresponds to the total investment required for the network, we make use of several hierarchical models and algorithms that are being introduced in the next subsections. Figure 2 details the connection between the different models and algorithms (pseudocodes) that are used to evaluate the fitness of the individuals. Finally, the algorithm stops when a maximum number of generations is reached.

Evolutionary Approach Pseudocode

Pseudocode 1
   Initialilization

1. Random size-N population selection.
2. Calculate the fitness of the initial population individuals given by the hub structure H'.

Steady state
   While {Iterations < MAX_ITER}

3. Random operator selection: Crossover/Mutation
   3.1. In case of crossover operator, random selection of two parents.
   3.2. In case of mutation operator, random selection of one parent.
4. Individual ranking based replacement. /*replacement hypergeometric rule*/
5. Calculate the fitness for the new individual.

The Conduit Installation, Demand Routing and Capacity Assignment Problem

For a given hub structure (H') we can formulate the conduit installation, demand routing and capacity assignment problem (**CI-DR&CA-P**). The model considers feasible links in the network, as well as those feasible paths verifying the delay condition. This condition means that paths including more hubs than certain maximum are forbidden since the switching process takes the main component in the delay function of fiber optic networks. So, the set P(k) of the model includes all these aspects.

$$\text{minimize} \quad \sum_{\{i,j\} \in A} f_{ij} y_{ij} + \sum_{e \in E} c_e \left(x_e \right) \qquad \boxed{\textbf{CI-DR\&CA-P}}$$

$$\text{s.t. :}$$

$$\sum_{h \in P(k)} p_{kh} = 1, \ \forall k \in K$$

$$\sum_{k \in K} \sum_{h \in P(k)} p_{kh} \delta_{ij}^{kh} \leq |K| y_{ij}, \ \forall \{i,j\} \in A$$

$$x_e = \sum_{k \in K_0} \sum_{h \in P(k)} \gamma_k \, p_{kh} \delta_e^{kh}, \ \forall e \in E$$

$$\sum_{h \in P(k)} p_{kh} \delta_j^h \leq \rho_k, \ \forall k \in K_0, \ \forall j \subset H'$$

$$p_{kh} \leq \rho_k, \ \forall h \in P(k), \ \forall k \in K_0$$

$$x_e \geq 0, \ \forall e \in E$$

$$p_{kh} \geq 0, \ \forall h \in P(k), \ \forall k \in K$$

$$y_{ij} = \{0,1\}, \ \forall \{i,j\} \in A$$

Being:

- y_{ij} binary variable indicating if the non-directed arc $\{i,j\}$ is part of the network infrastructure. Note that A is the set of arcs (conduits).
- p_{kh} continuous variable indicating the origin-destination pair demand fraction, k, through the path h.
- x_e continuous variable indicating the flow in each link e. Note that E is the set of transmission links.
- γ_k communication demand of each origin-destination pair, k.
- ρ_k maximum demand fraction of origin-destination pair k that can be transported in a communication path or switched in a hub.
- δ_{ij}^{kh} binary parameter, equal to one if the path h connecting pair k uses the arc $\{i,j\}$, and zero otherwise.
- δ_e^{kh} binary parameter, equal to one if the transmission link e is included in the path h that connects pair k, and zero otherwise.
- δ_j^h binary parameter, equal to one if the path h includes the node j and zero otherwise.
- f_{ij} civil engineering cost for arc $\{i,j\}$. It includes the ditch and conduit costs.
- $c_e(x_e)$ cost of the link e when the flow is x_e.

To make more tractable this *NP-Hard* problem, we have divided it into two different subproblems that are shown below.

The Conduit Installation Subproblem

The problem includes the most relevant design costs: the ditch and conduit network costs. This first subproblem looks for a κ_k-connected network where the costs of the arcs are the infrastructure costs. We denote by κ_k-connected network a survivability grade κ_k network. In fact, it is usual considering survivability level 2 for the inter-regional network and survivability level 1 for the intra-regional networks. Telecommunications network basic infra-structure problem can be adequately represented by means of the Steiner problem (**S-P**). Next model presents the typical coverage formulation for the well-known Steiner problem.

$$\text{minimize} \quad \sum_{\{i,j\} \in A} f_{ij} y_{ij} \qquad\qquad \boxed{\textbf{S-P}}$$

s.t. :

$$\sum_{\{i,j\} \in A} a_{m,(i,j)} y_{ij} \geq 1, \ \forall m = 1, 2, \ldots, q$$

$$y_{ij} = \{0, 1\}, \ \forall \{i, j\} \in A$$

Where, m represents every feasible cut of the graph leaving terminal nodes (in Steiner problem terms) at both parts of the cut. And $a_{m,(i,j)}$ depict the coverage matrix parameters.

But the Steiner problem provides a network without survivability conditions, and our conduit installation problem requires such type of network. In order to do so, we state the conduit basic infrastructure topology by means of 2-trees techniques with survivability conditions, combining Steiner trees as in [27]. The solution from this first stage (conduit installation problem) is the set of arcs (conduits of the physical network), as well as the feasible set of paths and transmission links in the logical network. The pseudocode for the basic infrastructure is:

Pseudocode 2

1. Calculate the Steiner trees corresponding to the low-speed networks.
   1.1. Terminal nodes are urban nodes in each region.
   1.2. Steiner nodes are passive nodes.
   1.3. Costs of the arcs are civil engineering costs (costs of ditches and conduits).
2. Calculate the Steiner tree corresponding to the high-speed network.
   2.1. Terminal nodes are the hubs of the problem.
   2.2. Steiner nodes are the rest of nodes of the graph.
   2.3. Costs of the arcs are civil engineering costs (costs of ditches and conduits). Arcs that were used in step 1 are set to null cost.
3. Calculate the survivable high-speed network
   3.1. After steps 1 and 2, all the hubs in a final position of the trees are set as terminals. The rest of the nodes are Steiner nodes.
   3.2. The cost of the arcs used in steps 1 and 2 are set to an upper bound (e.g., the summatory of all the arcs costs in the network). The costs of the rest of the arcs are the civil engineering costs (costs of ditches and conduits).
   3.3. Calculate the Steiner problem with these conditions.

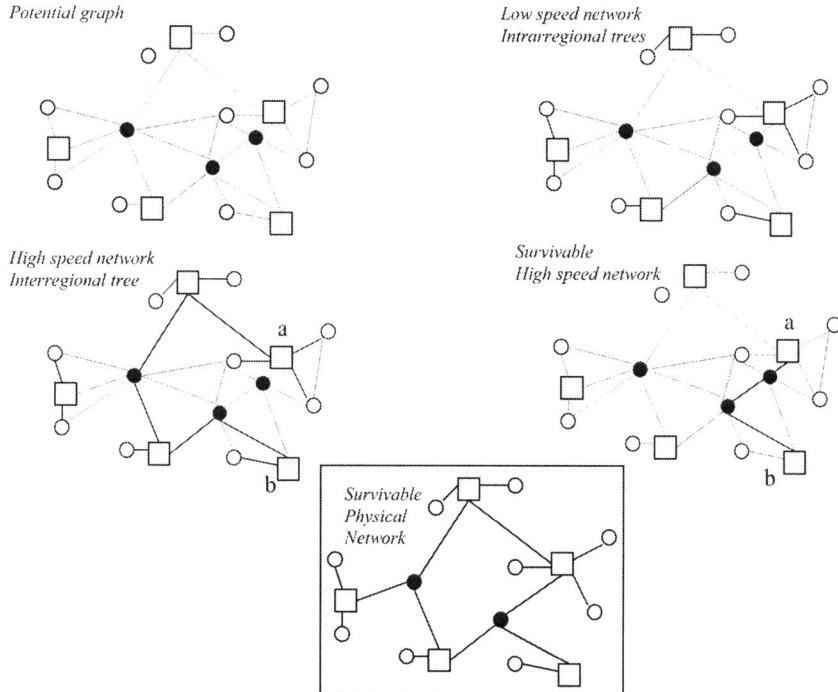

Fig. 3. Procedure that provides a survivable physical network (pseudocode 2)

3.4. Erase redundancies. After these 3-step procedure, redundancies could appear and it is known that nodes connected through more than four arcs are suboptimal [27].

Figure 3 details so described procedure with an example. Note that pseudocode step 3 constructs the Steiner tree between nodes a and b, because they are the only hubs with incidence arc set cardinality equal to one in the high-speed network.

The Demand Routing and Capacity Assignment Problem

Once the basic infrastructure has been stated from Sect. 3.1, the second stage is to route the traffic and to assign enough capacity to the links. Now, the routing problem searches for a set of transmission links with enough capacity to transport the demands. It is noted as **DR&CA-P**.

DR&CA-P problem has been reduced to a multicommodity flow problem with reliability conditions (equivalent to capacity constraints) and with a concave objective function. A concave minimization problem is *NP-Hard*, even in such special cases as that of minimizing a quadratic concave function over very simple polytopes such as hypercubes.

$$\text{minimize} \sum_{e \in E} c_e(x_e) \qquad \boxed{\textbf{DR\&CA-P}}$$

s.t. :

$$\sum_{h \in P(k)} p_{kh} = 1, \ \forall k \in K_0$$

$$x_e = \sum_{k \in K_0} \sum_{h \in P(k)} \gamma_k p_{kh} \delta_e^{kh}, \ \forall e \in E$$

$$p_{kh} \le \rho_k, \ \forall h \in P(k), \ \forall k \in K_0$$

$$x_e \ge 0, \ \forall e \in E$$

$$p_{kh} \ge 0, \ \forall h \in P(k), \ \forall k \in K_0$$

The objective function of DR&CA-P is a complex one. Formally $c_e(x_e)$ must be calculated depending on the type of fiber-optic technology used. The correct specification of the link cost is a key factor in the network planning. Melián et al. [26] proposes a cost function quite similar to ours, but our function brings some slight differences. To consider the link cost function we have managed two main options. The first option corresponds to high quality fiber optic link (without regenerators). It can be estimated as (1).

$$C_{ij}(D_{ij}^{sG}, C_{\max}) = \frac{P_{SDHo}}{n_o} + \frac{P_{SDHd}}{n_d} + P_{fo}^{sG} \cdot D_{ij}^{sG} + C_{oam} \cdot \left\lceil \frac{D_{ij}^{sG}}{D_{\max}} \right\rceil$$

$$+ 2C_{tr} \cdot \left\lfloor \frac{C_{\max}}{C_{10Gb}} \right\rfloor \qquad (1)$$

The second option corresponds to lower quality fiber optic link (with regenerators) that can be estimated as (2).

$$C_{ij}(D_{ij}, C_{\max}) = \frac{P_{SDHo}}{n_o} + \frac{P_{SDHd}}{n_d} + P_{fo} \cdot D_{ij} + C_{reg} \cdot \left\lceil \frac{D_{ij}}{D_{\max}} \right\rceil$$

$$+ 2C_{DWDM} \cdot \left\lfloor \frac{C_{\max}}{C_{10Gb}} \right\rfloor \qquad (2)$$

Where:

- P_{SDHo} and P_{SDHd} are the origin and destination SDH multiplexer costs.
- n_o and n_d are the number of links in the SDH equipment placed in the node. Each SDH multiplexer can serve diverse digital links.
- D_{ij}^{sG} is the arc distance that can be covered without regenerators.
- D_{ij} is the arc distance that must be covered with regenerators.
- P_{fo}^{sG} is the linear cost of a high quality cable of 24 monomode fibers plus the cost of connecting it.
- P_{fo} is the linear cost of a standard cable of 24 monomode fibers plus the cost of connecting it.

- C_{oam} is the cost of an optical amplification system.
- C_{reg} is the cost of a regenerator.
- D_{max} is the amplification maximum distance. Typically 140 km for option 1 and 40 km for option 2.
- C_{tr} is the cost of DWDM interface card (transmitter/receiver) with capacity up to $10 \, \mathrm{Gb \, s^{-1}}$.
- C_{max} is the capacity that must be equipped.

This cost function corresponds to a linear function with respect to the distance covered by the link, and a piecewise function respect to the flow in the link attending to the discrete steps of capacity for each specific value of capacity given by a set of cables. Each cable is characterized by a number of fibers and a binary regime for each fiber (STM-u) accordingly with (1) and (2). To be able to differentiate the cost function, we consider its exponential-like smooth cost approximation, $\bar{C}(p)$ (an in depth discussion on this technique can be found in Bertsekas [5]). Using this approximation, the cost function can be differentiated, and (3) depicts the global optimality conditions:

$$
\frac{\partial \bar{C}(p)}{\partial p_{kh}} - \mu_k = u_{kh} - v_{kh} \quad \text{being } \mu_k \text{ free}
$$

$$
(p_{kh} - \rho_k) \cdot v_{kh} = 0 \quad \text{being } v_{kh} \geq 0 \tag{3}
$$

$$
p_{kh} \cdot u_{kh} = 0 \quad \text{being } u_{kh} \geq 0
$$

These results lead us to consider three possible situations. First one is depicted by the set of equations 4.

$$
p_{kh} = 0 \Rightarrow u_{kh} \geq 0 \text{ and } v_{kh} = 0
$$

$$
\frac{\partial \bar{C}(p)}{\partial p_{kh}} - \mu_k = u_{kh} \geq 0 \tag{4}
$$

$$
\sum_{e \in E} [\bar{c}'_e(x_e)] \cdot \left[\gamma_k \cdot \delta_e^h\right] \geq \mu_k \Rightarrow \sum_{e \in E} \bar{c}'_e(x_e) \cdot \delta_e^h \geq \mu'_k = \frac{\mu_k}{\gamma_k}
$$

Equation 4 shows the case where the origin-destination pair k does not use the path h. In this situation, the cost variation sum associated to the links shaping path h is bounded from below by an indicator parameter, μ'_k, equal to the Kuhn–Tucker multiplier associated to the demand balance equation divided into the total demand volume of pair k.

The other three situations correspond to a case in which the origin-destination pair k uses the path h. Here two alternatives are presented: the transmission link transports all the flow allowed (according to the survivability constraint), (5), or the transmission link does not saturate the survivability constraint, (6).

$$
p_{kh} = \rho_k \Rightarrow u_{kh} = 0 \text{ and } v_{kh} \geq 0
$$

$$
\frac{\partial \bar{C}(p)}{\partial p_{kh}} - \mu_k = -v_{kh} \leq 0 \tag{5}
$$

$$\sum_{e \in E} [\bar{c}'_e(x_e)] \cdot [\gamma_k \cdot \delta^h_e] \le \mu_k \Rightarrow \sum_{e \in E} \bar{c}'_e(x_e) \cdot \delta^h_e \le \mu'_k = \frac{\mu_k}{\gamma_k}$$

$$0 < p_{kh} < \rho_k \Rightarrow u_{kh} = 0 \text{ and } v_{kh} = 0$$

$$\frac{\partial \bar{C}(p)}{\partial p_{kh}} - \mu_k = 0 \tag{6}$$

$$\sum_{e \in E} [\bar{c}'_e(x_e)] \cdot [\gamma_k \cdot \delta^h_e] = \mu_k \Rightarrow \sum_{e \in E} \bar{c}'_e(x_e) \cdot \delta^h_e = \mu'_k = \frac{\mu_k}{\gamma_k}$$

Equation 5 states that the cost variation sum associated to the links shaping path h is upper bounded by the indicator parameter. On the other hand, (6) shows how the cost variation sum associated to the links shaping path h is exactly equal to the indicator parameter, setting its value.

So, up to three different kinds of paths can be found. In monetary terms, they can be divided into cheap paths that must be flow saturated; expensive paths that must be flow empty; and one indicator path, which transports the rest of the flow to satisfy the demand equation problem. This indicator path plays the role of the frontier value between the cheap and expensive paths. Therefore the optimality conditions reduce to verify $\bar{c}'_{h_s}(p_{kh}) \le \bar{c}'_{h_i}(p_{kh}) \le \bar{c}'_{h_v}(p_{kh})$ for each path $h = \{h_s$ (saturated path), h_i (indicator path), h_v (empty path)$\} \in P(k)$.

The explanation of the Kuhn-Tucker multipliers can be deduced from (4–6). So, μ'_k what is associated to the demand balance equation, refers to the cost variation sum along the indicator path ($\bar{c}'_{h_i}$). Multiplier v_{kh}, what is associated to the survivability equation, refers to the difference between the cost variation sum along the indicator path and the cost variation sum along the saturated path ($\bar{c}'_{h_i} - \bar{c}'_{h_s}$). Multiplier v_{kh} must be positive in the optimum, otherwise the indicator path should be flow saturated and the saturated path should not be saturated, since it would be cheaper. Finally, multiplier u_{kh}, what is linked to the non-negative path variable equation, refers to the difference between the cost variation sum along the empty path and the cost variation sum along the indicator path ($\bar{c}'_{h_v} - \bar{c}'_{h_i}$). Multiplier u_{kh} must be positive in the optimum, otherwise the empty path should transport positive flow and the indicator path should be empty, since it would be cheaper. Based on the optimality conditions interpretation we have discussed, we propose the next algorithm to route the demand and assign capacity to the links.

Pseudocode 3
Initialisation

1. *Survivability in transmission links.* Survivability of grade(2) in transmission links between hubs is imposed by constructing two minimum spanning trees (MST) overlapped. Note that from Section 'The Conduit Installation Subproblem', there is a survivability network attending to the infrastructure (physical) network. Additionally we have to impose survivability in the logical network too. So, MSTs are constructed over the survivable network previously constructed. First MST spans over all the nodes in the

graph including terminals and hubs. Second MST spans over all the hubs with incidence arc set cardinality equal to one. Figure 3 describes the survivable logical network construction. First MST spans over all the nodes. And second MST spans only over hubs. It has to be noted that several hubs have already grade 2 of survivability, so the second MST must be expanded only over hubs with grade 1 of survivability. They are hubs a and b at this stage (Fig. 4).

2. *Indicator path.* For every pair of communication, k, search the path that communicates them incorporating a greater number of hub switching processes. This path is the indicator path. For example in Fig. 4, the case is the path showed in MST-1 for a communiction between hubs a and b. With this procedure, the remaining hierarchically most important costs are considered, i.e., the transmitter, receiver and multiplexer costs. The indicator path carries the remaining demand until satisfying the flow balance equation in **DR&CA-P**, and rest of paths are saturated.

3. If the optimality conditions are verified, the procedure finishes. Otherwise go to iterations.

Iterations

1. H(k) is the set of origin-destination pairs not satisfying the optimality conditions.

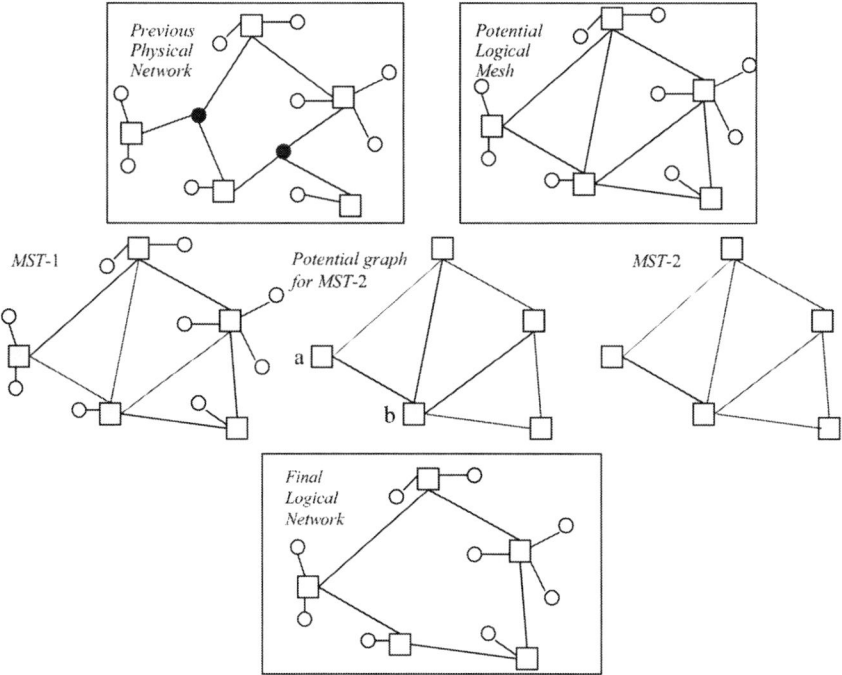

Fig. 4. Procedure that provides a survivable physical network (pseudocode 3)

2. Select the pair k^*, such as: $k^* = \max_{k \in H(k)} \left\{ \max_{h_s} \left\{ \left| \bar{c}'_{h_s} \left(p^0_{kh} \right) - \bar{c}'_{h_i} \left(p^0_{kh} \right) \right| \right\} \right\}$

3. For pair, k^*: a list of paths is generated and ordered by the value of $\bar{c}'_h \left(p_{k^*h} \right)$. The maximum demand allowed by the survivability constraint is assigned to the paths in the list while demand remains. They are the new paths saturated. The last path with a demand greater than zero (and less than the survivability constraint could allow) is the indicator path and it transports the remaining demand in order to satisfy the flow balance equation.

3.2 Computational Results

To check the suitability of the models and algorithms, we solve 600 trial problems. They are accessible via web in http://io.us.es/RedTeltrials/. Our EC approach is compared with a heuristic method used for telecommunication network planning in the real life (called 2-step heuristic, see Sexton and Reid [29], Chap. 13). Two-step heuristic proposes a basic conduit infrastructure by connecting directly nodes with higher traffic and reconnect isolated nodes to that node nearest by applying the shortest path algorithm. Nodes with a higher of links are set as hubs (this ends the first step). After that, the second step states the capacity of the links by applying the shortest path algorithm and rebalances oversized links onto undersized links. Table 2 shows the results for both methods.

For every case, EC approach provided better results than 2-step heuristic attaining a best result that improved in an 80.1% for instance 50_13. It is important to note that a substantial improvement is reached for the majority of the instances, so the average improvement is 36.2%. Only for five examples improvement less than 10% was achieved, and the worst case got an improvement equal to 1.8% for instance 70_10. Even more, for larger size networks EC approach obtained the better results.

Figure 5 provides a graphical comparison on the results obtained by the 2-steps and EC heuristics. EC approach always outperforms 2-steps one, even clearer when the problem size rises. As a direct result, the analysis with respect to the terminal and region parameters provides a much better performance for EC. On the other side, the results with respect to the nodes and arcs do not provide a so clear outperforming, though EC improvements are fairly better. For Table 2 costs can be taken in thousand of euros.

Two-step heuristic obtained faster computation time results for specific ranges because of processing simpler rules without making use of iterative calculus. Contrarily, EC requires iterations. Anyway, for 20 cases EC obtained results in less time than 2-step heuristic, although these instances corresponded to cases of networks with a low number of regions (these results are represented with negative figures in Table 2). For those cases with time consumption greater than 2-step heuristic we have to claim that the time

Table 2. Summary of results

Instance	Problem characteristics				2-step heuristic		Hierarchical heuristic		Comparison	
PROB	NODES	TERM	ARCS	REG.	COST	TIME	COST	TIME	% Better	Time increase
60_8	60	47	877	13	8251,3	52,6	6244,5	113,5	24,3%	60,9
60_9	60	32	735	9	5627,6	81,0	3743,2	49,8	33,5%	−31,1
60_10	60	32	1257	8	5236,3	56,0	3420,8	46,0	34,7%	−10,0
60_11	60	22	1411	6	3996,9	63,7	2719,0	16,3	32,0%	−47,4
60_12	60	36	440	17	10116,0	25,4	7885,2	203,6	22,1%	178,2
60_13	60	31	1545	11	7103,1	142,6	5455,0	84,0	23,2%	−58,5
60_14	60	31	740	10	7653,1	67,0	4504,3	57,2	41,1%	−9,8
60_15	60	28	562	14	10028,6	23,7	5685,5	94,5	43,3%	70,8
70_1	70	70	222	39	22292,2	11,3	17701,9	1702,4	20,6%	1691,0
70_2	70	68	894	33	15378,7	106,8	10778,6	993,9	29,9%	887,2
70_3	70	70	343	38	18510,6	27,2	12155,0	1462,6	34,3%	1435,4
70_4	70	27	772	12	7689,9	39,0	5732,7	105,0	25,5%	66,0
70_5	70	63	353	28	15625,5	26,2	11685,9	777,9	25,2%	751,6
70_6	70	62	482	30	15607,6	39,1	11013,2	825,7	29,4%	786,6
70_7	70	41	571	19	13566,9	31,7	6496,8	220,5	52,1%	188,8
70_8	70	23	526	10	6302,4	21,0	5366,5	57,1	14,8%	36,0
70_9	70	21	700	7	5345,0	25,0	3626,9	21,0	32,1%	−4,0
70_10	70	22	629	8	4778,6	15,2	4690,8	36,7	1,8%	21,5
70_11	70	7	865	3	2508,4	130,5	2005,2	0,9	20,1%	−129,6
70_12	70	38	514	18	9114,6	25,7	5588,0	190,4	38,7%	164,7
70_13	70	32	1050	13	9679,5	51,6	5042,4	108,3	47,9%	56,7
70_14	70	62	947	28	15120,6	100,7	9037,2	686,1	40,2%	585,4
70_15	70	66	207	24	15628,8	13,4	14141,8	638,0	9,5%	624,7
80_1	80	33	1526	12	7648,9	78,0	4117,6	105,1	46,2%	27,1
80_2	80	8	1167	3	4568,0	9,0	2109,1	1,4	53,8%	−7,6
80_3	80	14	1317	6	3118,0	28,7	2391,4	6,2	23,3%	−22,5
80_4	80	68	2624	34	15692,5	198,2	7645,0	1129,3	51,3%	931,1
80_5	80	48	599	16	13188,3	35,3	8334,0	229,7	36,8%	194,5
80_6	80	66	1617	30	13843,4	176,2	8079,3	769,4	41,6%	593,2
80_7	80	71	587	34	22077,6	68,6	12610,6	1281,5	42,9%	1212,9
80_8	80	80	1653	44	23576,2	297,6	10204,1	2281,3	56,7%	1983,7
80_9	80	63	278	31	21257,9	21,3	15483,6	1022,4	27,2%	1001,1
80_10	80	51	1696	23	17456,8	202,1	8082,2	420,8	53,7%	218,7
80_11	80	56	1589	26	12345,4	156,2	7203,9	504,0	41,6%	347,8
80_12	80	53	2435	25	12873,3	187,6	6907,8	540,3	46,3%	352,7
80_13	80	73	2487	31	11278,1	108,3	7683,4	926,3	31,9%	818,0
80_14	80	76	2480	38	14561,0	112,8	8888,2	1487,0	39,0%	1374,2
80_15	80	38	1654	10	9023,3	23,0	4469,0	72,1	50,5%	49,1
90_1	90	44	3237	12	6978,6	327,6	4689,9	101,9	32,8%	−225,7
90_2	90	65	2985	35	23125,6	87,6	12117,4	1722,0	47,6%	1634,4
90_3	90	66	3100	31	16745,9	78,9	11932,5	1218,5	28,7%	1139,6
90_4	90	69	1151	34	18956,0	56,7	10163,9	1212,8	46,4%	1156,1
90_5	90	90	3286	50	23657,0	187,5	11177,6	3489,1	52,8%	3301,6
90_6	90	66	2104	39	18674,9	100,7	8312,2	1651,3	55,5%	1550,6
90_7	90	86	413	43	26958,9	134,7	17299,7	2366,5	35,8%	2231,8
90_8	90	88	2796	44	14598,7	102,4	9793,2	2170,6	32,9%	2068,2
90_9	90	75	2845	20	13478,5	78,9	7404,6	379,0	45,1%	300,1

(continued)

Table 2. (continued)

Instance	Problem characteristics				2-step heuristic		Hierarchical heuristic		Comparison	
PROB	NODES	TERM	ARCS	REG.	COST	TIME	COST	TIME	% Better	Time increase
90_10	90	66	3233	33	16005,7	67,0	8001,0	1087,5	50,0%	1020,5
90_11	90	47	441	17	10333,4	41,0	8213,7	284,6	20,5%	243,5
90_12	90	70	2890	22	19987,9	104,2	9028,6	518,7	54,8%	414,5
90_13	90	68	863	24	16781,0	111,9	10317,2	578,5	38,5%	466,6
90_14	90	56	749	19	10986,9	140,2	8129,8	317,3	26,0%	177,1
90_15	90	29	1059	10	13925,3	72,2	5030,0	111,8	63,9%	39,5
100_1	100	29	688	13	13478,4	123,3	7337,8	231,2	45,6%	107,9
100_2	100	65	2213	23	17638,5	102,4	8177,4	466,1	53,6%	363,7
100_3	100	74	4711	20	17345,2	107,3	8321,8	403,3	52,0%	296,0
100_4	100	13	911	5	5789,0	21,0	3204,8	30,1	44,6%	9,1
100_5	100	27	3613	7	4832,0	45,0	3234,0	77,4	33,1%	32,4
100_6	100	96	998	32	17567,3	109,5	12048,5	1253,5	31,4%	1144,0
100_7	100	75	1551	27	14769,6	79,7	10236,5	867,1	30,7%	787,4
100_8	100	39	1966	14	7689,2	94,6	4942,7	144,8	35,7%	50,2
100_9	100	20	2743	7	4561,2	39,7	3561,0	41,1	21,9%	1,4
100_10	100	84	2833	25	17894,6	109,5	8727,1	591,9	51,2%	482,4
100_11	100	10	2069	5	5678,9	3,0	2110,6	2,2	62,8%	−0,8
100_12	100	100	2238	52	21568,2	367,8	13185,6	4038,0	38,9%	3670,2
100_13	100	100	1676	59	23817,6	435,4	12273,6	5119,5	48,5%	4684,1
100_14	100	46	3250	13	8751,2	98,8	5051,7	143,0	42,3%	44,2
100_15	100	90	2115	38	16789,4	101,1	11147,1	1679,8	33,6%	1578,7

execution was clearly feasible for the planning stage problem. So, the worst case corresponded to instance 100_13 with 5,119.5 s (4,684.1 s more than 2-step heuristic), and the average execution time was 487.0 s.

In brief, EC approach obtained substantial better results than 2-step heuristic considering the network deployment cost, and slightly worse results attending to execution time. However, the time results are completely assumable when dealing with a planning problem, and the so obtained improvement in terms of deployment cost compensates clearly it.

4 Conclusions

This paper has been divided into two parts. In its first one, main research carried out in telecommunications using evolutionary approaches has been presented. The history of successes and efficient applications shows telecommunications as a promising field of application for EC.

The second part is focused on one of the fields of applications previously described. The case corresponds to a realistic telecommunication network deployment planning problem that involves a huge number of well-known operations research problems such as hub location, conduit network design,

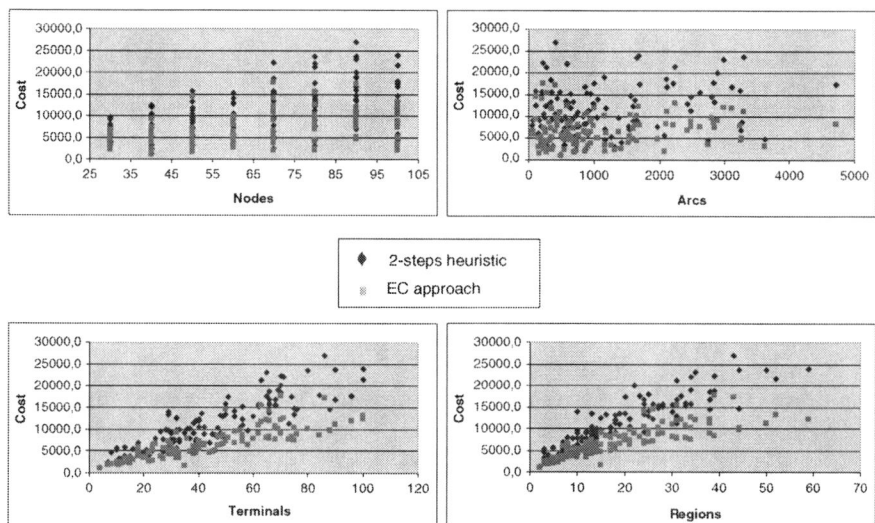

Fig. 5. Comparative analysis between 2-steps heuristic and EC approach

Steiner trees or multicommodity flow problems. To deal with such type of problems the operational researchers have been trying to adapt the reality to the corresponding well-known model more than trying to deepen into the real problem. Here, a finer successive formulation of the overall problem has been presented relating all these concepts. So, the differentiation between the physical and logical network special characteristics, and in their turn, among links, arcs and paths increases the complexity of the modelling effort. Historically, these details have not been dealt with in the telecommunications network topological design literature.

To solve the problem, our proposal is based on an evolutionary strategy that considers a hierarchical approach starting with the most relevant costs and falling into the rest of elements of the network. The formulation goes though different models evaluating the infrastructure costs, as well as the costs associated to the links. It also includes constraints of survivability and grade of service. To test the efficiency of the methodology we constructed a library of randomly generated problems. This EC method showed a much better behaviour than 2-step heuristic, outperforming it in all the analysed ranges.

References

1. Abuali FN, Schoenefeld DA, Wainwright RL (1994) Designing telecommunications networks using genetic algorithms and probabilistic minimum spanning trees. In: Proceedings of the 1994 ACM Symposium on Applied Computing, SAC 1994, ACM Press, Phoenix, pp. 242–246

2. Alba E, Chicano F (2005) On the behavior of parallel genetic algorithms for optimal placement of antennae in telecommunications. International Journal of Foundations of Computer Science 16: 343–359
3. Albiñana M, Escudero LF, Monge JF, Sánchez-Soriano J (2007) On the enrouting protocol problem under uncertainty. European Journal of Operational Research 181 (2): 887–902
4. Armony M, Klincewicz JC, Luss H, Rosenwein MB (2000) Design of stacked self-healing rings using a genetic algorithm. Journal of heuristics 6: 85–105
5. Bertsekas DP (1982) Constrained optimization and Lagrange multiplier methods, Academic Press, New York
6. Cardwell RH, Monma CL, Wu TH (1989) Computer aided design procedure for survivable fiber optic networks. IEEE Journal on Selected Areas in Communications 7: 1188–1197
7. Chen L, Xu ZQ (2004) An effective genetic algorithm for multicast routing tree with degree-delay-constrained. In: Yoo SJ Ben, Chang G-K, Li G, Cheung K-W (eds.) SPIE proceedings series International Society for Optical Engineering Proceedings Series, Vol. 5626 (2), pp. 793–804
8. Cortés P, Guerrero F, Canca D, García JM (2001) An evolutionary algorithm for the design of hybrid fiber optic-coaxial cable networks in small urban areas. Lecture Notes in Computer Science 2084: 749–756
9. Cortés P, Muñuzuri J, Larrañeta J, Onieva L (2002) A genetic algorithm based on cell loss for dynamic routing in ATM networks. In: Roy R, Köppen M, Ovaska S, Furuhashi T, Hoffman F (eds.) Soft Computing in Industry – Recent Applications, Spinger-Verlag, Berlin, pp. 627–640
10. Cortés P, Muñuzuri J, Onieva L, Larrañeta J, Vozmediano JM, Alarcón JC (2006) Andalucía assesses the investment needed to deploy a fiber-optic network. Interfaces 36 (2): 105–117
11. Din D-R (2004) Genetic algorithms for multiple multicast on WDM ring network. Computer Communications 27 (9): 840–856
12. Din D-R (2007) A genetic algorithm for solving virtual topology configuration transition problem in WDM network. Computer Communications 30 (4): 767–781
13. Elbaum R, Sidi M (1995) Topological design of local area networks using genetic algorithms. In: Proceedings of IEEE Conference on Computer Communications (INFOCOM'95), Boston, MA, pp. 64–71
14. El-Madbouly H (2005) Design and bandwidth allocation of embedded ATM networks using genetic algorithm. Transactions on Engineering. Computing and Technology 8: 249–252
15. Gódor I, Magyar G (2005) Cost-optimal topology planning of hierarchical access networks. Computers & Operations Research 32: 59–86
16. Grötschel M, Monma CL, Stoer M (1995) Design of survivable networks. In: Ball MO, Magnanti TL, Monma CL, Nemhauser GL (eds.) Network Models, North-Holland, Elsevier, pp. 617–672
17. Guimarães M, Rodrigues L (2003) A genetic algorithm for multicast mapping in publish-subscribe systems. In: Proceedings of the Second IEEE International Symposium on Network Computing and Applications, pp. 67–74
18. Hazem MM, Sayoud H, Takahashi K (2003) Hybridizing genetic algorithms for a survivability study of broadband communication networks: solution of the MSCC. In: APCC 2003. The 9th Asia-Pacific Conference. Vol. 3 (21–24): pp. 949–953

19. Huang R, Ma J, Frank Hsu D (1997) A genetic algorithm for optimal 3-connected telecommunication network designs. In: Proceedings of the International Symposium on Parallel Architectures, Algorithms and Networks (ISPAN '97), pp. 344–350
20. Karunanithi N, Carpenter T (1994) A ring loading application of genetic algorithms. In: Proceedings of the 1994 ACM symposium on applied computing, pp. 227–231
21. Kerner M, Lemberg HL, Simons DM (1986) An analysis of alternative architectures for the interoffice network. IEEE Journal on Selected Areas in Communications 4: 1404–1413
22. Kim S-S, Smith AE, Lee J-H (2007) A memetic algorithm for channel assignment in wireless FDMA systems. Computers & Operations Research 34 (6): 1842–1856
23. Lansard E, Frayssinhes E, Palmade JL (1998) Global design of satellite constellations: a multi-criteria performance comparison of classical walker patterns and new design patterns. Acta Astronautica 42 (9): 555–564
24. Le VT, Jiang X, Ngo SH, Horiguchi S, Inoguchi Y (2006) A novel dynamic survivable routing in WDM optical networks with/without sparse wavelength conversion. Optical Switching and Networking 3 (3–4): 173–190
25. Medhi D, Tipper D (2000) Some approaches to solving a multihour broadband network capacity design problem with single-path routing. Telecommunication Systems 13 (2–4): 269–291
26. Melián B, Laguna M, Moreno-Pérez JA (2004) Capacity expansion of fiber optic networks with WDM systems: problem formulation and comparative analysis. Computers & Operations Research 31: 461–472
27. Monma CL, Munson, BS, Pulleyblank WR (1990) Minimum-weight two-connected spanning networks. Mathematical Programming 46: 153–171
28. Pitsillides A, Stylianou G, Pattichis CS, Sekercioglu A, Vasilakos A (2000) Bandwidth allocation for virtual paths (BAVP): investigation of performance of classical constrained and genetic algorithm based optimisation techniques. In: Proceedings of the Nineteenth Annual Joint Conference of the IEEE Computer and Communications Societies INFOCOM 2000 3 (26–30), pp. 1501–1510
29. Sexton M, Reid (1997) A Broadband Networking. ATM, SDH and SONET. Artech House
30. Sheng-Ling Wang, Yi-Bin Hou, Jian-Hui Huang, Zhang-Qin Huang (2006) Adaptive Call Admission Control Based on Enhanced Genetic Algorithm in Wireless/Mobile Network. 18th IEEE International Conference on Tools with Artificial Intelligence (ICTAI'06) 3–9
31. Sherif MR, Habib IW, Naghshineh M, Kermani PA (1999) Generic bandwidth allocation scheme for multimedia substreams inadaptive networks using genetic algorithms. In: Proceedings of the Wireless Communications and Networking, Vol. 3, pp. 1243–1247
32. Wang S-L, Hou Y-B, Huang J-H, Huang Z-Q (2006) Adaptive call admission control based on enhanced genetic algorithm in wireless/mobile network. In: Proceedings of the 18th IEEE International Conference on Tools with Artificial Intelligence (ICTAI'06), pp. 3–9
33. Watanabe S, Hiroyasu T, Miki M (2001) Parallel evolutionary multi-criterion optimization for mobile telecommunication networks optimization. In: Proceedings of the EUROGEN 2001 Conference, pp. 167–172

34. Wille CG, Mellia M, Leonardi E, Ajmone M (2005) Topological design of survivable ip networks using metaheuristic approaches. In Third International Workshop on QoS in Multiservice IP Networks, Catania (Italy)
35. Wille CG, Mellia M, Leonardi E, Ajmone M (2006) Algorithms for IP networks design with end-to-end QoS constraints. Computer Networks 50 (8): 1086–1103
36. Yener A, Rose C (1997) Genetic algorithms applied to cellular call admission problem: local policies. IEEE Transactions on Vehicular Technology 46 (1): 72–79
37. Yoon M, Baek Y, Tcha D (1998) Design of a distributed fiber transport network with hubbing topology. European Journal of Operational Research 104: 510–520
38. Zhang O, Leung Y-W (1999) An orthogonal genetic algorithm for multimedia multicast routing. IEEE Transactions on EC 3 (1): 53–62
39. Zhou G, Gen M (2003) A genetic algorithm approach on tree-like telecommunication network design problem. Journal of the Operational Research Society 54: 248–254

Survivable Network Design with an Evolution Strategy

Volker Nissen[1] and Stefan Gold[2]

[1] Technical University of Ilmenau, Chair of Information Systems in Services,
  PF 10 05 65, D-98684 Ilmenau, Germany, volker.nissen@tu-ilmenau.de
[2] Technical University of Ilmenau, Chair of Information Systems in Services,
  PF 10 05 65, D-98684 Ilmenau, Germany, stefan.gold@web.de

Summary. In this chapter, a novel evolutionary approach to survivable network design is presented when considering both economics and reliability. It is based on a combinatorial variant of the evolution strategy and clearly outperforms existing benchmark results by genetic algorithms. The integration of domain-specific knowledge in the mutation operator and a repair heuristic raises the performance of an otherwise broadly applicable but less effective metaheuristic. Moreover, the results underline the important but so far often neglected potential of evolution strategies in combinatorial optimization.

1 Introduction to Survivable Network Design

This chapter describes an evolution strategy approach to combinatorial optimization. It is presented within the framework of optimal telecommunication network design when considering both economics and reliability. We abstract from the consideration of flows and capacities at the planning stage. Instead, the focus is on costs for connecting the nodes in the network, different technology options and reliabilities associated with these links, and, finally, reliability constraints assigned to the network as a whole.

Over the last years, business-to-business telecommunication has become a key success factor for many companies. Communication networks are the backbone of important business processes such as electronic commerce, customer relationship management or supply chain management. Moreover, they enable end user services like multimedia, IP telephony and mobile communication. Due to the availability of reliable high-performance communication networks it is possible to integrate business partners efficiently on a worldwide scale and implement new concepts such as service-oriented IT-architectures [17] and enterprise application integration [20]. Thus, the design of fault-tolerant networks is an issue of paramount importance.

The combinatorial problem of survivable network construction, as addressed here, asks for a network design that can be implemented with a minimum

V. Nissen and S. Gold: *Survivable Network Design with an Evolution Strategy*, Studies in Computational Intelligence (SCI) **92**, 263–283 (2008)
www.springerlink.com

of resources while a given network dependability is secured. The resulting complexity of the optimization problem requires adequate decision support methods in the design phase.

It is assumed that node locations are predetermined and different technology options to connect individual nodes exist. Network setup costs are a function of the distance of nodes connected in the network and the chosen technology options for connecting nodes. It is further assumed that nodes are perfectly reliable and do not fail, and that edges have two possible states – good or failed. Edges fail independently and repair is not considered. Installation and maintenance costs are also not considered. These are common assumptions among this family of problems [14, 19, 27, 38, 39].

Cost optimal network design under reliability constraints is an NP-hard combinatorial optimization problem [12, 23, 43]. For germany30, the largest problem presented here, the solution space has a size of $7.87 \cdot 10^{261}$. Heuristics are, therefore, an appropriate solution method.

In our approach we develop a combinatorial version of the evolution strategy, originally created by Rechenberg and Schwefel [36, 42], and today applied mainly to continuous parameter optimization problems. The combinatorial version of the evolution strategy is shown to be flexible and computationally effective on a test suite of diverse network design problems.

2 Formal Statement of the Problem

The network design issue addressed in this chapter can be treated as a graph theoretic problem. Each problem instance has a fixed number of nodes. A fully connected network is modeled as an undirected graph $G(V, E)$. V is the set of given nodes and E the set of possible edges in graph G, where $|V|$ denotes the number of nodes and $|E|$ the number of possible edges. Each edge $e_i \in E$ represents a bi-directional link between two nodes $v_i \in V$ of the graph G. Links can be realized through different technology options $l_k(k = 1...k_{max})$. The technology options differ w.r.t. cost $c(l_k(e_i))$ and reliability $r(l_k(e_i))$ of edges, where $l_k(e_i)$ is the currently chosen option for a particular edge. Generally, technology options with higher reliability also cause higher costs. Each link is associated with a state s_{e_i}, where $s_{e_i} = 1$ means "operative" and $s_{e_i} = 0$ means "not operative".

Figure 1 shows a sample graph with $|V| = 4$ nodes and $|E| = 6$ possible edges. The edges with dotted lines are not employed in this graph. The other edges are associated with different technology options $k = 1...3$.

The overall reliability of the network is of particular interest. It is measured by the all-terminal reliability (R_{all}) that can be defined as the probability that a connecting path exists between each pair of nodes in a probabilistic graph [11]. This is equivalent to the existence of a spanning tree that connects all nodes of the graph [12]. The all-terminal reliability is an adequate measure

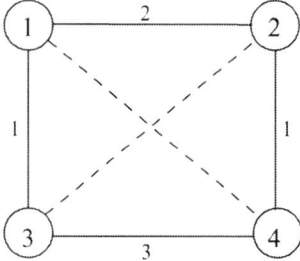

Fig. 1. Sample graph of a small network

of the robustness of a network design when individual connections fail [14]. To calculate the all-terminal reliability we follow the approach suggested in [39] which is based on Monte Carlo simulations similar to previous work [15,19,27].

Each solution to the network design problem is represented as a subgraph $G_N(V, E_N \subset E)$, where $C(G_N)$ is the total cost of subgraph G_N. The total cost is given by the sum of the cost of the edges $e_i \in E_N$. The cost of each edge $c(l_k(e_i))$ is given by the Euclidean distance of the two nodes multiplied with the unit cost of the connecting edge. If R_0 is a reliability constraint for the network design and $R_{all}(G_N)$ is the all-terminal reliability of the graph G_N then the objective function for the design of cost minimal reliable networks can be defined as

$$C(G_N) = \sum_{e_i \in E_N} c(l_k(e_i)) \to min \quad \text{with} \quad R_{all}(G_N) \geq R_0 \quad (1)$$

3 Related Work

There is some work on the evolution strategy in combinatorial and discrete optimization. Herdy [24] investigates discrete problems with some focus on neighborhood sizes during mutation. Rudolph [40] develops an evolution strategy for integer programming by constructing a mutation distribution that fits this particular search space. Bäck [6] discusses mutation realized by random bit-flips in binary search spaces. Nissen [33] modifies the coding and the mutation operator of the evolution strategy to solve combinatorial quadratic assignment problems. In [34] this approach is extended to solve facility layout problems. Schindler et al. [41] apply the evolution strategy to combinatorial tree problems by using a random key representation which represents trees with real numbers. Schwefel and Beyer [9] present permutation operators for the evolution strategy in combinatorial ordering problems. Li et al. [30] develop a mixed-integer variant of the evolution strategy that can optimize different types of decision variables, including continuous, normal discrete, and ordinal discrete values. However, continuous parameter optimization is still the dominant field of application for the evolution strategy, as the operators of the standard form are particularly well adapted to this type of problem.

Previous approaches to network design have either been enumerative, which is applicable only for small network sizes [2, 28], or heuristic-based, which can be applied to larger networks but does not guarantee optimal solutions. Previous heuristic approaches include tabu search [21], neural networks [1], genetic algorithms [13, 15, 22, 29], and simulated annealing [5, 16, 35, 44].

The issue of providing a reliable network design under a cost constraint is investigated in [10] using a simulated annealing approach. Different technology options are not considered, and a simpler measure than all-terminal reliability is used to secure network dependability. Liu and Iwamura [31] develop a genetic algorithm for minimal cost network design while also considering the k-terminal reliability of the network. A similar work is presented in [4]. Ghosh et al. [22] also develop a GA-approach for survivable network construction that discards invalid solutions instead of repairing them. Again, simpler measures than the all-terminal reliability are used. An overview of metaheuristics for the optimal design of computer networks is provided by Altiparmak et al. [3]. Soni et al. [44] present a simulated annealing approach to cost minimal communication access network design, but they do not use all-terminal reliability as a measure of network dependability. In [1] a neural network is used for planning survivable networks using the all-terminal reliability measure. However, different technology options for network links are not included. This also applies to Dengiz et al. [15] where an evolutionary approach is presented. Survivability conditions in networks are also addressed by [13].

The research that comes closest to our own is by [14, 19, 38, 39]. In all these contributions, cost minimal network design with reliability constraints is investigated while allowing for different technology options with associated unit costs. Thus, the final network may have a heterogeneous combination of differing edge reliabilities and costs. This relaxation over older contributions in the field, while greatly improving the relevance to real-world network design, also complicates the network reliability calculation and exponentially increases the search space [14].

A reference of special importance is [39]. In his PhD-dissertation, Reichelt develops two different genetic algorithms (GAs), termed LaBORNet and BaBORNet for the same suite of problems as used in this chapter.

In LaBORNet (Lamarckian Based Optimizer for Reliable Network Design Problems) solutions are coded as vectors of edges and technology options. Network topology and technology options are determined simultaneously by the GA. If the genetic operators create an invalid solution, a repair heuristic called CURE [37] is applied to determine a valid network design in the neighborhood of the invalid result. CURE is based on previous work of [18] and may be interpreted as a local search heuristic that is combined with the GA-approach. Only valid results are evaluated w.r.t. the objective function.

BaBORNet (Baldwin Based Optimizer for Reliable Network Design Problems) is a GA inspired by the natural Baldwin effect in evolution. The overall problem is decomposed in two subproblems: topology search and technology design. Each solution in the GA contains information only on the network topology. Hence, the code is a binary vector of edges. Each edge is initially associated with the least costly technology option. When the overall network reliability constraint is violated, the CURE heuristic is applied to change edges and technology options until the solution is valid. The invalid GA solution is then associated with the fitness value of the valid solution produced by CURE.

Both GA-approaches of Reichelt demonstrate good performance on the test suite and clearly outperform previous heuristics in the literature. Reichelt experiments with different population sizes, where $popsize = 400$ produces the best overall results on the test suite. For the problem deeter10, he improves the previous best known result (cost value 5,661) of [14] by 21% to 4,386 (LaBORNet with $popsize = 200$). In the case of turkey19, the previous best known solution of [19] with a cost value of 1,755,474 is improved by 9% to 1,591,110 (BaBORNet with $popsize = 400$). Reichelt also creates benchmark results for new and partly larger problem instances with numbers of nodes ranging from 15 to 30.

In a direct comparison, BaBORNet produces better solutions on large problem instances than LaBORNet while also requiring more CPU-resources. The two GAs by Reichelt are the best available heuristics so far for the class of network design problems investigated in this chapter. Consequently, the performance or our own Combinatorial Evolution Strategy will be evaluated against these benchmarks.

4 An Evolution Strategy Approach to Network Design

4.1 Some General Remarks

The evolution strategy (ES) was originally invented by Rechenberg and Schwefel [36, 42] and soon applied to continuous parameter optimization problems. Like genetic algorithms [25, 32], it belongs to the class of evolutionary algorithms that form broadly applicable metaheuristics, based on an abstraction of the processes of natural evolution [7,8]. However, ESs are less well known than GAs despite of their substantial optimization potential. Moreover, they are not often applied in combinatorial optimization. In short, the main differences between ESs and GAs are:

- A GA focuses on recombination (crossover) as the main search operator, while an ES puts the focus for search on mutation.
- In an ES λ offspring solutions are created from μ parents by copying and modifying the parents through mutation and, if desired, recombination. Common ratios of μ and λ are between 1:5 and 1:7. This differs from the generation cycle in a GA that generally keeps the population size constant. Moreover, selection is deterministic in the ES while it is frequently stochastic in a GA.

- Mutation in standard-ESs is performed using normally-distributed random variables so that small changes in a solution are more frequent than large changes. For GAs, many different mutation operators and problem representations have been developed that address continuous parameter optimization as well as combinatorial problems.
- The concept of self-adapting strategy parameters was introduced early in ES-research. In particular, the mutation step size in an ES can be made self-adaptive by incorporating it in the solution representation and applying a log-normally distributed random variable to mutate the step size before using this step size to mutate decision parameters.

ESs apparently have not been applied to the class of network design problems investigated here. This may be due to the fact that the original ES-concept is well-adapted to continuous parameter optimization but modifications are required for combinatorial problems. The ES suggested in this chapter is a combinatorial version that aims at transferring main concepts from the standard-ES while modifying necessary technical details for the specific application. Changes are mainly required for the mutation operator.

4.2 Coding and Initialization of the ES-Population

A network design problem is represented as a vector $\mathbf{g} \in Z^n$ of length $|E| = \frac{|V|(|V|-1)}{2}$. Each element g_i of $\mathbf{g}$ represents an edge of the graph G with technology option $l_k (1 \le k \le k_{max})$. If an edge is not employed in the graph then $g_i = 0$. Figure 2 shows the coding of the sample graph from Fig. 1, using a node connection matrix [26]. The gray shaded upper triangle of this symmetrical matrix is transferred line by line to the solution vector $\mathbf{g}$. Thus, the coding for the sample graph would be $\mathbf{g} = (2, 1, 0, 0, 1, 3)$.

An ES-individual consists of the vector $\mathbf{g}$ plus a strategy parameter "mutation step size" σ which is required during mutation. The population is initialized randomly. For each pair of nodes in each individual of the population, an edge is created with probability 0.4. If the edge is created, one of the available technology options is chosen with even probability. The mutation step sizes are initialized with $\sigma_0 = 3$, following the suggestion of [6].

v_i/v_j	1	2	3	4
1	-	2	1	0
2	2	-	0	1
3	1	0	-	3
4	0	1	3	-

Fig. 2. Node connection matrix of sample graph

4.3 Fitness Function and Constraint Handling

Since a fitness evaluation is required for each valid ES-individual, it is sensible to check the reliability constraint during the evaluation and start repair when the constraint is violated. The following pseudo-code demonstrates the fitness implementation of our ES-approach:

1. function $objective$ (G_N, R_0)
2. R_{all} ← All-Terminal Reliability (G_N)
3.   while $R_{all} < R_0$ do
4.   repair invalid G_N with CURE heuristic
5. R_{all} ← All-Terminal Reliability (G_N)
6.   end while
7.   fitness (G_N) ← calculate total cost $C(G_N)$
8.   return fitness
9. end function

The fitness evaluation receives as input the decoded subgraph G_N and the reliability constraint R_0. Then the all-terminal reliability is calculated, following an approach of [39]. If the constraint is violated ($R_{all} < R_0$) then the CURE repair heuristic [37] is called until G_N has been converted to a valid solution of the network design problem. The total cost of the repaired solution is calculated and returned as fitness value.

4.4 Selection

A (μ, λ)-selection is employed for all problem instances, meaning that in every generation λ offspring are generated from μ parent solutions. It is a deterministic, non-elitist selection method that prevents parents to survive to the next generation cycle. However, the best solution found during an experimental run is always stored and updated in a "golden cage". It represents the final solution of the run. Following suggestions for the standard-ES in the literature [6,9], the ratio μ/λ is set to 1/5–1/7 during the practical experiments.

4.5 Recombination

Recombination operates on two randomly chosen solutions from the parent population. According to [42], it is particularly useful for the self-adaptation of the mutation step sizes. As in the standard-ES, intermediate recombination of the strategy parameter and discrete recombination of the decision parameters (edge information) is performed to create an offspring from two parents. The new individual then undergoes mutation. In the combinatorial ES outlined here intermediate recombination of is always performed, but discrete recombination of decision parameters is optional. Experiments with and without discrete recombination were performed and results are reported in a later section of this chapter.

4.6 Mutation

To create a powerful optimization method from a broadly applicable meta-heuristic, such as the Evolution Strategy, it is useful to combine the evolutionary approach with domain-specific knowledge. Mutation is not only the main search operator in ESs but also the preferred place to integrate such domain knowledge. The design of mutation operators is, thus, problem-dependent [9]. The mutation applied in our combinatorial version of an ES has a domain-independent and a domain-dependent part. The first step, visualized in the pseudo code below has a strong resemblance to the standard-ES mutation.

1. procedure $mutate - 1(G_N, G)$
2. $\tau \leftarrow 0.1$
3. $\sigma' \leftarrow \sigma exp(\tau \cdot N(0,1))$
4. $drift \leftarrow N(0, \sigma')$
5. $count \leftarrow round(|drift|)$
6.   for $i \leftarrow 1$, count do
7.   chose an edge $e_i \in E$ with probability $p = \frac{1}{|E|}$
8.   generate an even-distributed random number $z = [0, 1]$
9.   if $z < 0.5$ then
10.   edge e_i will be part of G_N
11.   chose technology option $l_k(e_i)$ with probability $p = \frac{1}{k_{max}}$
12.   else
13.   edge e_i will not be part of G_N
14.   end if
15.   end for
16.   call procedure $mutate - 2(G_N, count)$
17. end procedure

Mutation is here controlled by a normally-distributed random variable that utilizes the mutation step size σ of each ES-individual. First, this strategy parameter is mutated using a log-normal distribution with the factor τ set to 0.1. This heuristic value for τ worked better in our combinatorial case than other suggestions for the standard-ES from the literature by [6,9]. Still, as in the basic ES-concept, small mutations are more likely than large ones, and mutation step size can vary over the course of a run through a self-adaptive mechanism, thereby tuning the step size to the fitness landscape topology.

Then, by creating a $(0, \sigma')$-normally-distributed random number ($drift$) the extent of the mutation of edges in the network design is determined. The rounded absolute value of $drift$($count$) decides how many positions (edges) in the integer string of an individual are randomly changed. $Count$-times an edge of the complete graph G is chosen, independently whether it is already part of G_N or not. With a probability of 0.5 the state s_{e_i} of this edge is flipped. If the edge is part of G_N then one of the technology options $l_k(1 \leq k \leq k_{max})$

is associated with the edge with even probability. Finally, the domain-specific mutation operator is called. It works as follows:

1. procedure $mutate - 2(G_N, count)$
2. $border \leftarrow 3$
3.   if $count > 0$ then
4.   mutation heuristic $(G_N, min(count, border))$
5.   end if
6. $count' = count - border$
7.   if $count' > 0$ then
8. $mutation\ heuristic(G_N, count', removeEdge)$
9.   end if
10. end procedure

Depending on the value of $count$ a certain number of edges in G_N are downgraded in their technology option (if possible) or completely removed. The edges are chosen by a domain-specific mutation heuristic that focuses on edges with either high costs or low contribution to the overall reliability of the network. The parameter $border$, set heuristically to a value of 3, represents a threshold. When the mutation should be intense, meaning $count > border$, then edges are removed from the solution, otherwise their technology option is downgraded.

The mutation heuristic used here is based on the concept of spanning trees. As was pointed out before, the all-terminal reliability is used to measure the reliability of a network design. It can be interpreted as the probability that at least on spanning-tree exists that connects all nodes of a probabilistic graph. [18] introduced a method (spanning tree count – STC) to quickly calculate the number of spanning-trees of a graph that was further refined by [39]. The spanning-tree based mutation heuristic used in the combinatorial ES can be thought of as determining a cost-benefit-ratio for the edges of the graph in a solution. It is explained in the following pseudo-code:

1. procedure $mutation\ heuristic\ (G_N, count, removeEdge)$
2. $L \longleftarrow \oslash$
3. $currentTrees \longleftarrow STC(G_N)$
4.   for all $e_i \in E_N$ do
5. $remainTrees_{e_i} = STC(G_N \setminus e_i)$
6. $\triangle STC_{e_i} = currentTrees - remainTrees_{e_i}$
7. $EdgeRatio_{e_i} = c(l_k(e_i))/ \triangle STC_{e_i}$
8. $L \cup \{e_i, EdgeRatio_{e_i}\}$
9.   end for
10.   sort L w.r.t. $EdgeRatio_{e_i}$ in descending order
11.   for $i \leftarrow 1, count$ do
12. $e_{worst} \leftarrow$ worst edge $e_i \in L$
13.   if $removeEdge$ then
14.   remove e_{worst} from G_N
15.   else

```
16.            if l_k(e_worst) > 0 then
17.                l_{k-1}(e_worst)
18.            end if
19.        end if
20.        L ← L \ (e_worst, EdgeRatio_{e_worst})
21.    end for
22. end procedure
```

First, the number of spanning-trees in the given solution is calculated. Second, each edge $e_i \in E_N$ is individually removed from the graph G_N. Third, the difference $\triangle STC_{e_i}$ of spanning-trees of this modified solution and the original solution is calculated. $\triangle STC_{e_i}$ measures the contribution of the edge to the overall reliability of the network design. Fourth, the cost $c(l_k(e_i))$ of the edge is set in relation to $\triangle STC_{e_i}$, giving the *EdgeRatio* of the current edge. Finally, when all *EdgeRatios* have been calculated, the edges are ranked in descending order according to their *EdgeRatios*, and those with the worst results are either removed from the solution or their technology option is downgraded.

4.7 Destabilization and Termination of the ES

In highly multi-modal fitness-landscapes there is always a risk for optimization techniques to converge prematurely on some suboptimum. Population-based methods, such as evolution strategies, are generally less affected by this risk than traditional methods. Nevertheless, an operator called destabilization is used that helps to escape from local optima. Destabilization was successfully applied in [33]. The general idea is to intensify mutation when the solutions do not seem to improve any more.

In our implementation, the mutation step size of each individual is set to a value $\sigma_{new} > \sigma$, allowing for larger movements in the search space to leave a local suboptimum. After some experiments, the value for σ_{new} is determined heuristically as

$$\sigma_{new} = \frac{2\sigma_{avg} + \sigma_{max}}{2} \tag{2}$$

where σ_{avg} is the mean of all mutation step sizes and σ_{max} is the largest step size in the parent population.

Destabilization is used when the best solution found during the run could not be improved for 10 generations. If no improvement occurs for the next 10 generations after destabilization, then the run is terminated. Thus, after 20 generations without success the ES terminates. This is in accordance with a termination criterion of [39] in his GA-approaches. A second termination criterion stops the run independently, if for 15 generations the fitness value of the best individuals in the population lies within a range of $\varepsilon = 0.01$.

4.8 Overall Outline of the ES

To summarize this section, an high-level overview of the combinatorial ES is given in pseudo code:

1. combinatorial ES-heuristic
2. $t \leftarrow 0$
3. create initial population P_0^μ
4. determine fitness of individuals in P_0^μ
5. repeat
6.   for $i = 1, \lambda$ do
7. $a_1, a_2 \leftarrow$ random selection from P_t^μ with prob. $1/\mu$
8. $\sigma \leftarrow$ intermediate recombination $(\sigma_{a_1}, \sigma_{a_2})$
9. $g \leftarrow$ discrete recombination $(\mathbf{g}_{a_1}, \mathbf{g}_{a_2})$ or replication of $\mathbf{g}_{a_1}$
10. $\sigma' \leftarrow \sigma \exp(\tau N(0,1))$
11. $g' \leftarrow$ domain-independent mutation of g
12. $g'' \leftarrow$ domain-specific mutation of g'
13.   generate new individual $a_z = (g'', \sigma')$
14. $P_t^\lambda \cup a_z$ (add new individual to offspring population)
15.   end for
16.   determine fitness of individuals in P_t^λ
17. $P_{t+1}^\mu \leftarrow$ the best μ individuals from P_t^λ (comma selection)
18. $t \leftarrow t + 1$
19.   if no success for 10 generations then destabilization
20. until termination criterion holds
21. output of best solution found during entire run

5 Test Suite

For benchmarking purposes an established set of network design problems under reliability constraints from the literature is employed. The distance matrices between nodes for each problem can be found in the appendix, as well as cost and reliability figures for the technology options.

The problem deeter10 was first proposed in [14]. It is characterized by a network of ten nodes where each edge has one of three technology options with different unit costs $c_u(l_1) = 8$, $c_u(l_2) = 10$, $c_u(l_3) = 14$, and reliabilities $r(l_1) = 0.7$, $r(l_2) = 0.8$, and $r(l_3) = 0.9$. The reliability requirement for the entire network is set at 0.95.

The next problem instance, turkey19 was also included in [14]. It has 19 nodes that represent universities in Turkey. In contrast to deeter10, this problem is characterized by much smaller differences in reliability between the three technology options: $r(l_1) = 0.96$, $r(l_2) = 0.975$, and $r(l_3) = 0.99$. The reliability requirement for this network is higher than in deeter10 and set at 0.99. Unit costs are $c_u(l_1) = 333$, $c_u(l_2) = 433$, $c_u(l_3) = 583$.

The problems germany15–30 were introduced in [39] with 15 to 30 nodes respectively. The nodes represent the largest towns in Germany. Technology options have unit costs $c_u(l_1) = 8$, $c_u(l_2) = 10$, $c_u(l_3) = 14$, and reliabilities $r(l_1) = 0.7$, $r(l_2) = 0.8$, and $r(l_3) = 0.9$. The reliability requirement for the networks is set at 0.95. Germany30 is the largest problem instance published so far.

6 Results

For each problem instance 30 independent runs with the ES were performed, starting from randomly initialized populations. Since the absolute population size frequently influences run-time requirements and success probability of an evolutionary algorithm on complex problems, experiments with different population sizes were conducted for the ES. Tables 1 and 2 present the results for (30,200)- and (10,50)-selection. Additionally, the (30,200)-ES was tested with and without the optional discrete recombination of decision parameters. All runs with the ES were performed on Pentium-4, 2.4 GHz workstations.

Table 1. Results for deeter10 and turkey19. C_{result} represents mean and std.dev. of cost values. Fitness eval. gives mean and std.dev. of fitness evaluations. $C_{best-all}$ is the best solution found by the respective heuristic in all runs. 30 runs were performed with the ES for each problem, starting from randomly initialized solutions. The new best known solutions are bold and underlined. GA-DeSm is the genetic algorithm approach published by [14]. GA-FCC is the parallel genetic algorithm proposed by [19]. LaBORNet and BaBORNet are two genetic algorithms by [39]. The benchmark results by Reichelt as well as Deeter and Smith were based on ten randomly initialized runs. For Flores et al. the number of runs performed and further details are unclear from the reference.

| Problem | Heuristic | C_{result} | | $C_{best-all}$ | Fitness eval | |
		mean	std.dev.		mean	std.dev.
deeter10	(10,50)-ES	4,514	57	4,433	1,675	446
	(30,200)-ES	4,466	33	4,403	5,223	1,198
	(30,200)-ES with recomb.	4,441	27	**4,354**	5,310	1,170
	GA-DeSm	6,033	n.a.	5,661	n.a.	n.a.
	LaBORNet	4,450	13	4,433	9,220	256
	BaBORNet	4,561	6	4,598	7,120	376
turkey19	(10,50)-ES	1,789,411	217,217	1,614,450	4,328	1,350
	(30,200)-ES	1,619,627	48,936	1,555,080	13,810	3,493
	(30,200)-ES with recomb.	1,588,156	25,117	**1,548,440**	11,450	1,677
	GA-DeSm	n.a.	n.a.	7,694,708	n.a.	n.a.
	GA-FCC	n.a.	n.a.	1,755,474	n.a.	n.a.
	LaBORNet	2,023,240	179,716	1,689,800	25,140	1,366
	BaBORNet	1,610,041	13,875	1,591,110	29,040	728

Table 2. Results for problems germany15–30. Experimental setup identical to Table 1. The new best known solutions are bold and underlined.

Problem	Heuristic	C_{result}		$C_{best-all}$	Fitness eval	
		mean	std.dev.		mean	std.dev.
germany15	(10,50)-ES	42,694	523	41,448	2,903	934
	(30,200)-ES	41,918	497	41,110	8,323	1,671
	(30,200)-ES with recomb.	41,432	382	**40,960**	9,510	1,690
	LaBORNet	43,248	867	42,440	23,480	970
	BaBORNet	43,730	336	43,150	21,580	780
germany20	(10,50)-ES	46,143	724	44,804	3,222	803
	(30,200)-ES	45,551	489	44,494	9,610	1,761
	(30,200)-ES with recomb.	44,886	334	**44,164**	9,910	1,482
	LaBORNet	48,868	2,245	46,014	29,180	1,525
	BaBORNet	47,683	872	46,754	30,780	882
germany25	(10,50)-ES	49,932	2,144	**47,120**	4,133	1,064
	(30,200)-ES	48,237	516	47,248	11,323	1,463
	(30,200)-ES with recomb.	47,632	245	47,136	11,257	2,026
	LaBORNet	64,275	3,184	59,638	39,240	209
	BaBORNet	54,569	715	53,102	36,560	352
germany30	(10,50)-ES	56,719	1,250	53,950	4,405	994
	(30,200)-ES	54,913	764	53,268	13,623	1,414
	(30,200)-ES with recomb.	53,739	406	**53,202**	12,643	1,648
	LaBORNet	123,678	3,811	116,470	38,820	327
	BaBORNet	89,315	18,097	77,962	35,080	1,016

Since it is difficult to compare CPU-time with results of other authors, the average number of fitness evaluations (valid solutions inspected during a run) is a more reliable measure to compare the computational requirements of our ES to the results of others. This applies in particular to [39] who uses an identical approach and evaluates the fitness only for valid solutions.

However, it is worth pointing out that for Reichelt's GAs as well as our own ES-heuristics the CPU-requirements on a normal workstation are high. For instance, even for the smallest problem deeter10 the average duration of a single run is 593 sec for the (10,50)-ES, 1,522 s for the (30,200)-ES without recombination, and 2,038 s for the (30,200)-ES with recombination. With a population size of 400 Reichelt's LaBORNet-GA requires 3,153 s and BaBORNet-GA 6,966 sec for a run on deeter10, using Pentium-4, 2.0 GHz workstations. This is due to the complexity of the problem and the additional effort for reparing invalid solutions by the CURE-heuristic.

Tables 1 and 2 summarize our results for the ES and compare them with the best results achieved by [39] with GAs of popsize = 400.

For some of our network design problems other published results exist and are also given in the tables. [14] propose a penalty-based GA-approach and report results for deeter10 and turkey19. In [19] a parallel GA-approach is described with an application to turkey19. All solutions reported in Tables 1 and 2 are valid with respect to the reliability constraints as defined for the problem instances in the test suite.

For all problem instances new best solutions were discovered with the combinatorial evolution strategy. The improvements over previous best solutions as given in [39] are 0.7 % for deeter10, 3.5 % for germany15, 2.7 % for turkey19, 4.0 % for germany20, 11.3 % for germany25, and 31.8 % for germany30. These results were achieved with considerably fewer fitness evaluations than in the benchmark GAs by Reichelt.

7 Discussion

The quality of results demonstrates the flexibility and effectiveness of the combinatorial ES with respect to network design problems. The problem instances consider different technology options for edges in the network graph and, thus, reflect an important aspect of practical network design. However, the technology options greatly increase the size of the search space and, consequently, the complexity of the optimization problem.

The combinatorial ES was tested in three variations. Overall, the (30,200)-ES with discrete recombination delivers the best solutions w.r.t. total network cost. The (10,50)-ES that operates with a smaller population size uses much less computational resources while still delivering good cost results. The (30,200)-ES without discrete recombination performs almost as good as the version with recombination, but also has high computational requirements. Clearly, a larger population helps to achieve a more cost-efficient network design by avoiding or delaying convergence on local suboptima. Destabilization should be regarded as an optional operator that is useful particularly for larger problem instances, where it helped to improve results in roughly half of all runs.

As a side note, our experiments also confirmed that the rather disruptive discrete recombination often increases the repair effort for invalid solutions and, thus, raises runtime-requirements when compared to an otherwise identical approach without recombination. However, the improved results seem to justify this extra effort.

It is worth noting that the combinatorial evolution strategy was not tuned for any individual problem instance. On a relative basis, the ES performs least good on turkey19 which is characterized by very small differences in reliability between the three technology options for edges. This in turn means that the fitness landscape here has less marked "hills" and "valleys" than in the other problem instances. The spanning-tree-based mutation heuristic used in the ES is apparently not quite as efficient for this problem instance.

It focuses mainly on creating the right edges in a graph. A specific tuning of the mutation would probably further improve the performance for turkey19-like problems. Also, a lower selection pressure might be helpful in this case. However, even for this difficult problem a new best solution was found, and the mean performance of the (30,200)-ES with discrete recombination is better than previous GA-approaches.

The ES outperforms the benchmark GAs by Reichelt not only in terms of effectiveness, with better average cost values, but also in terms of flexibility and efficiency. While LaBORNet seems to perform well for small problem instances, it is less applicable to the large problems. With BaBORNet, the opposite applies. The combinatorial ES performs well on all problem instances and in all three ES-variations tested. Moreover, when examining the results of germany15–30, the ES scales much better than LaBORNet and BaBORNet, increasingly improving over previous best solutions with problem size. Even the (10,50)-ES with its relatively low computational requirements generates better results on the problems germany15–30 than LaBORNet and BaBOR-Net. The (10,50)-ES is the most efficient heuristic presented here, while the (30,200)-ES with discrete recombination is the most effective and, thus, currently the best heuristic for this class of optimization problems. The superior performance of the ES can be attributed to the operators, not the coding, since LaBORNet uses the same solution coding as our ES.

The runtime variation, measured by the standard deviation of the number of fitness evaluations is higher for the ES than for Reichelt's GA-approaches. This may be associated with the use of destabilization in the ES that sometimes revitalizes an otherwise more or less converged population to continue the search.

Runtime requirements are frequently high for evolutionary approaches in optimization. As was indicated in the previous section, this also applies to the evolution strategy. However, since network design is no time-critical application and much money can be saved by creating a cost-minimal reliable network structure, the computational effort is definitely justified.

8 Conclusion and Future Work

The proposed combinatorial version of an evolution strategy performs better than benchmark approaches for this important type of network design problem. It clearly outperforms the best-so-far results by genetic algorithms and finds new best solutions for every problem instance. Moreover, the ES appears to scale better for large problems instances than the benchmark genetic algorithms. Therefore, it is fair to conclude that the combinatorial ES is a valuable aid to solve complex network design problems.

Moreover, the results underline the important but so far often neglected potential of evolution strategies in combinatorial optimization. As was shown here, it is possible to keep the overall working of a standard-ES, while simultaneously adapting mutation as the main search operator to the combinatorial problem domain. Finally, the integration of domain-specific knowledge in the mutation operator and a repair heuristic proves a useful measure to raise the performance of an otherwise broadly applicable but less effective metaheuristic.

Future research with the evolution strategy could address the design of cost minimal reliable networks as a multiobjective optimization problem. In multiobjective optimization, the network designer receives a set of paretooptimal solutions as output. Thus, the trade-off between the goals cost and reliability can be visualized, allowing for an informed decision that may vary depending on the relative importance of the two individual goals. More research is also needed on optimal settings for the algorithmic design parameters of the combinatorial ES. Finally, other practical constraints such as capacity and reaction time requirements for network edges could be integrated.

References

1. Aboelfotoh HMF, Al-Sumait LS (2001) IEEE Trans Reliab 50(4):397–405
2. Aggarwal KK, Rai S (1981) IEEE Trans Reliab 30(1):32–35
3. Altiparmak F, Dengiz B, Smith A (2003) J Heuristics 9(6):471–487
4. Arabas J, Kozdrowski S (2001) IEEE Trans Evol Comput 5(4):309–322
5. Atiqullah MM, Rao SS (1993) Microelectron Reliab 33(9):1303–1319
6. Bäck T (1996) Evolutionary algorithms in theory and practice. Oxford University Press, New York
7. Bäck T (2002) (ed) Handbook of evolutionary computation. Institute of Physics Publishing, Bristol
8. Bäck T, Fogel D, Michalewicz Z (2000) (eds) Evolutionary computation, Vols 1 and 2. Institute of Physics Publishing, Bristol
9. Beyer H-G, Schwefel H-P (2002) Nat Comput 1:3–52
10. Cancela H, Urquhart M (1995) Simulated annealing for communication network reliability improvement. In: Proceedings of the XXI Latin American conference on informatics (PANEL '95). ACM Press, New York, pp 1413–1424
11. Chen Y, Li J, Chen J (1999) A new algorithm for network probabilistic connectivity. In: Proceedings of military communications conference (MILCOM '99) Vol 2. Atlantic City, pp 920–923
12. Colbourn, CJ (1987) The combinatorics of network reliability. Oxford University Press, Oxford
13. Cortes P, Munuzuri J, Onieva L, Larraneta J, Vozmediano JM, Alarcon JC (2006) Interfaces 36(2):105–117
14. Deeter D, Smith AE (1998) IIE Trans 30:1161–1174
15. Dengiz B, Altiparmak F, Smith AE (1997) IEEE Trans Reliab 46:18–26
16. Dengiz B, Alabap C (2001) A simulated annealing algorithm for design of computer communication networks. In: Callaos N (ed): Proceedings of world multiconference on systemics, cybernetics and informatics, SCI 2001. IIIS, Orlando, pp 188–193

17. Erl T (2006) Service-oriented architecture: concepts, technology and design. 5. print. Prentice-Hall, Uppder Saddle River, NJ
18. Fard N, Lee T-H (2001) Comput Commun 24:1348–1353
19. Flores SD, Cegla BB, Caceres DB (2003) Telecommunication network design with parallel multi-objective evolutionary algorithms. In: IFIP/ACM Latin America networking conference. ACM Press, New York, pp 1–11
20. Fowler M (2006) Patterns of enterprise application architecture. 10. print. Addison-Wesley, Boston
21. Glover F, Lee M, Ryan J (1991) Ann Oper Res 33:351–362
22. Ghosh L, Mukherjee A, Saha D (2002) Design of 1-FT communication network under budget constraint. In: IWDC '02 Proceedings of the 4th international workshop on distributed computing, mobile and wireless computing. London, UK. Springer, Berlin Heidelberg New York, pp 300–311
23. Grötschel M, Monma CL, Stoer M (1995) Design of survivable networks. In: Ball, MO (ed) Network models. Elsevier, Amsterdam, pp 617–672
24. Herdy M (1990) Application of the 'evolutionsstrategie' to discrete optimization problems. In: Schwefel H-P, Männer R (eds) Parallel problem solving from nature. Springer, Berlin Heidelberg New York, pp 188–192
25. Holland JH (1992) Adaptation in natural and artificial systems. 2. print. MIT, Cambridge/MA
26. Horowitz E, Sahni S (1983) Fundamentals of data structures. Computer Science Press, New York
27. Jan R-H (1993) Comput Oper Res 20(1):25–34
28. Jan R-H, Hwang F-J, Chen S-T (1993) IEEE Trans Reliab 42(1):63–70
29. Kumar A, Pthak RM, Gupta YP, Parsaei HR (1995) Comput Ind Eng 28(3):659–670
30. Li R, Emmerich MTM, Bovenkamp EGP, Eggermont J, Bäck T, Dijkstra J, Reiber JHC (2006) Mixed integer evolution strategies and their application to intravascular ultrasound image analysis. In: Rothlauf F (ed) Applications of evolutionary computation (LNCS 3907). Springer, Berlin Heidelberg New York, pp 415–426
31. Liu B, Iwamura K (2000) Comput Math Appl 39:59–69
32. Michalewicz, Z (1999) Genetic algorithms + data structures = evolution programs. 3., corrected print. Springer, Berlin Heidelberg New York
33. Nissen V (1994) Solving the quadratic assignment problem with clues from nature. IEEE Trans Neural Netw 5(1):66–72
34. Nissen V, Krause M (1994) Constrained combinatorial optimization with an evolution strategy. In: Reusch B (ed) Fuzzy Logik. Theorie und Praxis. Springer, Berlin Heidelberg New York, pp 33–40
35. Pierre S, Hyppolite M-A, Bourjolly J-M, Dioume O (1995) Topological design of computer communication networks using simulated annealing. Eng Appl Artif Intell 8(1):61–69
36. Rechenberg I (1994) Evolutionsstrategie '94. Frommann-Holzboog, Stuttgart
37. Reichelt D, Rothlauf F (2004) CURE: Eine Reparaturheuristik für die Planung ökonomischer und zuverlässiger Kommunikationsnetzwerke mit Hilfe von heuristischen Optimierungsverfahren. Working Paper 10/2004 (in German), University of Mannheim, Department of Business Administration and Information Systems
38. Reichelt D, Rothlauf F (2005) Int J Comput Intell Appl 5(2):251–266

39. Reichelt D (2006) Kommunikationsnetzwerkplanung unter Kosten- und Zuverlässigkeitsgesichtspunkten mit Hilfe von evolutionären Algorithmen. (PhD-Diss.) http://www.db-thueringen.de/servlets/DocumentServlet?id=5635 (in German). Technical University of Ilmenau

40. Rudolph G (1994) An Evolutionary algorithm for integer programming. In: Davidor Y, Schwefel H-P, Männer R (eds): Parallel problem solving from nature - PPSN III (LNCS 866). Springer, Berlin Heidelberg New York, pp 139–148

41. Schindler B, Rothlauf F, Pesch E (2002) Evolution strategies, network random keys, and the One-Max Tree problem. In: Applications of Evolutionary Computing: EvoWorkshops 2002 (LNCS 2279). Springer, Berlin Heidelberg New York, pp 29–40

42. Schwefel H-P (1995) Evolution and optimum seeking. Wiley, New York

43. Smith AE, Dengiz B (2000) Evolutionary methods for the design of reliable networks. In: Corne DW (ed): Telecommunications optimization: heuristic and adaptive techniques. Wiley, Chichester, pp 17–34

44. Soni S, Narasimhan S, LeBlanc LJ (2004) Telecommunication access network design with reliability constraints. IEEE Trans Reliab 53(4):532–541

A

Appendix: The Problem Instances

Table A.1. deeter10: cost and reliability of technology options

option (l_k)	reliability	cost per distance unit
0 (not connected)	0	0
1	0.7	8
2	0.8	10
3	0.9	14

Table A.2. deeter10: distance matrix

	2	3	4	5	6	7	8	9	10
1	47.3995	41.7519	24.999	48.876	54.7605	33.7527	34.6182	43.7072	27.3028
2		20.3195	28.9359	39.9203	31.7823	65.4767	68.7027	32.1302	62.2745
3			34.7425	19.8154	15.4418	49.4191	53.0689	11.8351	64.1167
4				50.814	50.1819	54.7977	56.7332	43.0894	33.3708
5					13.6208	41.8359	45.8073	8.2105	75.0041
6						54.1884	58.1259	12.0843	78.7647
7							3.97178	42.6274	58.0398
8								46.5197	57.6986
9									68.8805

Table A.3. turkey19: cost and reliability of technology options

option (l_k)	reliability	cost per distance unit
0 (not connected)	0	0
1	0.96	333
2	0.975	433
3	0.99	583

Table A.4. turkey19: distance matrix

	2	3	4	5	6	7	8	9	10	11	12	13	14	15	16	17	18	19
1	111	126	120	122	115	116	132	346	968	343	344	106	107	105	454	613	828	1261
2		15	15	17	5	6	243	458	1079	454	456	10	11	5	565	724	939	1342
3			15	17	13	14	258	473	1094	469	471	25	26	23	580	740	954	1357
4				2	3	6	248	460	1082	456	457	12	13	15	570	730	943	1353
5					8	9	251	463	1085	459	460	15	16	18	573	733	946	1355
6						1	246	457	1080	454	455	10	9	12	568	728	940	1350
7							245	456	1079	453	454	9	8	11	567	727	939	1351
8								384	383	380	381	235	236	240	322	542	831	1301
9									766	3	4	450	451	453	580	542	487	920
10										763	764	1074	1075	1077	1345	1307	972	624
11											1	450	451	453	582	544	489	921
12												449	459	452	583	545	490	922
13													1	4	560	720	932	1337
14														3	561	721	933	1338
15															563	723	934	1340
16																469	898	1424
17																	553	1079
18																		526

Table A.5. germany15-30: cost and reliability of technology options

option (l_k)	reliability	cost per distance unit
0 (not connected)	0	0
1	0.7	8
2	0.8	10
3	0.9	14

Table A.6. germany15–30: distance matrix

	2	3	4	5	6	7	8	9	10	11	12	13	14	15	16	17	18	19	20	21	22	23	24	25	26	27	28	29	30
1	232	467	536	355	430	414	528	429	496	496	112	273	362	131	430	430	486	232	342	475	136	430	429	415	486	537	497	232	205
2		561	432	133	303	362	560	302	352	352	304	130	451	262	303	303	466	361	175	393	131	303	302	362	467	433	353	172	112
3			427	595	442	312	149	442	491	492	363	468	112	342	442	442	248	235	494	368	445	442	442	312	248	428	492	671	451
4				360	130	143	312	131	110	110	507	305	377	438	130	131	182	429	261	70	413	131	132	142	183	1	110	603	356
5					232	334	559	232	261	261	408	130	491	352	233	232	444	432	111	340	231	232	231	334	445	360	261	286	175
6						133	363	1	69	69	421	178	362	350	1	1	234	371	131	111	299	1	1	132	235	131	70	475	236
7							233	133	180	180	371	234	243	304	132	133	111	286	222	72	306	133	133	1	111	144	180	522	263
8								363	396	397	441	444	184	397	363	363	133	313	449	266	466	363	364	233	132	312	397	698	450
9									70	70	420	178	362	350	1	1	234	371	130	131	298	1	1	132	234	132	70	474	235
10										1	420	236	420	419	69	69	264	438	177	131	363	69	70	180	264	110	236	474	298
11											490	235	420	419	69	69	264	439	177	131	363	69	70	180	265	111	1	524	298
12												299	264	70	420	420	421	130	364	439	178	421	420	371	421	508	490	41	236
13													362	235	178	178	341	305	69	263	137	178	177	234	342	306	236	299	68
14														234	362	362	218	134	394	310	333	362	362	243	218	378	420	560	341
15															350	350	362	110	298	371	130	350	350	304	362	439	420	333	177
16																1	233	371	131	110	299	1	1	132	234	131	70	475	236
17																	234	371	130	111	298	1	1	132	235	131	70	474	236
18																		310	333	133	395	234	234	111	1	183	264	620	362
19																			357	358	232	371	371	287	311	430	439	443	262
20																				232	205	130	129	222	334	262	177	348	137
21																					357	111	112	72	134	71	131	560	306
22																						299	298	306	395	414	364	232	69
23																							299	133	235	131	264	474	236
24																								133	235	132	439	473	235
25																									111	143	180	522	263
26																										183	265	621	363
27																											110	604	357
28																												525	298
29																													259

Evolutionary Computations for Design Optimization and Test Automation in VLSI Circuits

A.K. Palit, K.K. Duganapalli, K. Zielinski, D. Westphal, D. Popovic, W. Anheier, and R. Laur

University of Bremen, FB1/ITEM, Otto-Hahn-Allee 1, 28359 Bremen, Germany, palit@item.uni-bremen.de

Summary. Evolutionary Computation (EC) mimics the process of natural evolution, and particularly its versions known as Genetic Algorithms (GAs) and Differential Evolution (DE) help engineers and researchers to optimize their designs using the Darwin's principle of "survival of the fittest". These optimization principles are very frequently used in electronics for design of complex electronic systems such as VLSI circuit design and for their automated test.

The present chapter of the book should intimate the reader with the optimization requirements in VLSI circuit design and testing and also demonstrate how the optimization could be achieved using evolutionary computation methodology. This will also be demonstrated on some specific application examples of design optimization and post-production test automation of VLSI circuit.

1 Introduction

In modern semiconductor fabrication processes the need to design and manufacture integrated circuits with a small budget and hard time limits but still with a high performance increases more and more. Because circuit design is a complex and time-consuming task, designers have to be assisted by software tools which provide them with optimized results in reduced time compared to the intuitive approach. In this chapter it is shown how problems arising in the production of VLSI circuits can be tackled using optimization algorithms.

In the last decades, evolutionary algorithms (EAs) attracted a lot of attention from researchers and engineers because of their advantages over classical optimization methods and because of their capability to solve complex optimization problems. Due to their population-based nature the algorithms have the ability to find the global best solution of optimization problems, and they do not require building of any derivatives of the objective function for this purpose as several classical optimization algorithms do. Therefore, they can be flexibly used for solving a large range of problems.

A.K. Palit et al.: *Evolutionary Computations for Design Optimization and Test Automation in VLSI Circuits*, Studies in Computational Intelligence (SCI) **92**, 285–311 (2008)
www.springerlink.com   © Springer-Verlag Berlin Heidelberg 2008

A problem of the majority of EAs is that their performance is dependent on control parameter settings which are usually problem-dependent. Experienced users of these algorithms may sometimes be able to guess which settings may be good for a given optimization problem, but in general no reliable guidelines are available. The approach that is normally used in Evolutionary Computation (EC) for selection of control parameters consists of executing several optimization runs with different parameter settings in order to determine the best setting by trial-and-error method. This approach takes considerable computational time which is especially disadvantageous when dealing with real-world problems with computationally expensive constraint and objective functions. Therefore, a method for adaptive adjustment of control parameter settings based on methods from Design of Experiments (DoE) is presented in this chapter. A further advantage is that parameter values can be changed over time using this approach. As the task changes during an optimization run from exploration in the beginning (where a large area should be covered) to exploitation during the end (where the search should concentrate on the vicinity of the best solution to fine-tune it to the global best solution), it is natural to assume that parameter values should reflect these changes [7].

Because real-world problems like the ones arising in circuit design require the fulfillment of various specifications, there is a great need for efficient multi-objective optimization algorithms. Multi-objective optimization is usually more difficult than single-objective optimization, because the goals are generally conflicting so not one single best solution but the Pareto-optimal front has to be generated that consists of solutions which cannot be improved in one goal without deteriorating another goal. Thus, new problems arise like maintaining enough diversity in the population to achieve a well-distributed Pareto-optimal front (meaning that the extent is as large as possible and the distribution is as uniform as possible, thus giving the decision maker the largest freedom of choice).

This chapter is organized as follows: In Sect. 2 the professional need for optimization in VLSI fabrication processes is explained. Section 3 gives an introduction to the optimization algorithms used here which are Differential Evolution (DE), Particle Swarm Optimization (PSO) and Genetic Algorithm (GA), and also explains an approach for adaptive parameter control. In Sect. 4 the application of DE and PSO to the design of an operational amplifier is shown, while a GA is used in Sect. 5 to solve problems arising in test automation of VLSI circuits. The chapter ends in Sect. 6 with conclusions.

2 Professional Need and Motivation for Optimization

In contemporary semiconductor fabrication processes, with rising technical feasibilities and complexity, the dimensions of the basic components of integrated circuit elements are scaled down more and more using the latest lithography and deposition technologies. This allows the reduction of both the lateral and, even more important, the vertical dimensions, e.g. of transistors. If the strength of the electrical field over the oxide layer in a new process is similar to the electrical field used in the mature technology, the transistors' threshold voltage can be reduced and accordingly also the supply voltage of the process.

The down-scaling of the dimensions of circuit elements has two important effects on the commercial exploitation of semiconductor devices. As the dimensions of single circuit elements are reduced, the size of the entire circuit decreases, enabling the fabrication of more circuits on a single wafer in the same fabrication steps. This also drastically reduces the fabrication costs of the single circuit. Furthermore, the supply voltage reduction accompanied with the related current reduction enables circuit production for battery powered applications like cell phones, Bluetooth devices, wireless sensor networks, medical sensors, RFID systems, PDAs, digital cameras, MP3 players, game stations, home automation and so on.

The diversity of usable integrated circuits allows the designer of board level systems to make his choice from a pool of devices from various manufacturers and types to meet his expectations. From the point of view of a manufacturer, this means an increased pressure to develop his (potentially innovative) products faster than his competitors, and also with lower production costs and with more functions per chip.

Under these circumstances the designer of integrated circuits has to solve more complex problems with more complex constraints, with a smaller budget and within hard time limits. The achievement of these restrictions is accomplished by various modern CAD tools on all design levels, including simulators for a wide range of special problems during the design process, like DC behavior, frequency analysis, crosstalk analysis, temperature behavior and profile, parasitic analysis, field simulation, process simulation, post-layout simulation and so on.

During a typical design process the designer gets his specifications as boundary conditions for the circuit to be developed. These can be the maximum target size of the circuit, a maximum current consumption, the specified function of the circuit, a designated I/O bus topology, a specified temperature range for a safe operation, frequency specifications, the expected precision of the circuit functions and noise constraints, etc. In addition, the designer has to take into consideration some parasitic effects, to integrate some noise suppression features and to minimize the influence of some typical technical effects like short channel effects in MOS transistors, temperature effects, input and output resistance of single circuit stages, circuit noise, supply noise, breakdown effects, and layout effects like crosstalk between power and signal lines, substrate crosstalk, temperature profile caused by power losses, signal integrity, ESD protection, electromagnetic stability, high frequency signal routing through the signal lines, isolation of single basic circuit elements (e.g. bipolar transistors) and the compliance of the technology specific design rules.

After receiving the circuit design specifications the designer has to create a basic design, i.e. to split the global function over the entire circuit topology that might be hierarchical, assigning to the individual functional stages some adequate specific local functions. This includes decisions regarding which functions are handled by a digital part or by analogue stages, divided into sub-stages like power supply, signal references, front-ends, back-ends, filter, amplifier, converters, etc.

After the circuit's topology has been established the individual macros, made by basic circuit elements that implement the functions of the individual stages, have to be specified. If there are no useable macro cells or blocks from some earlier designs satisfying the required performance, like in most cases of ASIC design, the designer has to continue the design by calculating the circuit elements at transistor level, e.g. in operational amplifiers, reference voltages and current sources. To verify the resulting performance the designer simulates the resulting macro and compares the simulation results with the expected performance parameters. In most cases these first simulation results will show the expected principle function of the sub-circuit, but also an extensive difference to the given specifications. This is caused due to the above described complexity of today's circuit specifications near the technological limits supported by the complexity of the latest simulation models, where the designer is not able to handle all technological influences without CAD assistance anymore. So, with his fundamental knowledge and experience about the behavior of the basic circuit elements, the designer makes systematic changes on the design parameters to reach the expected performance. This is an iterative process of simulations until the defined performance is matched as good as possible by the circuit macro. The design is finished, if all circuit stages have been designed in this way and the resulting global circuit meets the specified performance.

Simulation is an effective economic way to design an integrated circuit in contrast to a trial-and-error approach during various hardware processing rounds, which is cost and time intensive. But as mentioned above, as short as possible time to market is expected for a commercial success today. This is only possible if the designer is assisted by further software tools, which are able to help optimizing the circuits or the individual circuit stages in a reduced time with possibly better results than the intuitional approach. Thus, for a practical relevance of optimization tools the complex constraints to be met have to be inserted into the optimization tool. The optimization tools must be capable of generating optimal design results on different design levels, e.g. basic stages like operational amplifiers or higher level stages like entire filters with potentially included pre-optimized basic macros. Also, it has to be flexible to be used with the special simulation tools in the particular design step or level.

3 Optimization Algorithms

In the following the optimization algorithms used in this chapter are introduced. First, some principal differences will be emphasized before explaining each algorithm in detail. A commonality of DE and PSO is that they both use real-coded individuals instead of the binary representation employed in the GA (real-coded GAs have also been developed but the original GA as well as the one used here have binary representation). Using real-valued individuals, the need for coding parameters and objectives is obsolete, and the parameters are used as floating-point numbers as they appear, e.g. in design optimization problems. However, for problems like the ones arising in test automation, binary representation is better suited because only some discrete values exist for the parameters.

3.1 Differential Evolution

The process of Differential Evolution is similar to other evolutionary algorithms: After a random initialization of the population, the individuals are evolved using the evolutionary operators mutation, recombination and selection until a stopping criterion is satisfied.

A weighted sum based on three randomly chosen, mutually different population members $\vec{x}_{r1}$, $\vec{x}_{r2}$ and $\vec{x}_{r3}$ is computed in the mutation process:

$$\vec{v}_i = \vec{x}_{r1} + F \cdot (\vec{x}_{r2} - \vec{x}_{r3}), \tag{1}$$

where F is one of three control parameters of DE. The mutated vector $\vec{v}_i$ as well as the target vector $\vec{x}_i$ are used in the recombination process to build the trial vector $\vec{u}_i$ ($i \in \{0, \ldots, NP - 1\}$ where NP is the number of individuals which is another user-defined control parameter).

$$u_{i,j} = \begin{cases} v_{i,j} & \text{iff } rand_j \leq CR \text{ or } j = k \\ x_{i,j} & \text{otherwise} \end{cases} \tag{2}$$

where CR is the third DE control parameter that decides in comparison with a random number $rand_j$ that is newly computed for each vector component $j \in \{0, \ldots, D - 1\}$ (where D is the dimension or the number of objective function parameters) if components should be copied from the target vector $\vec{x}_i$ or the mutated vector $\vec{v}_i$. At least one component is taken from $\vec{v}_i$ by randomly choosing $k \in \{0, \ldots, D - 1\}$ for each individual, so $\vec{u}_i \neq \vec{x}_i$ is ensured. This happens due to the fact that every target vector is compared to the corresponding trial vector during selection.

Different DE schemes have been developed that vary in the mutation and recombination operators. They are usually given in the notation DE/x/y/z, where x denotes the mutated vector, y is the number of difference vectors and z is the crossover scheme. With the given notation, the variant used here can be described as DE/rand/1/bin.

For selection in unconstrained single-objective optimization the target vector and the corresponding trial vector are compared, and the vector that yields the smaller objective function value is chosen for the next generation. As no deterioration with regard to the objective function value is possible, the DE selection scheme is called greedy [25]. Because this property includes that the best individual will always be retained, the DE selection scheme can also be described as elitist.

For constrained optimization the selection scheme has to be modified. When a vector $\overrightarrow{a}$ is compared to a vector $\overrightarrow{b}$, $\overrightarrow{a}$ is considered better if:

- Both vectors are feasible, but $\overrightarrow{a}$ yields the smaller objective function value.
- $\overrightarrow{a}$ is feasible and $\overrightarrow{b}$ is not.
- Both vectors are infeasible, but $\overrightarrow{a}$ results in the lower sum of constraint violations (where scaled constraints are used).

Feasibility means that all constraints are satisfied. No additional parameters are needed for this constraint-handling technique. Information about the amount of constraint violations is utilized to lead the individuals to feasible regions. In case of an unconstrained problem the replacement procedure is the same as for the original DE.

For multi-objective optimization the selection scheme has to be modified again, so the dominance relation is used if two feasible individuals are compared. A vector $\overrightarrow{f}(\overrightarrow{x})$ dominates a vector $\overrightarrow{f}(\overrightarrow{y})$ if:

$$\forall i \in \{0, \ldots, M - 1\} : f_i(\overrightarrow{x}) \leq f_i(\overrightarrow{y}), \tag{4}$$

$$\exists i \in \{0, \ldots, M - 1\} : f_i(\overrightarrow{x}) < f_i(\overrightarrow{y}), \tag{5}$$

where M is the number of objective functions.

During selection the target vector is replaced by the trial vector if the trial vector dominates the target vector. If the trial vector is dominated by the target vector, it is discarded. If both vectors are non-dominated, the trial vector is added to the generation, and it will be determined later (after generating trial vectors for each vector of the current generation) depending on domination and crowding distance which vectors will be part of the next generation. This procedure is similar as in NSGA-II (Non-dominated Sorting Genetic Algorithm II) [6]: First, a fast non-dominated sorting is done that orders the population into non-dominated fronts $\mathcal{F}_0 \ldots \mathcal{F}_l$, so that members of one front are non-dominated by each other, and the best non-dominated solutions are stored in $\mathcal{F}_0$. The next generation is filled beginning with members of $\mathcal{F}_0$, and subsequently adding members of following fronts. If not all members of a front can be added because otherwise NP would be exceeded, the decision about which vectors are kept is made based on crowding distance [6] that is a distance measure considering the nearest neighbors of a solution. In this work, vectors of all fronts which contribute to the next generation are considered during calculation of crowding distance [31]. The objective function values are scaled to the interval $[0, 1]$ before crowding distances are calculated to avoid undesired effects due to different scaling of the objective functions.

Several alternatives exist for the handling of boundary constraints, for example, the position that exceeds a limit can be newly generated or it can be replaced by a random number in the feasible region. In this work the position of the respective individual is set to the middle between the old position and the limit.

3.2 Particle Swarm Optimization

PSO is derived from the behavior of social groups like bird flocks or fish swarms. Although the "survival of the fittest" principle is not used in PSO, it is usually considered as an EA because of the co-operative behavior and self-organization of the individuals [9].

Optimization is achieved by giving each individual in the search space a memory for its previous successes, information about successes of a social group and providing a way to incorporate this knowledge into the movement of the individual. Therefore, each individual (called particle) is characterized by its position $\vec{x_i}$, its velocity $\vec{v_i}$, its personal best position $\vec{p_i}$ and its neighborhood best position $\vec{g_i}$. Several neighborhood topologies have been developed [16]. In this work the *von-Neumann* topology is used for single-objective optimization as it showed promising results in the literature, e.g. in [10]. Because of the high computational effort that would be needed for multi-objective optimization with the *von-Neumann* neighborhood, the *gbest* model is used in that case.

The dynamic behavior of PSO is generated by the update equations for velocity and position of the particles:

$$\vec{v_i}(t + 1) = w \cdot \vec{v_i}(t) + c_1 r_1 \left[\vec{p_i}(t) - \vec{x_i}(t) \right] + c_2 r_2 \left[\vec{g_i}(t) - \vec{x_i}(t) \right] \qquad (6)$$

$$\vec{x_i}(t + 1) = \vec{x_i}(t) + \vec{v_i}(t + 1) \qquad (7)$$

Due to these equations the particles are drawn towards their personal best position and their neighborhood best position. Furthermore, the velocity of the previous iteration is kept weighted with the inertia weight w that is a control parameter of PSO. Other parameters are c_1 and c_2 which influence the impact of the cognitive and social component, respectively. To add a stochastic element to the movement of the particles, the numbers r_1 and r_2 are chosen randomly from the interval $[0,1]$ in each iteration. Further parameters of PSO are the population size NP and the maximum velocity V_{max} that is used for preventing oscillations with increasing magnitude [9].

The same handling for boundary constraints is used as described for Differential Evolution in Sect. 3.1. Thus, if a particle goes beyond a border, it is reset to the middle between the old position and the violated limit for the respective parameter. Note that this approach is debated in the PSO community as several researchers feel it is better to let the particles go beyond bounds [11].

Handling of constraint functions is also similar to the procedure described in Differential Evolution but it is applied in a different situation [29, 30]. In DE the constraint-handling is applied during selection when trial vector and respective target vector are compared. In PSO the particles are not compared with each other, but particles are compared to their personal and neighborhood best positions. In unconstrained single-objective optimization a personal or neighborhood best position is replaced if the current position has a lower objective function value (for minimization problems). For constrained single-objective optimization this rule is altered as described for DE. Similarly, for unconstrained problems the modified PSO algorithm is exactly the same as the original PSO.

As for DE, the comparison of two feasible individuals has to be modified again for multi-objective optimization so that it is based on the dominance relation. In this work one personal best position is allowed for each particle, as it is the case in single-objective optimization. If the current position dominates the personal best position, the personal best position is replaced. If the current position is dominated by the personal best position, it is discarded. If both are non-dominated, it is randomly decided which to keep. In contrast, several global best solutions are saved with a maximum number of 100 global best solutions. If the maximum number of solutions is exceeded, the ones with the largest crowding distances are kept. As for DE, the objective function values are scaled to the interval $[0, 1]$ before calculating crowding distances.

3.3 Genetic Algorithm

Genetic Algorithms (GAs) are random but yet directed search methods based on the laws of natural selection and genetics. GAs are particularly much superior to pure random search algorithms as they from the very beginning search for the relatively "prospective" regions in the search space [19]. Typically, GAs are characterized by the following features: Genetic representation, i.e. encoding of solutions of optimization problems, a population of encoded solutions or chromosomes, a fitness function that evaluates the optimality or quality of each solution, genetic operators that help in generating a new population from the existing population and control parameters. A typical execution of GA involves the following:

- Random generation of initial population $Y(k) := (y_1, y_2, \ldots, y_N)$ with N chromosomes at generation $k = 0$
- Computation of fitness value of each chromosome y_l, $l = 1, 2, \ldots, N$, in the current population $Y(k)$
- Checking whether termination condition is met (i.e. when k reached the maximum value)
    1. If YES, then pick up the best chromosome, i.e. one with highest fitness value and stop the search process.

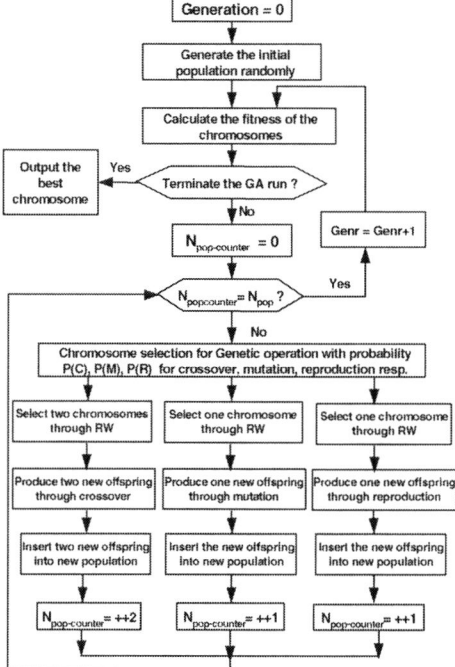

Fig. 1. GA flow-chart for test generation of VLSI circuits (source [19])

2. If NO, then create new population $Y(k + 1)$ with N new chromosomes applying the selection, crossover, mutation and reproduction
   genetic operators from the current population $Y(k)$ and start the new
   iteration with a fitness computation.

The GA implemented for our test generation purposes is based on the
flowchart in Fig. 1 (source [19]). Furthermore, the description of various
genetic operators used for GA implementation can also be found in [15, 19].
In our algorithm Roulette Wheel Selection mechanism followed by separately
crossover, mutation and reproduction operations were performed.

Elitist GA implies that the chromosome with best fitness from the previous generation is always copied into the next generation of population and
thereby, further eliminating the possibility of losing the so-far-obtained best-fit
chromosome in the future generation of GA.

3.4 Adaptive Parameter Control

Because optimization problems like the ones occurring in design optimization include computationally expensive objective and constraint functions, a
method for adaptive control of parameter settings is shown here. The adaptive
control for both DE and PSO is based on the same principle and therefore,

it is described for both algorithms in this section (for GAs a similar procedure would be possible but because adaptive control is not necessary for
the test automation problems reported here, this is not regarded further).
A Design of Experiments (DoE) technique named evolutionary operation
(EVOP) is employed because it is especially suited for continuous monitoring
and improvement of processes with varying performance [17]. In this case,
the process equals the optimization run that changes from exploration in the
beginning to exploitation towards the end, so there is a varying performance. It
can be determined statistically if changing a certain parameter yields a significant effect, or if performance variations are caused by randomness. Not only
can the influence of individual parameters be analyzed but also the interaction of parameters which is an important advantage [7]. Due to sophisticated
designs (settings of input variables) less test runs are necessary as if only
one of the parameters is regarded at a given time. Here, a two-level factorial
design is chosen, meaning that for every factor (parameter) two settings are
examined at a given time. It is a relatively simple design that nevertheless
allows detecting significant effects of individual parameters and also interaction effects. Performance differences are evaluated using a statistical method
called analysis of variance (ANOVA). It is calculated if the effect of changing
parameter settings is significant, associated with a confidence coefficient α.

For DE the parameters F and CR, which influence the mutation and
recombination processes, are subjected to adaptive control. For PSO parameters w, c_1 and c_2, which have direct effect on the update equations, are
chosen to be adaptively controlled. In each generation two different settings
of each parameter are applied, leading to $2^2 = 4$ different parameter combinations for DE and $2^3 = 8$ parameter combinations for PSO. Each parameter
combination is applied to the same percentage of population members which
are randomly chosen in each generation.

To be able to evaluate performance differences, a performance measure
has to be found that mirrors the success of the individuals. Because in single-
objective and multi-objective optimization different selection schemes are
used, different performance measures have to be employed.

For single-objective optimization the percentage of individuals which
resulted in a better objective function value than the parent (meaning that
for DE the trial vector is better than the target vector, and for PSO a new
personal best position was generated), definitely makes a statement about
success. However, if this percentage is used for the calculation of effects, successful individuals are rewarded regardless of the amount of change. As a
result parameter values might be favored that induce small improvements
of many individuals. Other parameter settings might lead to larger improvements but for fewer individuals. Because these settings might be essential to
escape from local optima, the ratio of successful vectors to all vectors with
the same parameter combination is weighted with the average improvement
for the respective parameter combination.

For multi-objective optimization a performance value is introduced that is increased by 2 if the challenging individual dominates its competitor, and it is increased by 1 if the challenging individual and its competitor are non-dominated (again, the challenging individual equals the trial vector for DE and a current particle for PSO, and the competitor is the target vector for DE and the personal best position for PSO).

After each generation it is checked based on the performance measure if a significant effect has emerged, so each generation equals a replicate of the DoE analysis. In case of a significant effect, the respective parameter is varied accordingly. In the present implementation of adaptive control it is concluded from the statistical analysis which parameter should be changed but no statements are derived about the preferable amount of change. Instead, the parameters are changed in fixed step sizes of 0.05 and 0.1: If, e.g. the DoE analysis suggests that the higher setting w_+ is better than w_-, the higher setting is increased by 0.05 and the lower setting is increased by 0.1. Thereby, the lower setting w_- is adapted towards better settings, and it is tested if larger values than w_+ give even better results without moving too far from an already discovered good setting.

If many replicates have been conducted without detecting a significant effect, the DoE analysis is restarted because the process performance may have changed. In [17] it is suggested to start a new DoE analysis if no significant effects occurred after 5–8 replicates. As this recommendation is rather strict, it is varied here so the DoE analysis is restarted if the average performance measure of the current generation is below 1% of the performance measure averaged over all generations belonging to the current DoE analysis. As suggested in [17], the parameter settings are changed when restarting the DoE analysis. Based on a random decision, the parameters are increased or decreased using the fixed step sizes.

4 Optimization Example: Operational Amplifier Design

Single-objective versions of DE and PSO have already been used in design optimization, e.g. in [26] and [27]. In this section the application of both single- and multi-objective variants of DE and PSO to the operational amplifier design shown in Fig. 2 is discussed. Adaptive parameter control is employed for DE as well as PSO. There are other works based on Genetic Programming where the topology of the circuit is evolved also [12] but here it is assumed that the topology is known, thus optimization of parameter values is sufficient. The model of the operational amplifier can be simulated using the tool **Spectre** which is a part of the design framework distributed by Cadence. The interface is realized via script files. The optimization tool based on DE and PSO implementations writes parameters to a script file, which is afterwards called to start a simulation. The output of the simulation (i.e. the values of the objective and constraint functions) is written to a file that is read by

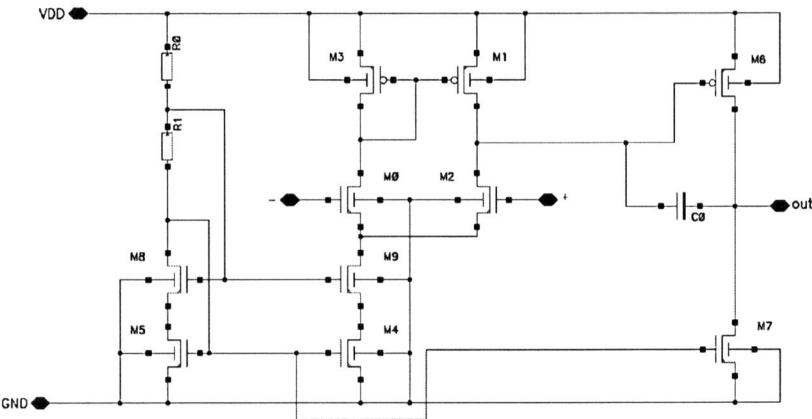

Fig. 2. Operational amplifier

the optimization tool, which processes the information and generates new parameter values depending on the optimization algorithm used.

Many objectives can be selected for being optimized (see Sect. 2). However, for the present exemplary test case only some of the most important characteristics of the operational amplifier have been chosen which are overall amplifier gain, bandwidth and current consumption, where the gain and the bandwidth are maximized and the current consumption is minimized. The gain (in dB) is evaluated by measuring the output voltage at the frequency 1 Hz (the operational amplifier has low-pass characteristics). For an assessment of the bandwidth the 3 dB cut-off frequency is considered, and the current consumption is computed as the difference of the current at the output node and at the input node.

As parameters the widths and the lengths of the transistors, the resistance values of the resistors and the capacitance value of the capacitor have been selected. A priori knowledge about current mirrors is utilized by combining the widths and lengths of the correspondent transistors into one parameter, respectively. Thereby, the number of parameters is reduced as it is also done in other circuit design applications [12]. As a result, the optimization problem has a dimension of $D = 13$.

One essential demand for the operational amplifier is stability. It can be checked by observing the phase margin, thus the phase margin is included as a constraint in the optimization problem.

The output signal of the operational amplifier should have sinusoidal form similar to the input signal. However, solutions generated during an optimization run often exhibit distorted signals. Therefore, the form of the signal must also be taken into account in the optimization problem. In preliminary tests no single parameter was found that is able to correctly reflect the form of the signal, therefore a combination of several parameters is used. Fourier transform of the output signal is performed by the simulator, and the magnitude of the result at the frequencies 0 and 10 kHz should be maximized (which equals the constant component as well as the frequency of the desired sinusoidal signal). All other frequencies should be suppressed, i.e. their magnitude should be minimized. In addition, the integrated difference between the output and the input signal is minimized. To keep the example simple, these variables are not included as objective functions but as constraints into the optimization problem.

As the indicator of optimization quality, an original design specification is regarded as reference for which the parameters have been optimized by an experienced designer based on his knowledge. The main task is to show how the optimization algorithms can be used for the same task and to check whether they are able to provide better results. Thus the constraints are formulated in a way that the results must be at least as good as the ones from the original design. For the objectives the reference values are as follows: A gain value of 52.66 dB, bandwidth of 3.031 kHz and a current consumption of 14.9 μA (in the original design, the current consumption was only 4.9 μA, but later it was decided to allow larger currents).

Because the objective functions are computationally very expensive, only few optimization runs were carried out. It is to be noted that these runs are examples, and the performance might vary for different runs because of the randomness involved in the evolutionary process.

Because the application of single-objective optimization methods is easier, it is worth a thought if problems with several objectives can be converted into single-objective optimization problems. One possibility is building a weighted sum of the objective functions, but because the selection of the weights is difficult, and minimal variations may lead to very different results [5], this approach is discarded. For circuit design frequently the boundary values for certain properties exist, e.g. the bandwidth of the circuit must not be below a given value. Therefore, instead of optimizing all properties of a circuit, it is also possible to take these boundary values as constraints into the optimization problem and select only one property that is of particular importance for being optimized, thus generating a single-objective optimization problem (ε-constraint method). In literature also approaches can be found where the optimization run is started with many objectives and the goals are changed during the search depending on intermediate results and user preferences [8], e.g. goals can be excluded from the search, or transformed into constraints. However, this approach requires considerable interaction with the user, so it is not regarded further in this work.

4.1 Experimental Settings

Although adaptive parameter control is used for most of the control parameters of DE and PSO, there are still some parameters to be set. For DE the number of individuals was chosen to be $NP = 20$ for single-objective optimization while $NP = 100$ was used for multi-objective optimization [28, 32].

Initial settings and limits are needed for the parameters which are subjected to adaptive control. The initial settings of DE are chosen based on guidelines from literature [14, 25, 28]: $F_- = 0.5$, $F_+ = 0.9$, $CR_- = 0.2$, and $CR_+ = 0.9$. The limits $F \in [0.2, 2]$ and $CR \in [0.1, 1]$ are used because in literature often ranges of $[0,1]$ or $[0,2]$ are given for F, and CR must be from $[0,1]$. Settings near 0 are not allowed because they usually do not perform well [28].

For PSO the number of individuals has to be divisible by eight because eight different parameter combinations are considered. Because it is assumed that effects are less correctly reflected with a smaller number of individuals for every parameter combination, $NP = 40$ is used for single-objective PSO, and $NP = 80$ was chosen for multi-objective optimization. V_{max} is assigned to one half of the search space in every dimension.

For the adaptively controlled parameters the limits $w \in [0.4, 1]$, $c_1 \in [0.4, 4]$ and $c_2 \in [0.4, 4]$ are used. The initial settings $\vec{w} = \{0.6, 0.8\}$, $\vec{c_1} = \{0.8, 1.5\}$ and $\vec{c_2} = \{0.8, 1.5\}$ have been chosen because mostly values of $0.5 \le c_1, c_2 \le 2$ are used [21], and for the inertia weight often values around $w = 0.7$ produce good results [10]. A confidence coefficient of $\alpha = 0.1$ is used for both DE and PSO.

4.2 Results

When optimizing the operational amplifier design with the single-objective DE, the gain value of 71.17 dB was achieved, this is almost 20 dB more than in the original design. With the single-objective PSO an even higher value of 73.68 dB was reached. Interestingly, even after 200 generations when the neighborhood best positions have already converged, there are many infeasible particles in the population, indicating that the optimum is located at or near the border between feasible and infeasible space. Unfortunately, nothing is known about the shape of this border, there might even be several disjoint feasible regions. It might be the ability of the particles to get into feasible space that resulted in the better gain value found by PSO in contrast to DE. Both DE and PSO are basically able to cross infeasible space but DE individuals can do it only if vector differences of the current population allow reaching a feasible point because a feasible individual is not allowed to become infeasible. In contrast, PSO individuals can become infeasible at any time so they are able to cross an infeasible region in a different way than DE individuals.

The Pareto-optimal fronts reached by the multi-objective DE and PSO algorithms are shown in Fig. 3. The original solution depicted by a circle shows that the algorithm was able to find solutions of similar quality. Additionally, the Pareto-optimal fronts include a large variety of other trade-off solutions from which a decision maker can choose an appropriate design.

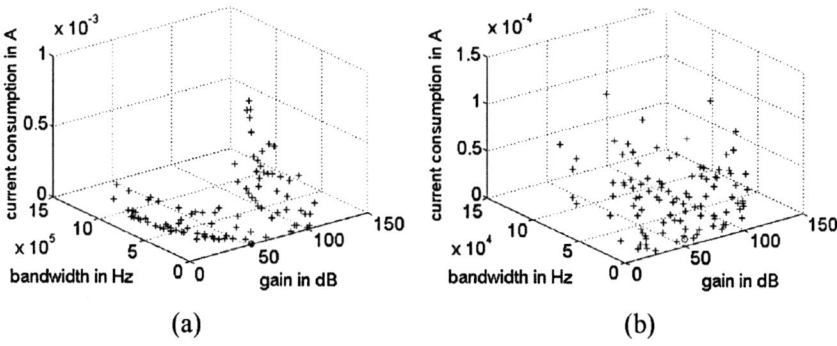

Fig. 3. Pareto-optimal front for (**a**) DE with $NP = 100$ and (**b**) PSO with $NP = 80$

5 Test Automation Applications

In this section we will discuss how GAs can be used for automated test generation. We begin with an overview of test generation problem and then discuss how test vectors can be generated in a GA framework.

5.1 Overview of Test Generation Problem

Once a chip is designed and fabricated, it must be tested to ensure that it functions as designed, while the assumption is made that the design is correct. The objective of automated test generation is to obtain a set of test vectors (input stimuli) that should be applied to the primary input(s) of fabricated IC to detect all the defects that could occur during the manufacturing process. However, covering all potential defects in the chips would be very difficult and would require an enormous number of input patterns or test vectors. Therefore, automated test generators operate on an abstract representation of defects referred to as *faults* and model a subset of the potential faults. A fault represents the logical effects of a defect on the chip. For instance, stuck-at-0 or s-a-0 and stuck-at-1 or s-a-1 faults at the gate input or output terminal can occur when the concerned input or output terminal is tied up with the power supply ground (V_{ss}) and supply high (V_{dd}) voltage, respectively. Therefore, due to the presence of fault the output response of the defective chip must be different from the correct output response of the good chip (golden device), while suitable test vectors are applied to the device under test (DUT). Thus, the defective chips can be sorted out through the post-production test, so that they are not shipped to customers.

Usually, various deterministic tests based on path oriented decision making (PODEM) and fan-out (FAN) algorithms are used for generating suitable test sequences for combinational circuits. However, those tests are mostly limited to stuck-at-faults (SAF), bridging faults and delay fault testing, etc. GAs have been recently also used for such kind of tests generation. For instance,

Krstic et al. [13] have also demonstrated the successful application of GA in digital circuit testing, where the fault models used are far more complex than simple stuck-at-fault principle. In the former application dynamic and intermittent faults like crosstalk-induced delay faults are detected by suitable tests based on GA. However, the authors in the previous example excluded the test generation for crosstalk-induced hazard tests. The success of GAs for several such fault models also motivated us to apply the same for similar complex, dynamic and intermittent faults like crosstalk-induced hazards test [20].

5.2 GA Framework for Crosstalk Test Generation

This section will describe a GA framework for crosstalk test generation. It is important to note that crosstalk-induced (coupling) noises are the major source of signal integrity faults and therefore, test generation of crosstalk faults can be considered as sub-set of test generation of signal integrity faults in high speed deep sub-micron (DSM) chips.

Overview of Crosstalk Fault Testing

In the DSM technologies process variation not only exists from die-to-die but also across a single die. Furthermore, aggressive scaling of active device dimensions and lower supply voltage, the increase of chip size and switching speed, high density of devices and interconnects, and the mixing of devices with different driving strengths in the DSM chips give rise to a situation in which crosstalk (coupling) effects are induced between the neighboring circuit wires. These problems are again aggravated by process variations in the DSM circuits. Crosstalk (coupling) noise, i.e. the leakage of signal between adjacent interconnect lines can lead to significant intermittent or permanent logic errors and timing violations/signal integrity losses, and thereby, creating major functionality problems of the integrated circuits. Hence, rigorous manufacturing tests for crosstalk-induced noise errors should be applied to ensure the quality of the produced chips. Therefore, both simulation-based verification and manufacturing test require high quality test vectors to be generated efficiently for integrated circuits. The quality of the vectors is determined by their ability to activate a large noise effect and propagate the same effect to primary outputs or to memory elements like flip-flops.

With the above objective, several authors have come up with different test generation procedures for crosstalk faults [1–4, 13, 22–24]. However, these works focused their attentions primarily on crosstalk-induced delays [4, 13]. Although, test generations for crosstalk-induced pulses were covered in [22, 23] earlier, they have used typically seven valued logic and the modified PODEM algorithm for this purpose. The efficiency of the latter was also found satisfactory for a few benchmark circuits. Unlike previous authors, in this section we investigate the efficacy of GAs for generation of test vectors particularly for crosstalk-induced pulses. The GAs have been applied earlier in different engineering disciplines as potentially good optimization tools [19] and also by [13,15,18] for various applications in VLSI Design, layout, EDIF digital system testing and also for test automation, particularly for stuck-at-faults and crosstalk-induced delay faults. However, the previous works paid little attention to the test problems related to crosstalk-induced pulse, although suitable tests for faults induced by crosstalk pulse were equally important as the tests for crosstalk-induced delay faults. Crosstalk pulse/glitch with considerable height and duration, in the best case, may cause reliability concerns in the VLSI circuits and will increase the dynamic power dissipation, whereas the same, in the worst case, will cause false transitions/logic errors. This was actually the main motivation for us to develop a suitable test algorithm based on GA, particularly for the logic faults caused due to crosstalk-induced positive or negative glitch. For this purpose, we have formulated the problem of test generation for crosstalk faults as a search of optimal solutions from the binary valued search space. Furthermore, an elitist GA has been developed and applied initially for the test pattern generation of stuck-at-faults and subsequently for crosstalk-induced logic faults as described in the following. A few benchmark and industrial circuit examples are considered and the test generation procedure for various fault models with GAs is illustrated. The generated test vectors from such GA-based method will be useful for both design simulation test and also for post-production tests of concerned digital circuits for stuck-at-faults and crosstalk faults.

Test Generation for Crosstalk-Induced Faults Viewed as a Finite Space Search Problem

The test generation for crosstalk-induced pulse can be viewed as a search of the n-dimensional 0–1 state-space of primary input patterns of an n-input combinational logic circuit. Consider the combinational circuit of Fig. 4, in which g and h are two internal nets, where g is a victim (affected) net and h

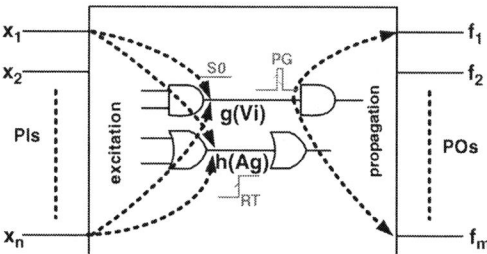

Fig. 4. Combinational circuits used for formulating test generation of crosstalk faults

AND	S0	S1	RT	FT	PG	NG	DR	DF
S0	S0	S0	S0	S0	S0	S0	S0	S0
S1	S0	S1	RT	FT	PG	NG	DR	DF
RT	S0	RT	RT	S0	PG	DR	DR	PG
FT	S0	FT	S0	FT	S0	FT	S0	FT
PG	S0	PG	PG	S0	PG	S0	S0	PG
NG	S0	NG	DR	FT	S0	NG	DR	FT
DR	S0	DR	DR	S0	S0	DR	DR	S0
DF	S0	DF	PG	FT	PG	FT	S0	DF

(a)

OR	S0	S1	RT	FT	PG	NG	DR	DF
S0	S0	S1	RT	FT	PG	NG	DR	DF
S1	S1	S1	S1	S1	S1	S1	S1	S1
RT	RT	S1	RT	S1	RT	S1	RT	S1
FT	FT	S1	S1	FT	DF	NG	NG	DF
PG	PG	S1	RT	DF	PG	S1	RT	DF
NG	NG	S1	S1	NG	S1	NG	NG	S1
DR	DR	S1	RT	NG	RT	NG	DR	S1
DF	DF	S1	S1	DF	DF	S1	S1	DF

(b)

XOR	S0	S1	RT	FT	PG	NG	DR	DF
S0	S0	S1	RT	FT	PG	NG	DR	DF
S1	S1	S0	FT	RT	NG	PG	DF	DR
RT	RT	FT	S0	S1	DR	DF	PG	NG
FT	FT	RT	S1	S0	DF	DR	NG	PG
PG	PG	NG	DR	DF	S0	S1	RT	FT
NG	NG	PG	DF	DR	S1	S0	FT	RT
DR	DR	DF	PG	NG	RT	FT	S0	S1
DF	DF	DR	NG	PG	FT	RT	S1	S0

(c)

Fig. 5. (**a**) Truth table for AND gate, (**b**) Truth table for OR gate and (**c**) Truth table for XOR gate using eight valued logic

is an aggressor (affecting) net. The objective is to generate a test for crosstalk positive glitch on the victim net g due to aggressor net h. The states of g and h can be expressed as two Boolean functions of the primary inputs $(x_1, x_2, \ldots, x_n)$.

Similarly, each primary output $(f_j, \ j = 1, 2, \ldots, \ m)$ can be expressed as a Boolean function of the states on net g and h as well as primary inputs $(x_1, x_2, \ldots, x_n)$. Note that for crosstalk glitch test the gate inputs connected directly to primary inputs $(x_1, x_2, \ldots, x_n)$ can have only four states such as static 0, i.e. $(0 \to 0)$ or $S0$, static 1, i.e. $(1 \to 1)$ or $S1$, rising transition, i.e. $(0 \to 1)$ or RT and falling transition, i.e. $(1 \to 0)$ or FT. These are shown in the truth tables of AND, OR and XOR gates by dark-colored inner rectangular block (see Fig. 5a–c). On the other hand, states of internal nets, or inputs and outputs of internal gates can be any one of the following eight valued logic: $S0$, $S1$, RT, FT, PG (positive glitch), NG (negative glitch), DR

(delayed rising) and DF (delayed falling), respectively (see the truth tables of basic gates).

Let us suppose that

$$g = G(x_1, x_2, \ldots, x_n) \tag{8}$$

$$h = H(x_1, x_2, \ldots, x_n) \tag{9}$$

$$f_j = F_j(g, h, x_1, x_2, \ldots, x_n) \tag{10}$$

where $j = 1, 2, \ldots, m$ and x_i can be $S0$, $S1$, RT and FT for $i = 1, 2, \ldots, n$.

The problem of test generation for crosstalk fault (positive glitch) on victim net g due to aggressor net h can be stated as solving a set of Boolean equations. In order to excite the victim with positive glitch we have to set static 0 ($S0$, i.e. $0 \rightarrow 0$) signal on the victim net g, whereas the aggressor net h has to be set as ($0 \rightarrow 1$) rising transition or RT. If the strong crosstalk (capacitive) coupling exists between the nets h and g, it will induce a crosstalk positive glitch (PG) on the victim net g. Thereafter, the generated fault effect (PG) has to be propagated to any (or all) primary output(s).

This can be done by setting the following:

$$G(x_1, x_2, \ldots, x_n) = S0 \tag{11}$$

$$H(x_1, x_2, \ldots, x_n) = RT \tag{12}$$

$$F_j(PG, RT, x_1, x_2, \ldots, x_n) \text{ XOR } F_j(S0, RT, x_1, x_2, \ldots, x_n) = 1 \tag{13}$$

Equation 13 must be satisfied for at least one j (or primary output), with $1 \leq j \leq m$ and x_i is either one from $S0$, $S1$, RT, FT for $1 \leq i \leq n$. Otherwise, fault is declared as untestable. Note that in the above equations the state of net h is RT implies that at t_1 instant of time h net assumes the logic value 0 and thereafter, at the time instant t_2 the net h has a logic value 1, which are of course dependent on the input patterns presented at the primary inputs $(x_1, x_2, \ldots, x_n)$ at t_1 and t_2 instants of time.

For testing of crosstalk negative glitch (NG) on victim g due to same aggressor h the set of equations from (11) through (13) will be changed slightly as follows:

$$G(x_1, x_2, \ldots, x_n) = S1 \tag{14}$$

$$H(x_1, x_2, \ldots, x_n) = FT \tag{15}$$

$$F_j(NG, FT, x_1, x_2, \ldots, x_n) \text{ XOR } F_j(S1, FT, x_1, x_2, \ldots, x_n) = 1 \tag{16}$$

Note that all the above equations can be evaluated with the help of the truth tables of four basic gates, i.e. AND, OR, XOR and NOT. In the following (Sect. 'Test Generation using Elitist GA'), GA actually defines the fitness function partly based on the above constraints.

Following the similar procedure one can also derive the set of Boolean equations for stuck-at-fault. For stuck-at-0 fault, i.e. s-a-0 on the net g of Fig. 4, for instance, the modified Boolean equations will be:

$$G(x_1, x_2, \ldots, x_n) = 1 \tag{17}$$

$$F_j(1, x_1, x_2, \ldots, x_n) \text{ XOR } F_j(0, x_1, x_2, \ldots, x_n) = 1 \tag{18}$$

Here also, equation (18) must be satisfied for at least one j (or primary output), with $1 \le j \le m$ and x_i is either 1 or 0 for $1 \le i \le n$. Otherwise, the s-a-0 fault on net g is declared as untestable fault.

Parameter Settings of Elitist GA

In order to develop the GA suitable for test generation for both stuck-at-fault and crosstalk fault the initial population of size N (say, $N = 40$ for crosstalk fault), was generated randomly. Here, each chromosome, i.e. y_l is of length n bits (for stuck-at-faults) where n bits correspond to actually n primary inputs (PIs) $(x_1, x_2, \ldots, x_n)$, whereas for crosstalk faults the chromosome y_l is of length $2n$ bits. In the latter each PI, i.e. x_i corresponds to two consecutive bits, representing any one of the four states from $S0$, $S1$, RT, FT. Thereafter, based on the fault excitation and fault propagation capabilities of individual chromosome fitness functions have been defined which will be given in the following.

Typically, probabilities of crossover, mutation and reproduction were set to 0.7, 0.05 and 0.25, respectively. Furthermore, for new generations 60% of the population was generated by crossover operation, 30% of the population was generated by mutation and the remaining 10% was generated by reproduction operator.

Test Generation using Elitist GA

In the following, we describe the simulation experiments of test generation carried out with elitist GA. Initially, GA was applied to a few simple circuits for stuck-at-fault generation that revealed its potential in finding the test patterns for stuck-at-fault.

For instance, GA was initially applied for the test generation of s-a-1 fault on net Y in circuit shown in Fig. 6. The fitness value was calculated in this case as follows:

Fitness = (1 if fault is excited else 0) + (fraction of inputs A-J set to 0) + (1 if fault is propagated to Z else 0) + (fraction of inputs K-S set to 0).

The corresponding results obtained from the GA run for the same circuit are shown in Table 1. Note that in order to demonstrate real search capabilities of GA, in this case initial population was deliberately created with only "1"s (not randomly and not with even single "0") so that the fitness values of all chromosomes in the 0th generation were always forced to 0 and

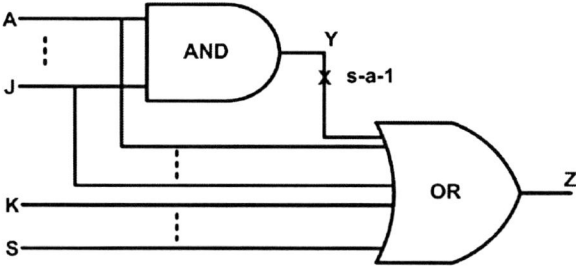

Fig. 6. GA-based test generation for stuck-at-fault

Table 1. Performance of GA in test generation for stuck-at-fault of circuit in Fig. 6

Generation	Best Fitness	Best Vector
0	0	1 1 1 1 1 1 1 1 1 1 1 1 1 1 1 1 1 1 1
1	1.51053	0 1 1 0 1 1 1 1 1 0 1 1 1 1 1 1 1 1 0
2	1.61053	0 1 0 1 1 1 0 0 1 1 1 1 1 1 1 1 1 1 1
3	1.81579	0 1 1 0 1 0 1 0 1 0 1 1 1 1 0 1 1 1 1
4	2.17895	0 0 0 1 0 1 0 0 1 1 0 0 1 1 1 0 0 0 1
5	2.17895	0 0 0 1 0 1 0 0 1 1 0 0 1 1 1 0 0 0 1
6	2.48421	0 0 0 1 0 0 0 0 1 0 0 0 1 1 1 1 0 0 0
7	2.48421	0 0 0 1 0 0 0 0 1 0 0 0 1 1 1 1 0 0 0
8	2.58421	0 0 0 0 0 0 0 0 1 0 0 0 1 1 1 1 0 1 0
9	2.68947	0 0 0 1 0 0 0 0 0 0 0 0 1 1 0 1 0 0 0
10	2.68947	0 0 0 0 0 0 0 0 1 0 0 0 1 0 1 1 0 0 0
11	2.79474	0 0 0 1 0 0 0 0 0 0 0 0 0 0 0 1 0 0 0
12	2.84211	0 0 0 0 0 0 0 0 0 0 0 0 1 0 1 1 0 0 0
13	2.94737	0 0 0 0 0 0 0 0 0 0 0 0 0 0 0 1 0 0 0
14	2.94737	0 0 0 0 0 0 0 0 0 0 0 0 0 0 0 1 0 0 0
15	2.94737	0 0 0 0 0 0 0 0 0 0 0 0 0 0 0 1 0 0 0
16	4	0 0 0 0 0 0 0 0 0 0 0 0 0 0 0 0 0 0 0

therefore, except for mutation operator all other operators, i.e. crossover and reproduction were practically inactive in the 0th generation. However, the GA run in the next generation (i.e. first generation here) itself has improved the best fitness to 1.5105 and quickly attained the desired fitness of 4.0 only in seventeen (i.e. 16th) generations. The test patterns corresponding to best fitness values are shown also in the right most column of Table 1. In this case desired input test pattern consists of 19 zeros, and the corresponding output for good circuit is 0, whereas faulty output at Z is 1. Similarly, test patterns for the stuck-at-fault on other nets can also be determined easily by similar GA runs.

In the following, GA performance was once again tested for stuck-at-fault in the error correction and translation (ECAT) circuit of Fig. 7. ECAT circuit is well known to the test engineers as the original D-algorithm failed to generate the test for stuck-at-fault in the same because of the presence of several

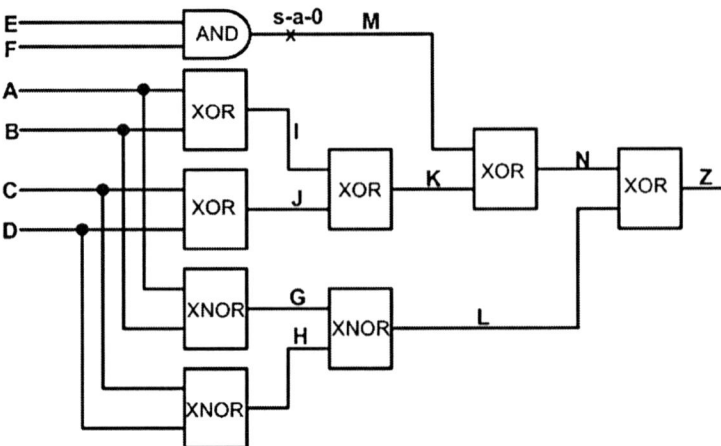

Fig. 7. GA-based test generation for stuck-at-fault in ECAT circuit

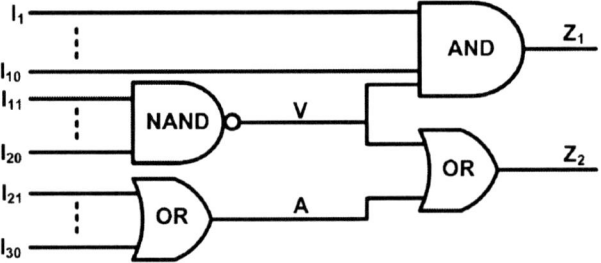

Fig. 8. GA-based test generation example circuit for crosstalk fault

XOR and XNOR gates. In this case also the test pattern for the stuck-at-fault (s-a-0) on the net M was quickly found by GA run in less than five generations and the corresponding test pattern is "011011" and the good circuit output at Z is 0 and for the faulty one it is 1.

As the GA was found to perform extremely well for both circuits in Figs. 6 and 7, we were further motivated to use the same for test generation of crosstalk-induced faults. For this purpose another circuit as shown in Fig. 8 has been selected. Here, the objective is to find the suitable test vectors for the crosstalk positive glitch (PG) generated on the victim net V due to crosstalk coupling effect from aggressor net A. Obviously, the input patterns, which will excite the fault effect on the victim net V, must set the logic state $S0$ on the victim and rising transition (RT) on the aggressor net A. Thereafter, pattern should facilitate the fault effect (PG) to be propagated either on the Z_1 or Z_2 primary output (PO). As mentioned earlier, here, each chromosome has bit-length $= 2 \cdot 30 = 60$ bits, as the number of inputs in this case is 30. Fitness function for this crosstalk fault is defined as follows:

Table 2. Performance of GA in test generation for crosstalk fault of circuit in Fig. 8

Generation	Best Fitness	Best Vector
0	1.45	1 3 2 3 0 1 1 0 1 2 3 3 1 2 1 3 3 1 2 2 1 2 0 2 1 1 0 3 2 3
1	1.9	1 2 0 1 1 1 1 0 1 2 1 1 0 1 3 1 1 1 2 0 1 0 1 0 2 2 3 1 1 3
2	1.9	1 2 0 1 1 1 1 0 1 2 1 1 0 1 3 1 1 1 2 0 1 0 1 0 2 2 3 1 1 3
3	2.1	1 2 1 1 1 1 1 0 1 2 1 1 0 1 1 1 1 2 0 1 0 1 0 2 2 3 1 1 3
4	2.25	1 1 1 1 1 1 1 0 1 2 1 1 1 1 3 1 1 1 2 0 1 0 1 0 2 2 3 0 1 3
5	2.35	1 2 1 1 1 1 1 1 1 2 1 1 0 1 1 1 1 1 1 0 1 0 1 0 2 2 2 1 1 3
6	2.45	1 1 1 1 1 1 1 1 1 2 1 1 1 1 3 1 1 1 2 0 0 0 1 0 2 2 2 0 1 3
7	2.55	1 2 1 1 1 1 1 1 1 1 1 1 0 1 1 1 1 1 1 1 1 0 1 0 2 2 2 1 1 3
8	2.65	1 1 1 1 1 1 1 1 1 2 1 1 1 1 3 1 1 1 1 0 0 0 1 0 2 2 2 0 0 2
9	3.75	1 2 1 1 1 1 1 1 1 1 1 1 1 1 1 1 1 1 1 1 1 1 0 0 0 2 2 2 1 1 2
10	3.75	1 2 1 1 1 1 1 1 1 1 1 1 1 1 1 1 1 1 1 1 1 1 0 0 0 2 2 2 1 1 2
11	3.85	1 1 1 1 1 1 1 1 1 1 1 1 1 1 1 1 1 1 1 1 1 1 0 0 0 2 2 2 1 1 2
12	3.85	1 1 1 1 1 1 1 1 1 1 1 1 1 1 1 1 1 1 1 1 1 1 0 0 0 2 2 2 1 1 2
13	3.95	1 1 1 1 1 1 1 1 1 1 1 1 1 1 1 1 1 1 1 1 1 0 0 0 0 2 2 2 1 0 2
14	3.95	1 1 1 1 1 1 1 1 1 1 1 1 1 1 1 1 1 1 1 1 1 1 0 0 0 2 2 2 0 0 2
15	6	1 1 1 1 1 1 1 1 1 1 1 1 1 1 1 1 1 1 1 1 1 1 0 0 0 0 2 2 2 0 0 2

Fitness = (1 if V is set to S0 else 0) + (fraction of inputs $I_{11}-I_{20}$ set to S1) + (1 if A is set to RT else 0) + (0.5 if any input of $I_{21}-I_{30}$ is set to RT else 0) + (0.5* fraction of inputs $I_{21}-I_{30}$ set to RT or S0) + (1 if fault is propagated to Z_1 else 0) + (fraction of inputs I_1-I_{10} set to S1)

In this case, population size $N = 40$ was selected, and as usually Roulette Wheel Selection mechanism followed by other genetic operations with fixed probabilities were used. The corresponding result of GA runs is depicted in Table 2. Note that in the Table 2 under *Best Vector* (test pattern) the following notations were used to indicate eight valued logic: 0, 1, 2 and 3 indicate *S0*, *S1*, *RT*, and *FT*, respectively. Similarly 4, 5, 6 and 7 indicate *PG*, *NG*, *DR* and *DF*, respectively, which are used by the inputs and outputs of internal gates (not by primary inputs).

Table 2 shows that best chromosome in the first (0th) generation has the fitness value of 1.45, which was gradually improved up to the desired level of 6 only in 16 (i.e. 15th) generations. So, it is obvious that GA could find the suitable test vectors very quickly for the crosstalk-induced logic faults for the example circuit of Fig. 8.

Similarly, GA runs for the crosstalk test generation were conducted for another example circuit (benchmark circuit c17) of Fig. 9. Combinational circuit c17 consists of six NAND gates with five primary inputs (PIs) designated as 1GAT(0), 2GAT(1), 3GAT(2), 6GAT(3) and 7GAT(4) and two primary outputs (POs) designated as 22GAT(10) and 23GAT(9) (see Fig. 9). Figure 9 also depicts the fault-free response and faulty response for the applied test vectors generated through GA for crosstalk fault, particularly for the internal nets 19GAT(7) and 16GAT(8) used respectively as aggressor net and victim

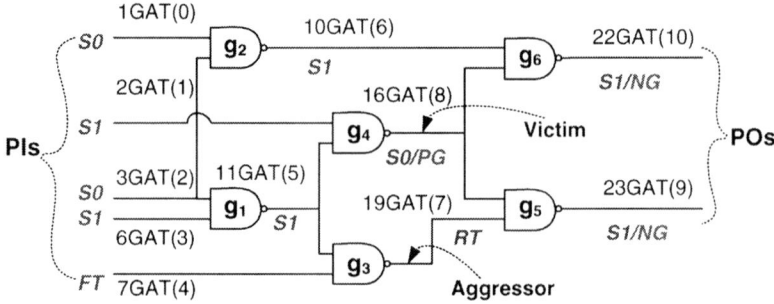

Fig. 9. GA-based test vectors of crosstalk fault for c17 benchmark circuit

net. In order to generate such test pattern for crosstalk positive glitch (PG) on the victim net 16GAT(8), at first victim net must be excited with static "0" or $S0$ signal and aggressor net 19GAT(7) must be excited with $0 \rightarrow 1$ rising transition or RT. Required signal excitation on the aggressor and victim nets can be performed by applying the $S1$, $S0$, $S1$ and FT transitions respectively on the 2GAT(1), 3GAT(2), 6GAT(3) and 7GAT(4) PIs. Now, because of the existence of any strong (capacitive) coupling between the two nets, aggressor will induce a positive glitch (PG) on the 16GAT(8) victim net. Thereafter, this fault effect (PG) needs to be propagated to at least one PO or both POs. Therefore, fault excitation on victim, and fault effect (PG) propagation to both POs can be simultaneously performed by applying $S0$, $S1$, $S0$, $S1$ and FT transitions, respectively, on the PIs 1GAT(0), 2GAT(1), 3GAT(2), 6GAT(3) and 7GAT(4). The corresponding desired responses for fault-free circuit from both POs are $S1$ and $S1$, whereas in presence of crosstalk the faulty responses at both POs are negative glitch (NG) and negative glitch (NG), respectively. GA run has quickly found these test patterns (01013; 11; 55) using the suitable fitness function in less than 10 generations. Here, "01013" respectively represent $S0$, $S1$, $S0$, $S1$ and FT transitions on the PIs (from top to bottom) and desired response "11" at the POs (from top to bottom) represent respectively $S1$ and $S1$ transitions at the POs, whereas faulty responses at the POs (top to bottom) "55" represent NG and NG, respectively.

Further simulations were performed with c17 circuits considering the different combinations of aggressor and victim nets and the results obtained through GA runs are illustrated in Table 3. Note that as illustrated in the first row of Table 3 different test patterns/test vectors (Input Transitions: FT, $S1$, $S0$, RT, FT; Output Responses (fault-free/faulty): $S1/NG$, $S1/NG$) are possible for same crosstalk fault induced by the same pair of aggressor-victim nets.

Furthermore, as shown in row 4 of Table 3 crosstalk fault may be occasionally untestable if the aggressor-victim nets cannot be excited with suitable transitions and/or the fault effect (here, NG) induced on the victim cannot be

Table 3. GA-based test generation's result for crosstalk fault of c17 circuit

Aggressor	Signal on Aggressor	Victim	Signal on Victim	Test Patterns
19GAT(7)	*RT*	16GAT(8)	*S0*	*FT S1 S0 RT FT; S1 S1; NG NG*
19GAT(7)	*FT*	16GAT(8)	*S1*	*FT S0 S0 FT RT; S0 RT; PG RT*
16GAT(8)	*RT*	19GAT(7)	*S0*	*FT FT S0 S0 S1; FT S1; FT NG*
16GAT(8)	*FT*	19GAT(7)	*S1*	NOT TESTABLE
11GAT(5)	*FT*	10GAT(6)	*RT*	*FT S0 S1 RT S1; FT FT; DF FT*
11GAT(5)	*RT*	10GAT(6)	*FT*	*RT S0 S1 FT S0; RT S0; DR S0*

propagated to PO(s) through the sensitized path(s). With slight modification of the existing GA, test patterns for crosstalk-induced delay fault are also generated and shown in the row 5 and row 6 of Table 3 (fifth column). Similarly, GA runs were conducted for several other benchmark circuits like s27, and the circuit shown in [22]. It is needless to mention that for all the considered circuits GA was found to be very successful in finding suitable test vectors. Similarly, the GA performance can further be investigated for medium and large (ISCAS85) benchmark circuits like c432, c499, c3540 and c5315 etc.

6 Conclusions

In this chapter optimization tools using DE and PSO have been developed, which are able to assist designers in the task of tuning design parameters to reach a specified functionality. Depending on the formulation of the optimization problem, different objectives can be emphasized. Multi-objective optimization techniques offer trade-off solutions and give freedom for the final design decision. The use of adaptive parameter control simplifies the application of optimization algorithms even for users who are inexperienced concerning optimization. This holds particularly for PSO because it is often difficult to adjust parameters, and depending on parameter settings the results may not be competitive with DE.

Apart from the same, the chapter also addresses the GA framework for test generation algorithm, which is developed initially for stuck-at-faults and thereafter, by slight modification, the same algorithm has been utilized for suitable test vectors generation for logic faults caused due to crosstalk-induced glitch and delay faults. The algorithm has been tested for a few small benchmark circuits for both stuck-at-faults and crosstalk-induced glitch and delay faults. It has been observed that, in all cases, with properly defined fitness function the GA has extraordinary capability to extract the suitable test vectors very quickly from randomly generated initial patterns (population). The developed test generation algorithm can further be investigated for medium and large benchmark circuits, e.g. c432, c499, c5315 and so on. The correctness of the generated test vectors/test patterns can also be verified through Verilog/VHDL simulation of the circuit file, where crosstalk glitch/delay fault is injected into the concerned aggressor-victim pairs using a suitable Verilog/VHDL hardware module. Besides the test generation part, the algorithm should accommodate also the test pattern compaction mechanism, which is currently left as future perspective of the work.

References

1. Aniket and Arunachalam R, A novel algorithm for testing crosstalk induced delay faults in VLSI circuits, Int. Conf. on VLSI Design, 2005, pp. 479–484.
2. Bai X, Dey S, and Krstic A, ATPG for crosstalk using hybrid structural SAT, Proc. of ITC, 2003, pp. 112–121.
3. Chen WY, Gupta SK, and Breuer MA, Test generation in VLSI circuits for crosstalk noise, Proc. of ITC, 1998, pp. 641–650.
4. Chen WY, Gupta SK, and Breuer MA, Test generation for crosstalk induced delay in integrated circuits, Proc. of ITC, 1999, pp. 191–200.
5. Deb K, Multi-objective optimization using evolutionary algorithms, Wiley, 2001.
6. Deb K, Pratap A, Agarwal S, and Meyarian T, A fast and elitist multiobjective genetic algorithm: NSGA-II, IEEE Transactions on Evolutionary Computation, vol. 6, no. 2, 2002, pp. 182–197.
7. Eiben AE, Hinterding R, and Michalewicz Z, Parameter control in evolutionary algorithms, IEEE Transactions on Evolutionary Computation, vol. 3, no. 2, 1999, pp. 124–141.
8. Fleming PJ, Purshouse RC, and Lygoe RJ, Many-objective optimization: an engineering design perspective, Proc. of the Third Internat. Conf. on Evolutionary Multi-Criterion Optimization, 2005.
9. Kennedy J and Eberhart RC, Swarm intelligence. Morgan Kaufmann, San Francisco, 2001.
10. Kennedy J and Mendes R, Population structure and particle swarm performance. In David B. Fogel, Mohamed A. El-Sharkawi, Xin Yao, Garry Greenwood, Hitoshi Iba, Paul Marrow, and Mark Shackleton, editors, Proc. of the IEEE-CEC '02, Honolulu, HI, USA, 2002, pp. 1671–1676.
11. Kennedy J, Some issues and practices for particle swarms, Proc. of the IEEE Swarm Intelligence Symposium, 2007.
12. Koza JR, Jones LW, Keane MA, Streeter MJ, and Al-Sakran S, Toward automated design of industrial-strength analog circuits by means of genetic programming, in Genetic Programming Theory and Practice II, Una-May O'Reilly, Rick L. Riolo, Gwoing Yu, and William Worzel, Eds. Kluwer, Dordecht, 2004, pp. 121–142.
13. Krstic A, Liou JJ, Jiang YM, and Cheng KT, Delay testing considering crosstalk-induced effects, Proc. of ITC 2001, pp. 558–567.
14. Lampinen J and Storn R, Differential evolution, in New Optimization Techniques in Engineering, GC. Onwubolu and B. Babu, Eds. Springer-Verlag, Berlin, Heidelberg, New York, 2004, pp. 123–166.
15. Mazumder P and Rudnick EM, Genetic algorithm for VLSI design, layout and test automation, Prentice-Hall, Eaglewood Cliffs, NJ, 1999.

16. Mendes R, Kennedy J, and Neves J, The fully informed particle swarm: simpler, maybe better. IEEE Trans. EC 8(3) 2004: 204–210.
17. Myers RH and Montgomery DC, Response surface methodology – process and product optimization using designed experiments. Wiley, New York, 2002.
18. Oommen Tharakan KT, et al., EDIF digital system testing using genetic algorithm – an experimental study, Proc. of Int. Conf. on Cognition and Recognition, 2005, pp. 403–410.
19. Palit AK and Popovic D, Computational intelligence in time series forecasting, theory and engineering applications, Springer-Verlag, London, 2005, Chapter 5, pp. 197–202.
20. Palit AK, Duganapalli KK, and Anheier W, GA based test generation for crosstalk induced logic faults, Proc. of the ITG-TUZ 2007, Erlangen, 11–13th March 2007.
21. Parsopoulos K and Vrahatis M, Particle swarm optimization method for constrained optimization problems, in Intelligent Technologies – Theory and Applications: New Trends in Intelligent Technologies, P. Sincak, J. Vascak, V. Kvasnicka, and J. Pospichal, Eds. IOS Press, 2002, pp. 214–220.
22. Rubio A, Sainz JA, Kinoshita K, Test pattern generation for logic crosstalk faults in VLSI circuits, Circuits, Devices and Systems, IEE Proceedings G, vol. 38, Issue 2, April 1991, pp. 179–181.
23. Sainz JA, Rubio A, Aspects of noise in VLSI circuits: crosstalk and common impedance coupling, PATMOS 1993, pp. 197–212.
24. Sinha A, Gupta SK, Breuer MA, Validation and test generation for oscillatory noise in VLSI interconnects, ISBN: 0-7803-5832-X/99, 1999 IEEE.
25. Storn R and Price K, Differential evolution – a simple and efficient heuristic for global optimization over continuous spaces, J. Global Optimiz. 11 1997: 341–359.
26. Storn R, Designing digital filters with differential evolution, in New Ideas in Optimization, McGraw-Hill, London, Eds. D. Corne, M. Dorigo and F. Glover, 1999, pp. 109–125.
27. Xie XF, Zhang WJ, and Bi DC, Optimizing semiconductor devices by self-organizing particle swarm, in Proc. of IEEE-CEC, 2004.
28. Zielinski K, Weitkemper P, Laur R, and Kammeyer KD, Parameter study for differential evolution using a power allocation problem including interference cancellation, in Proc. of the IEEE-CEC, 2006.
29. Zielinski K and Laur R, Constrained single-objective optimization using differential evolution, Proc. of the IEEE-CEC, 2006.
30. Zielinski K and Laur R, Constrained single-objective optimization using particle swarm optimization, Proc. of the IEEE-CEC, 2006.
31. Zielinski K and Laur R, Variants of differential evolution for multi-objective optimization, Proc. of the IEEE Symp. on Comput. Intellig. in Multicriteria Decision Making (MCDM '07), 2007, Honolulu, HI, USA.
32. Zielinski K and Laur R, Differential evolution with adaptive parameter setting for multi-objective optimization, Proc. of the IEEE-CEC, 2007.

Evolving Cooperative Agents in Economy Market Using Genetic Algorithms

Raymond Chiong

School of IT and Multimedia, Swinburne University of Technology
(Sarawak Campus), State Complex, 93576 Kuching, Sarawak, Malaysia

Summary. This chapter seeks to follow Axelrod's research of computer simulations on the Iterated Prisoner's Dilemma (IPD) game to investigate the use of Genetic Algorithms (GA) in evolving cooperation within a competitive market environment. We use an agent-based economy model as the basis of our experiments to examine how well GA could perform against the IPD game strategies. We also explore the strategic interactions among the agents that represent firms in a coevolving population, and study the influence of the genetic operators on GA to evolve cooperative agents.

1 Introduction

Genetic algorithms (GA) have been used to mimic natural evolution and solve many problems in a wide variety of fields. Among them, GA have aided the design of turbine blades for jet engines, the design of fiber-optic networks, controlling and managing of gas pipelines, scheduling, productions, criminal identification with Faceprints system, financial predictions, computer chip manufacturing, etc. [1, 2]. This chapter aims to investigate the use of GA in evolving cooperative agents within a competitive market environment using the Iterated Prisoner's Dilemma (IPD), a well-known non-zero-sum game. Our study seeks to follow Axelrod's research of computer simulations of the IPD game which is generally regarded as a benchmark for the studies on evolution of cooperation. In Axelrod's original work [3], two computer tournaments based on IPD were undertaken, and many strategies encoded with pure logical programming codes were being submitted for the competitions. The results of these two tournaments found that from an evolutionary stand point a strategy called Tit-for-Tat (TFT) was the eventual winner. Due to critics on the lack of robustness and dynamics of the program-based strategies, Axelrod subsequently used GA [4] to overcome the limitations of his initial work.

Axelrod's tournaments have literally laid the framework for much of the current research in IPD. Nevertheless, we are of the opinion that his work was

R. Chiong: *Evolving Cooperative Agents in Economy Market Using Genetic Algorithms*, Studies in Computational Intelligence (SCI) **92**, 313–326 (2008)
www.springerlink.com

a little restrictive and lack of a genuine real-world component. As such, we report on an agent-based simulation study that attempts to bridge the gap by applying GA to an economy model we previously built based on IPD [5]. In this chapter, we use the IPD economy market as the basis of our experiments to examine how well GA could perform against the IPD strategies. We also explore the strategic interactions among the agents that represent firms in a coevolving population within the simulated market, and study the influence of the genetic operators on GA to evolve cooperative agents.

The rest of the chapter is organised as follows: Sect. 2 introduces the classical Prisoner's Dilemma (PD) story. Following which, we describe the IPD tournaments conducted by Axelrod. Ensuing section presents the methodology used in our work, from the basic structure of our model, to the economy paradigm of the model, the strategy representation, and the fitness function. Our experimental results are then presented, and some related discussion is drawn. Finally, the Sect. 7 concludes our studies.

2 Prisoner's Dilemma

The classical PD is a non-zero-sum game that has been studied extensively since its inception in 1950 [6]. It presents a scenario where two men have been arrested for committing crime. They are being locked up in separated cells and are being interrogated. Due to insufficient evidence for placing a charge against them, the two prisoners have been offered a deal each, and both are aware that the deal offered is identical to the other. According to the deal, should both prisoners refuse to confess and remain silent, both would receive only small punishment. On the contrary, should any of them choose to confess and accuse the other while the other one remains silent, the prisoner who confesses will be freed and full punishment will be given to the prisoner who remains silent. Under circumstance whereby both prisoners confess and accuse the other, both would receive a heavy punishment but slightly less severe than the full punishment. Throughout the entire decision making process, both prisoners are neither allowed to communicate with one another nor given to know the decision of the other prior to their final decision.

From the above scenario we see a dilemma among the prisoners on making their choices of whether to accuse or remain silent. It is obvious that the rational choice from the viewpoint of each prisoner would be to accuse, because one would be freed if the other party remains silent. However, should both of them decide to accuse, they are worse off than if they remain silent. The nature of the PD game is therefore presented as a paradox between individual and group rationality. In fact, the most impressive aspect of the game is when the two prisoners put their act together for a common gain and forsake their own selfishness, the outcome of the game can be much better.

3 Axelrod's Tournament

The classical PD game is a one-shot scenario, as the two prisoners make their choice against each other once for all. We have seen in Sect. 2 that when the game is played only once, the most compelling choice for each prisoner is to selfishly accuse one another. A renewed interest in the game in early 1980s has urged political scientist Robert Axelrod to extend the one-shot game into a repeated form, which he called the IPD game. When the game becomes iterative, the prisoners are allowed to form strategies based on previous actions instead of using mere rationality. With that in mind, Axelrod pioneered the attempts of using computer-based approach to simulate the strategies in the game in order to study the emergence of cooperation and the process of strategy change within the simulated evolutionary framework [4]. He conducted two famous IPD tournaments and invited many experts from diverse areas to create and submit various strategies to play the game.

It is necessary to note that Axelrod had enforced some rules that all the participants must understand and consider before submitting their strategies to the tournaments. Generally, each strategy must decide its moves, which are composed solely with the combination of two actions – cooperate and defect, without knowing the opponent's action until the game takes place. As the only information available in deciding whether to cooperate or defect is the previous moves of other strategies in the game, each strategy has to remember the entire history of the game.

In the first tournament, there were 14 strategies submitted by game theorists from diverse fields in economics, biology, political science, mathematics, psychology and sociology [7]. Most people produced their strategies by capturing some behavioural assumptions of the sequence of moves for each strategy and specifying them accordingly to generate the current moves. All the strategies were arranged in a round robin format competing against one another, against itself, and also against an additional Random strategy that randomises its moves. The calculation of the scores for each strategy was based on a payoff table where the outcome for a particular pair of strategies pitted together was constant. Table 1 summarises the payoff table used by Axelrod for the tournament.

Based on the payoff structure in Table 1, all the 14 strategies submitted and a Random strategy were pitted against each other and itself for 200 rounds. The tournament was repeated for five times, and Axelrod confirmed

Table 1. The payoff table of the IPD tournament

Strategy 2

	Cooperate	Defect
Cooperate	*3, 3*	*0, 5*
Defect	*5, 0*	*1, 1*

Strategy 1

that the winner was a simple strategy submitted by Anatol Rapoport called TFT. TFT scored an average of 504 points to win the tournament with simply cooperating on the first move and thereafter repeating the preceding move of the other strategy. In overall, the tournament concluded that cooperative strategies had done better than those non-cooperative strategies, as eight of the top scoring strategies were those called by Axelrod as "nice" strategies which have never defected on their first move.

Subsequently, a second tournament was carried out by Axelrod after circulating the results of the first tournament publicly. In the second tournament, 62 strategies were submitted from six countries [8], where the participants were more widely ranged from computer hobbyists to professors of various expertises. Instead of playing 200 rounds for each strategy in the first tournament, this time the game was randomly terminated. As a result, an average of 151 rounds was played for each strategy.

Different tournament, but with the same old story – TFT was again being submitted by the same person, and again won the tournament in style. All the "nice" strategies were again fared better, except for one strategy called Tit-For-Two-Tat (TFTT). TFTT keeps cooperating unless the other strategy defected twice consecutively. It did well in the first tournament, but did poorly in the second. A thought that comes to mind is that – be nice, but not too nice (not until opponent defected twice)! Axelrod explained that some participants might have taken into account the lessons learned from the first tournament, and thus capitalised on the weaknesses of some strategies in order to exploit them.

Due to the lack of robustness and dynamics in the simulation of his two tournaments, Axelrod decided to explore further the evolution of cooperation in IPD using GA [4]. The purpose was to variegate the simulated environment and allow new strategies to be discovered based on the genetic operators of GA such as crossover and mutation. With this new way of implementation, Axelrod again conducted a computational simulation in 1987. In contrast to his previous works, Axelrod imposed fewer restrictions to the game strategies. He considered a set of strategies that were deterministic, where each strategy's decision on its current move was based on the outcome of its previous three moves rather than the entire history.

In this simulation, 20 random strategies had been included, plus another eight best performing strategies selected from his second tournament. At the conclusion of the simulation, Axelrod found that GA evolved most of the ordinary strategies. However, he specifically noted that TFT was one of the strategies that were able to match up with any other strategies in terms of the accumulated payoffs.

Axelrod then removed the eight specially selected strategies and repeated the simulation with only the 20 random strategies. He noticed that the level of complexity in the simulation intensified due to the concurrent evolution nature of the strategies. The average score accumulated by these evolving strategies gradually increased as the game repeated, hence implied that the strategies were slowly able to differentiate between the cooperators and the defectors (see [4] for more details).

Axelrod's contribution to the research of IPD has been significant, and the results from his tournaments have been considered by many as the definitive solution to the problem of cooperation in the PD game. In this chapter, we use Axelrod's works as the framework to conduct our experiments. Axelrod is interested in the types of strategies that could generate success for individuals, and at the same time could allow cooperation among groups. He used GA mainly to evolve other strategies and search for cooperators in the population. We extend his works by applying GA to a simulated agent-based economy market, for we believe that only with some genuine real-world components, where the agents are selfish and the resources are limited, the nature of IPD can be fully reflected. Instead of following Axelrod's way of using the fixed rates for the genetic operators in GA, we examine what happens when the rates are varied, in order to find a way to improve the performance of GA.

4 Methodology

In this section, we aim to describe the research approach we adopt and the methodology we use in conducting our experiments. For the purpose of simulation, we use an economy model and encode the IPD strategies to the spatially non-restrictive environment.

4.1 Basic Elements of the Model

The IPD has become a common modelling tool for economic analysis and understanding in recent years. The theory embedded in the game allows us to highlight and exemplify many of the strategic interactions among the trading parties. It thus helps to illuminate the strategic aspects of economics, including market competition which we are interested to simulate in this chapter. Based on [9], we have drawn up the following elements to form the basic structure of our model:

- The *Agents*: an agent is simply a firm that trades with other firms.
- The *Environment*: the environment is a simulated market that consists of many agents where each agent will trade its products or resources with other agents.
- *Actions*: an agent's action is the possible ways it reacts within the simulated environment, which could be trade (cooperate) or do not trade (defect).
- *Strategies*: a strategy is formed by the sequence of actions each agent decides to take under all possible circumstances in the market.
- *Payoffs*: a payoff is the earning an agent gets from each trade based on its action.

- *Outcomes*: the outcome is resulted from a combination of strategies in the form of total profits since the interaction between agents will generate action combinations.
- An *Equilibrium* Concept: the equilibrium concept implies stability that comes when there is a lack of individual incentives for an agent to change from its current action in response to the given actions of others.
- A *Solution* Concept: the solution concept aims to make refinements and eliminate equilibrium that does not make sense.

With these basic elements, we conduct experiments that simulate a fixed number of firms trading in a single market based on the selection of their own strategies. As each firm aims to maximise its own profit from the trading environment, immense competition will develop and thus the market will exhibit a variety of interesting behaviours or outcomes. The following section describes the model we use which consists of all the elements mentioned above.

4.2 The Economy Model

The economy model uses bit strings to represent natural resources and products created from those natural resources. Agents, which can represent firms, operate on these strings using grammar rules. A grammar rule consists of an input/output pair of bit strings. When a grammar rule is applied to a bit string S, then if the input string is contained within S as a substring, it is replaced with the output string. Only one instance of the input string within S, which is chosen at random, will be replaced if S contains more than one copy of the input string.

Each firm in the simulated market can produce its own products from the resources it has and can exchange the products or even the resources in its original form with other firms during the trading process. As such, the economy model maintains two "pools" of bit strings, the resource pool and the product pool (see Fig. 1). The resource pool contains a set of randomly generated initial strings which are always available. When a grammar rule acts on a resource string to change it into a new string, the original resource string is still preserved. The product pool is initially empty, and any new strings created from resource strings are placed in it. When a grammar rule acts on a product string, the original is removed and replaced with the new string. Thus resource strings are in infinite supply but product strings are not.

Activities in the simulated model are monitored in a sequence of generations. With a 0.5 probability, either a resource string or a product string is selected at random at each generation. The selected string is passed to an agent, which is also chosen at random. The agent selects one of its grammar rules, and applies it to the selected string. If the grammar rule matches, the string is transformed.

If the selected string was a resource string, it remains in the string pool when the transformed string is added. If the selected string was a product

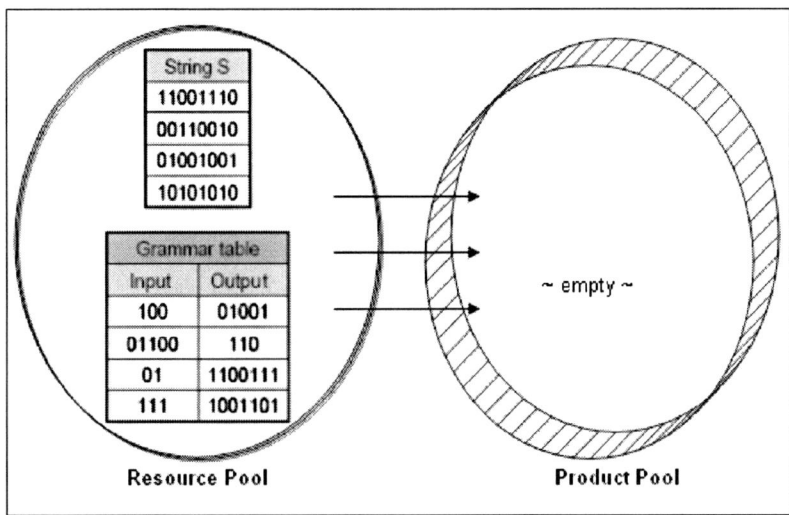

Fig. 1. Two pools of bit strings in the economy model with random grammars

string, however, the original string is removed from the string pool. This makes all the initial resource strings always available in the pool.

The model also adopts a simple graphical user interface that helps to visualise the outcomes of market activities during the trading process through real-time animation. The animation allows us to see the simulation dynamics while the agents interact within the market. In addition, an output screen that displays the simulation results is used to report the performance of agents in graph form. The graph measures the accumulated profit of each agent during the trading process. The results at the end of the simulation can be saved and read from a text file.

The screen-shot in Fig. 2 shows the environment of a simulated market where agents are represented as circles in graphical form. The colour of a firm can vary from brown to dark green. If an agent has more resources than products, it is brown in colour. As the agent started to produce more products from the resources it has, its colour changes from brown to dark green. The size of the agent, which is determined by the radius of the circles, represents how active an agent is involved in the market activities. The size increases if the agent is active in its trading but not necessarily representing higher profit. Over time, the size decreases if the agent did not manage to trade its products or resources with other agents for profits, thus signifying that the products or resources expired after certain period of time if there is no trade. Trade interaction between two agents is represented by drawing a line between them. The thickness of the line increases when there is more than one trade interaction between the two agents. It decreases when the trade interaction declines. Therefore, the thicker the line is, the more frequent the transfer of

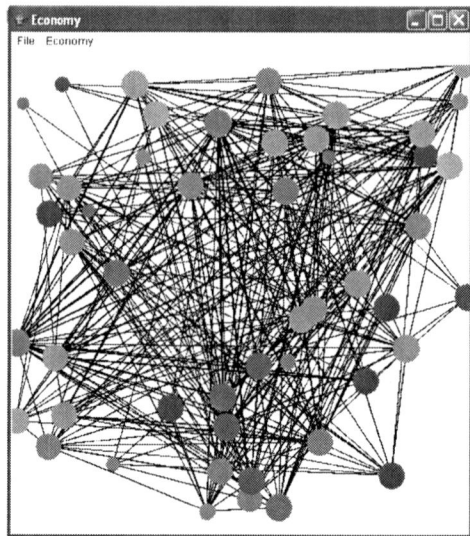

Fig. 2. Graphical representation of the simulated market

products or resources between the two agents, and thus more profits are shared between them.

4.3 Strategy Representation

The strategies in IPD game can be represented in many different ways. As the main goal of our experiments is to examine the performance of GA, a good representation that is sufficient to simulate the evolution of GA is important. We therefore adopt the representation developed by Axelrod in [4] for our model.

Under this representation, bit strings of 0s and 1s, which can be referred as "chromosome", are used to encode different game strategies. A 0 is therefore corresponded to cooperation and 1 to defection. An agent in the game has to keep the history of the past three rounds of play to decide on a current move. Each of these three rounds is recorded as a bit pair, thus a 6-bit string is produced for a three-round memory. For instance, 11 11 11 means both agents defected for the past three rounds, 00 00 00 means both agents cooperated for the past three rounds, 01 01 01 means the first agent cooperated while the second agent defected for the past three rounds, and so on. In deciding on a current move, a strategy must contain all the possible actions for the previous three rounds, resulting in $2^6 = 64$ possibilities, and thus a 64-bit string is required. Since there is no memory of previous rounds at the beginning of the game, an additional 6-bit is required to be the initial memory. This culminated in a strategy being encoded with the representation of a 70-bit string.

Table 2. The eight selected strategies used in this chapter

Strategies	Description
All-D	Always defects
Random	Randomly plays, cooperates and defects with probability 1/2
Pavlov	Cooperates on the first move and then cooperates only if the two firms made the same move
Prober	Begins by playing [cooperate, defect, defect], then if the trading partner cooperates on the second and the third move continues to defect, else plays TFT
Gradual	Cooperates on the first move and then plays [defect, cooperate, cooperate] if the trading partner defects; [defect, defect, cooperate, cooperate] if the trading partner defects again; [defect * the number of times the trading partner defects, cooperate, cooperate] when the trading partner defects for the third time and so forth
Mistrust	Defects, then plays trading partner's previous move
TFT	Cooperates on the first move and then plays what its trading partner played on the previous move
All-C	Always cooperates

After devising a way to represent the game strategies, an initial population of agents with different strategies needs to be determined. It is almost impossible for us to include all the game strategies from Axelrod's tournaments in our simulation experiments, therefore only six distinctive strategies are chosen, namely All-Defect (All-D), Random, Prober, Mistrust, TFT and All-Cooperate (All-C). In addition, Pavlov [10] and Gradual [11], two outstanding strategies claimed to be better than TFT, are also included. Table 2 below summarises the eight selected strategies used in this chapter.

We incorporate these eight IPD strategies to the economy model we have just described. Every agent thus has a choice of two actions to either defect or cooperate during each interaction based on the selection of strategies.

Various colours are used to represent different strategies selected by different agents. The colours replace the original colours of the agents based on the strategy each agent has chosen. When an agent decides to adopt another strategy from a more successful agent in order to increase its profit during a trading interaction, the colour of the agent will change, thus reflecting the new strategy of its choice. Table 3 shows the representation of colours for different strategies.

4.4 Fitness Function

The success or failure of a strategy is measured based on the fitness of the agents adopting that particular strategy. A fitness function, $F(agent_i)$, is therefore needed to calculate the agents' fitness by mapping the agents' chromosome, or bit strings, into the scores. The scores for our simulation are simply

Table 3. Strategy representation with different colours

	Colours	Strategies
	Red	All-D
	Green	Random
	Yellow	Pavlov
	Blue	Prober
	Magenta	Gradual
	Cyan	Mistrust
	Gray	TFT
	Black	All-C

For n generations:
  For each $agent_i$ of the population:
    Calculate and store $F(agent_i)$;
  Reproduce the fitter agents of the population, with better strategies inhabiting the weaker strategies. Discard the current population, and treat the reproductions as new population;

Fig. 3. Algorithms for reproduction of strategies

the scaled number of points a strategy accumulates based on the payoff table of Axelrod's tournaments in Table 1.

To see how well each strategy performs, the fitness of agents of a given strategy is studied based on its accumulated scores over an n number of generations. We determine the fitness of each agent by assuming the agent with highest score to be the fittest. Vice-versa, agents with lower scores are the weaker ones. This implies that the fitness function will determine which agent will adopt the strategy of another agent who is deemed to be fitter. By doing this, we will be able to see the evolution of stronger strategies within the given generations. Figure 3 below shows the algorithms of the reproduction of better strategies.

5 Experiments and Results

The aim of our experiments is to examine how well GA could perform against other IPD strategies. In doing so, we populate 50 agents to the economy model with the eight distinctive strategies mentioned in Table 2 being assigned to each of the agents. The experiments are designed for each agent to play the IPD game against one another iteratively for 100 generations. Scores are being calculated in every generation. Strategies with scores below the average scaled payoff in a given generation will be replaced by the more successful strategy from which they have interacted with. Before the simulation starts, the eight strategies are distributed randomly to different agents, as depicted in Fig. 4.

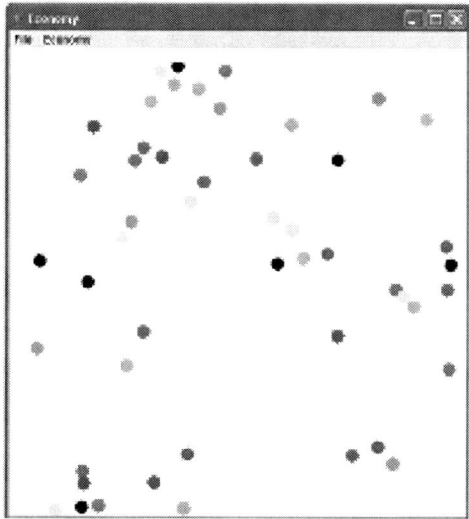

Fig. 4. Initial distribution of the simulated market

We conduct our experiments incrementally by first using the same crossover and mutation rate for GA at 0.1. We then increase and decrease the rate to examine the effects of changing the crossover and mutation. Finally, we turn off the crossover and mutation alternately to see how each genetic operator affects the performance of GA.

When the rates of crossover and mutation were fixed at 0.1, we noticed that although GA performed quite well and started to dominate from the 30th generation onwards, GA did not manage to evolve the cooperative strategies to optimality. Three strategies co-existed at the end of 100 generations as depicted in Fig. 5, namely the TFT, All-D and Pavlov. Based on the observation, we can conclude that with crossover and mutation rates set to 0.1, the performance of GA is unstable.

Next, we altered the rates of crossover and mutation by gradually increasing them from 0.1 to 0.5. We noticed that the increment caused disorder to the population, and increment beyond 0.25 brought randomness among the strategies. As we decreased the rates, the environment showed no improvement until they were below 0.1. When the rates reached 0.08, we observed that GA could learn better and manage to find the global optimum at times. When the rates were 0.05, GA literally has no problem in evolving the environment to full cooperation. Figure 6 shows the result with TFT fully populated the simulated market.

We repeated the same simulation for several times to observe any possible variation to the result of our experiments. From the repeated simulations, we noticed that GA generally did well when the crossover and mutation rates were set to around 0.05 or below, except on some occasions when GA ceased to evolve in the early stage.

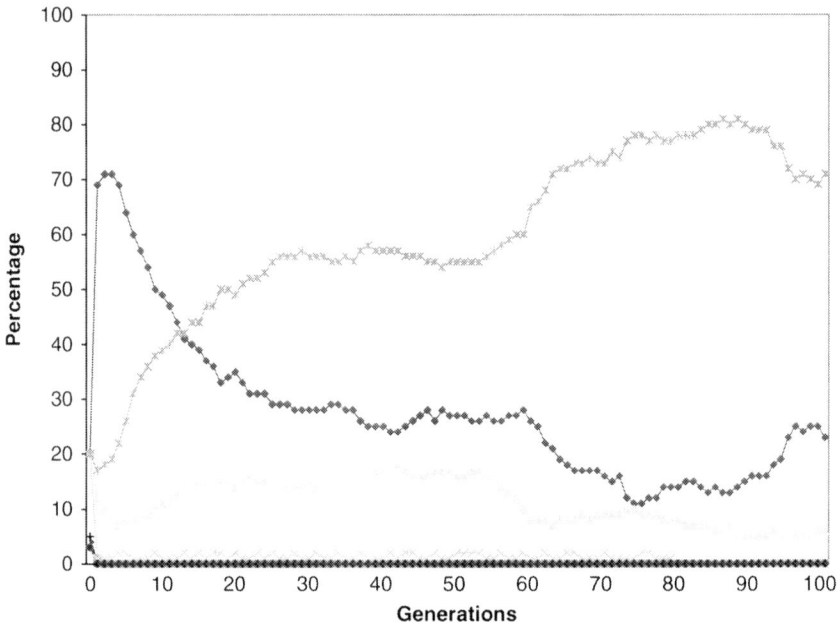

Fig. 5. Evolving strategies over 100 generations

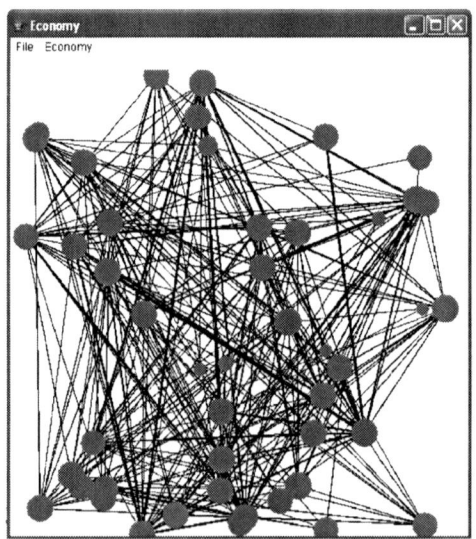

Fig. 6. GA evolved TFT to full cooperation

We have so far conducted our simulations with both crossover and mutation being carried out. In the final part of our experiments, we turned off the crossover and mutation rates alternately to examine how each of these genetic operators affects the performance of GA. We first turned off the crossover rate by setting it to 0. With mutation rate alone at 0.1, we noticed some random behaviour among the population and GA did not perform well. As we increased the mutation rate gradually, the random behaviour became more obvious. We then decreased the mutation rate to below 0.1, and we began to see the normal evolution of GA in learning and evolving the better strategies. However, with mutation rate below 0.01 or close to 0, GA simply replaced the weaker strategies with stronger ones without forming any new strategies.

Next, we turned off the mutation rate by setting it to 0. With crossover rate alone at 0.1, we observed a similar result as to the one showed in Fig. 4. When we increased the crossover rate gradually, we noticed that the pattern did not change much until the rate was beyond 0.5. As we decreased the rate, the performance of GA was significantly improved especially when the crossover rate was below 0.1. When the rate was close to 0, however, the performance of GA suddenly suffered a dip, and defective strategies immediately took the advantage to conquer the population.

6 Discussion

Our experiments demonstrated that GA could be highly potential in evolving cooperative strategies in the IPD game, although at times the performance was unstable. With the rates of crossover and mutation at 0.1, GA performed particularly well from the 60th generation to the 90th generation (see Fig. 5), but started to exhibit randomness thereafter. This showed that GA cannot effectively copy the successful strategies at the same crossover and mutation rate of 0.1. When the rates were beyond 0.1, more randomness occurred, resulting in worse performance for GA. We believe the reason behind this was that when the crossover and mutation rates were too high, the new strategies formed by GA were nothing like the successful strategies GA has copied, thus losing the winning traits from those good strategies.

Our experiments have also demonstrated that with the crossover and mutation rates below 0.08 (especially around 0.05), GA worked extremely well except on some exceptional occasions where GA ceased to evolve before the 10th generation. This often led to defective strategies dominating the population.

The final part of our experiments examined the effects when crossover and mutation were turned off alternately. We demonstrated that the random behaviour in the environment was largely due to inappropriate mutation rate. With small rate of mutation, GA were able to evolve as normal. We also showed that crossover was actually the key to GA's success. With reasonable crossover rate, GA could effectively perform well without the mutation.

7 Conclusion

In this chapter, we investigated the performance of GA against the IPD strategies. We adopted an empirical approach to examine how well GA could work against other common IPD strategies based on an agent-based economy model. Our work showed that cooperative strategies could be evolved through GA, although at times the performance of GA could be unstable. We conclude that the GA could effectively be enhanced when the genetic operators are set to reasonable values around 0.05. However, the exact crossover and mutation rates for GA to work perfectly are probably related in some way to the experimental settings such as the size of the population, generation of play, length of the bit strings, and so on. With proper tuning of these settings and the parameters used, we believe that GA would be extremely useful for optimising the outcome of an economy market. Further studies are needed to generalise our results.

References

1. Goldberg DE (1994) Genetic and evolutionary algorithms come of age. Communications of the ACM 37(3): 113–119
2. Drake AE, Marks RE (1998) Genetic algorithms in economics and finance: forecasting stock market prices and foreign exchange – a review. (AGSM Working Paper 98-004)
3. Axelrod R (1984) The Evolution of Cooperation. Basic Books, New York
4. Axelrod R (1987) The evolution of strategies in the iterated Prisoner's Dilemma. In: Davis L (ed.) Genetic Algorithms and Simulated Annealing. Morgan Kaufmann, Los Altos, CA, pp. 32–41
5. Chiong R, Wong DML, Jankovic L (2006) Agent-based economic modeling with iterated Prisoner's Dilemma. In: Proceedings of the IEEE International Conference on Computing and Informatics (ICOCI). Universiti Utara Malaysia, Kuala Lumpur, Malaysia
6. Poundstone W (1992) Prisoner's Dilemma. Doubleday, New York
7. Axelrod R (1980) Effective choice in the prisoner's dilemma. Journal of Conflict Resolution 24: 3–25
8. Axelrod R (1980) More effective choice in the prisoner's dilemma. Journal of Conflict Resolution 24: 379–403
9. Morrison WG, Harrison GW (1997) International coordination in a global trading system: strategies and standards. In: Flatters F, Gillen D (eds.) Competition and Regulation: Implications of Globalization for Malaysia and Thailand. Queens University: John Deutsch Institute for the Study of Economic Policy
10. Nowak MA, Sigmund K (1993) A strategy for win-stay, lose-shift that outperforms tit-for-tat in the Prisoner's Dilemma game. Nature 364: 56–58
11. Beaufils B, Delahaye JP, Mathieu P (1997) Our Meeting with Gradual: A Good Strategy for the Iterated Prisoner's Dilemma. In: Artificial Life V, Proceedings of the fifth International Workshop on the Synthesis and Simulation of Living Systems (ALIFE-96). The MIT Press, Cambridge, MA, pp. 202–209

Optimizing Multiplicative General Parameter Finite Impulse Response Filters Using Evolutionary Computation

Jarno Martikainen[1] and Seppo J. Ovaska[2]

[1] Helsinki University of Technology, Department of Electrical and
Communications Engineering, Otakaari 5 A, FI-02150 Espoo, Finland
[2] Utah State University, Department of Electrical and Computer Engineering,
4120 Old Main Hill, Logan, UT 84322-4120, USA

Summary. Evolutionary algorithms have been studied in the context of optimization problems for decades. These nature-inspired methods have lately received increasing amount of attention especially among demanding practical optimization tasks. This Chapter describes how evolutionary algorithms can be used to increase the speed and robustness of the design process of a digital filter used in power systems instrumentation. Designing such filters is a challenging optimization problem due to a discrete and exhaustive search space and time-consuming evaluation of the objective function. This chapter introduces new approaches to evolutionary computation using adaptive parameters, hierarchical populations, and the fusion of neural networks and evolutionary algorithms to tackle the bottlenecks of this challenging design process.

1 Introduction

In 50 Hz power systems instrumentation, predictive lowpass and bandpass filters play a crucial role. These signal processing tasks are delay-constrained so that the distorted line voltages or currents should be filtered without delaying the fundamental frequency component. In addition, the line frequency can vary typically up to ±2%, so the filter should be able to adapt to the changing input frequency. For this purpose, Vainio et al. introduced the multiplicative general parameter (MGP) finite impulse response (FIR) filtering scheme in [1, 2]. The aim of the MGP-FIR is to predict the signal value p steps ahead to compensate for delays introduced earlier in the system while simultaneously filtering out noise and harmonic components from the input signal. Additionally, the filter is designed so that the number of multiplications is minimized and the filter is thus computationally very efficient. Computational efficiency in MGP-FIRs is obtained by assigning the filter coefficients from a discrete set of $\{-1,\ 0,\ 1\}$, and, thus, the multiplication of a signal value

J. Martikainen and S.J. Ovaska: *Optimizing Multiplicative General Parameter Finite Impulse Response Filters Using Evolutionary Computation*, Studies in Computational Intelligence (SCI) **92**, 327–354 (2008)
www.springerlink.com

with a filter coefficient becomes an addition or subtraction. This cost effectiveness makes the MGP-FIR convenient, e.g., for Field-Programmable Gate Array (FPGA) implementations. However, the use of discrete filter coefficient set excludes derivative based design methods. Additionally, enumeration and random search are likely to take too long a time to produce competitive results since for a single filter with 40 coefficients there are 3^{40} different possibilities to select from. However, evolutionary algorithms [3–5] are widely known to excel optimization problems with discrete and extensive search spaces. Digital filters have been designed before using evolutionary computation, e.g., in [6] and [7], but the approaches have been mainly application oriented employing standard evolutionary mechanisms. This Chapter, however, introduces enhanced evolutionary schemes to produce competitive design performance. The use of adaptive variation operators and the use of multiple solution populations are shown to add to the performance of the standard evolutionary algorithms. Moreover, the fusion of neural networks and evolutionary computation offers the means to decrease the time required by the evaluation of a time-consuming objective function.

In the second section of this chapter the MGP-FIR filter is described in detail. Section 3 discusses the use of evolutionary computation in the design process of MGP-FIR filters and examines novel modifications to the standard evolutionary algorithms to improve the design process. Section 4 illustrates a scheme fusing evolutionary computation and neural networks in order to decrease the time required to evaluate a demanding objective function related to designing MGP-FIR filters. Section 5 concludes the chapter.

2 Multiplicative General Parameter Finite Impulse Response Filter

The multiplicative general parameter FIR filter is a cost effective filter to be used in power systems instrumentation for harmonics cancellation, noise reduction, and signal prediction. In a typical MGP-FIR, the filter output is computed as explained in (1).

$$y(n) = g_1(n) \sum_{k=0}^{N-1} h_1(k)x(n-k) + g_2(n) \sum_{k=0}^{N-1} h_2(k)x(n-k) \qquad (1)$$

where $g_1(n)$ and $g_2(n)$ present the adaptive MGP parameters, and $h_1(k)$ and $h_2(k)$ are the fixed coefficients of an FIR basis filter. Thus, the coefficients of the composite filter are $\theta_1(k) = g_1(n) h_1(k)$, $k \in [0, 1, \ldots, N-1]$, for the first MGP, and, $\theta_2(k) = g_2(n) h_2(k)$, $k \in [0, 1, \ldots, N-1]$, for the second MGP. An example of MGP-FIR with $N = 4$ is shown in Fig. 1. Here N denotes the filter length.

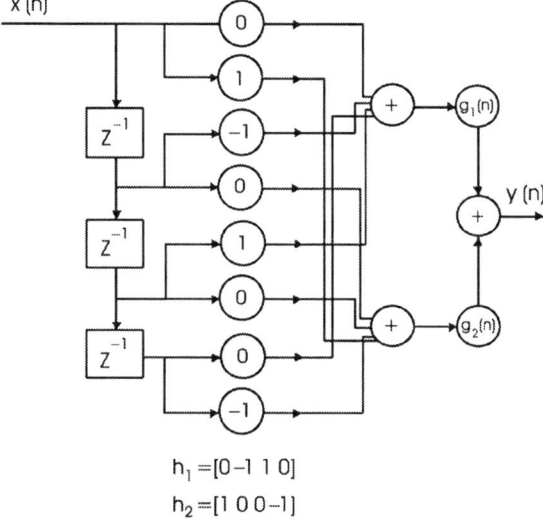

$$h_1 = [0{-}1\ 1\ 0]$$
$$h_2 = [1\ 0\ 0{-}1]$$

Fig. 1. An example of MGP implementation, where $N = 4$. Signal values $x(n-1)$ and $x(n-2)$ are connected to the first MGP and values $x(n)$ and $x(n-3)$ are connected to the second MGP with filter coefficients -1, 1, 1, and -1, respectively

The adaptive coefficients, $g_1(n)$ and $g_2(n)$, are updated online according to (2) and (3).

$$g_1(n+1) = g_1(n) + \mu e(n) \sum_{k=0}^{N-1} h_1(k)x(n-k) \qquad (2)$$

$$g_2(n+1) = g_2(n) + \mu e(n) \sum_{k=0}^{N-1} h_2(k)x(n-k) \qquad (3)$$

μ is the adaptation gain factor and $e(n)$ is the prediction error between the filter output and the training signal $s(n)$, i.e., $x(n) - y(n-p)$, p being the prediction step. The MGP-FIR has two adaptive parameters to adapt only to the phase and amplitude of the principal frequency. More degrees of freedom would allow the filter to adapt also to undesired properties, such as the harmonic frequencies. The training signal $s(n)$ is defined as

$$s(n) = \sin(2 \cdot \pi \cdot 49 \cdot n) + \sum_{m=3,5,7,\ldots}^{15} 0.1 \cdot \sin(2 \cdot \pi \cdot m \cdot 49 \cdot n)$$
$$+ 0.004 \cdot r(n), \ 0 < n \le 300 \text{ samples}$$

$$s(n) = \sin(2 \cdot \pi \cdot 50 \cdot n) + \sum_{m=3,5,7,\ldots}^{15} 0.1 \cdot \sin(2 \cdot \pi \cdot m \cdot 50 \cdot n) \qquad (4)$$
$$+ 0.004 \cdot r(n), \ 300 < n \le 600 \text{ samples}$$

$$s(n) = \sin(2 \cdot \pi \cdot 51 \cdot n) + \sum_{m=3,5,7,\ldots}^{15} 0.1 \cdot \sin(2 \cdot \pi \cdot m \cdot 51 \cdot n)$$
$$+ 0.004 \cdot r(n), \quad 600 < n \le 900 \text{ samples}$$

$r(n)$ denotes noise, i.e., a uniformly distributed random value between -1 and 1. The same training signal is used throughout the optimization process. The duration of the training signal is 900 samples and this signal is divided into three parts, each of which contains 300 samples. These training signal blocks correspond to frequencies of 49, 50, and 51 Hz. The training signal constitutes thus of the nominal frequency sinusoid, odd harmonics up to the 15th with amplitudes 0.1 each, and white noise. The training signal is similar to that used in [1] and [2] and it mimics the line signal (voltage/current) with varying fundamental frequency, harmonics, and noise.

3 Genetic Algorithms in the Design of MGP-FIRs

Evolutionary algorithms are computational models mimicking the evolution of species, individuals, or genes depending on the implementation. The well-known no free lunch theorem [8] states that no optimization algorithm is superior when considering all the possible optimization problems; rather, their average performance is the same. So, when selecting an algorithm for a specific task it is useful to try to find a match between the algorithm and the problem structure. Of all evolutionary computation methods genetic algorithms are chosen for optimizing MGP-FIR basis filter since the structure of the solutions can be conveniently modified using the mutation and crossover operators typical for GA implementations.

The filter tap cross connections and coefficients are modeled as chromosomes in the following way. There are two separate vectors in one chromosome: the first vector, $h_1(n)$, of the chromosome represents the weights of the input samples connected to the first MGP, and the second vector, $h_2(n)$, corresponds to the weights of the input samples connected to the second MGP. The weights can be -1, 0, or 1. -1 and 1 in a vector indicate that this particular input value is associated to the corresponding MGP. 0, on the other hand, determines that a particular input value is not associated with the MGP in concern; rather it is connected to the other MGP. An example of a chromosome is presented in Table 1.

Table 1. Example of a chromosome, $N = 40$

h_1	1 1 −1 0 0 1 −1 0 1 1 0 −1 1 0 1 0 −1 −1 0 1 0 0 0 1 1 0 −1 0 −1 0 1 −1 0 0 0 −1 0 1 1 0
h_2	0 0 0 1 −1 0 0 1 0 0 −1 0 0 1 0 −1 0 0 −1 0 1 −1 1 0 0 1 0 1 0 −1 0 0 1 1 1 0 −1 0 0 −1

The value of an individual solution can be determined based on (5).

$$fitness = \frac{K}{\max(NG49\ NG50\ NG51) \cdot (ITAE49 + ITAE50 + ITAE51)} \quad (5)$$

K is a scaling factor and it is assigned a value of 10^6. Terms $NG49$, $NG50$, and $NG51$ represent the noise gain at a specific stage of the test input signal. These terms were added to the fitness function to attenuate noise. The noise gain is calculated as in (6).

$$NG(n) = \sum_{k=0}^{N-1} [g_1(n) \cdot h_1(k)]^2 + \sum_{k=0}^{N-1} [g_2(n) \cdot h_2(k)]^2 \quad (6)$$

h_1 and h_2 associate the input signal values to the first and second MGPs, respectively. $g_1(n)$ and $g_2(n)$ represent the first and second MGP at the end of a certain frequency period of the test signal, respectively, whereas $h_1(n)$ and $h_2(n)$ denote the filter coefficients associated to the corresponding MGPs. In other words, $NG49$ is calculated using the MGP values after 300 samples, $NG50$ and $NG51$ are calculated after 600 and 900 samples, respectively, the frequency of the training signal changing every 300 samples.

$ITAE49$, $ITAE50$, and $ITAE51$ stand for the Integral of Time-weighted Absolute Error (ITAE) for each of the three signal parts, respectively. These terms were added to the fitness function to smoothen the adaptation of the filter to the varying input signal characteristics. The ITAE is calculated as given in (7).

$$ITAE = \sum_{n=1}^{M} n \cdot e(n) \quad (7)$$

The basic idea of MGP-FIR filters is that all the samples of input delay line should be connected either to the first or to the second MGP, and no value should be left unused. Generally, the MGP-FIRs are designed so that the basis filters $h_1(k)$ and $h_2(k)$ are optimized first. Then, the optimized filter is applied to the application and the basis filter is kept unchanged. The online adaptation is carried out by (2) and (3). In this chapter the length of the filters studied was 40. Population size of 80 individuals was found to be suitable through extensive testing.

In the following sections different types of genetic algorithms are studied. First, a simple genetic algorithm is modified using adaptive operators and seeding. Then, a two-population genetic algorithms scheme is discussed for further improving the design process of these filters.

3.1 Reference Genetic Algorithm

To create a reference point, we implemented a simple genetic algorithm without any fine tuning to solve the filter design problem. The algorithm starts by creating the initial population. The probabilities of assigning -1, 0, or 1 to a particular gene in the chromosome related to the first MGP are $1/3$ for all. In the case of assigning -1 or 1 to a gene in $h_1(n)$, 0 is assigned to this same gene in $h_2(n)$. This is because one input sample can be connected to only one MGP, not both. If 0 is assigned to a gene in the vector related to the first MGP, -1 or 1 is assigned to a gene related to the second MGP with the probability of $1/2$ for both -1 and 1.

After the generation of the initial population, the chromosomes are arranged in descending order according to their fitness values. Mating is based purely on the rank of the chromosomes: the best and the second best chromosomes act as parents for two offspring. The offspring are created using single-point crossover, the crossover point of which is selected randomly. Also, the third and fourth best chromosomes constitute parents for two offspring, and so on. Altogether, 40 best chromosomes produce 40 offspring. Thus the 40 chromosomes with the lowest fitness scores are replaced by the offspring of the 40 best chromosomes. Also the parents remain within the population thus keeping the population size constant at 80 chromosomes. Mutation probability of 5% and μ value 0.002 were used in this GA.

3.2 Adaptive Genetic Algorithm

Optimal parameters for a GA may change during the iteration process and thus it is usually difficult to choose initial parameters that produce satisfactory results. Adaptive GA (AGA) parameters have been studied by Srinivas et al. in [9]. In their approach, the mutation probability, p_m, and the crossover probability, p_c, are adjusted as follows.

$$p_c = \frac{k_1 (f_{\max} - f')}{f_{\max} - f_{ave}}, \ f' \geq f_{ave} \tag{8}$$

$$p_c = k_3, \ f' < f_{ave} \tag{9}$$

$$p_m = \frac{k_2 (f_{\max} - f)}{f_{\max} - f_{ave}}, \ f \geq f_{ave} \tag{10}$$

$$p_m = k_4, \ f < f_{ave} \tag{11}$$

k_1, k_2, k_3, and k_4 represent coefficients with values equal or less than one. f_{max} denotes the maximum fitness value of a generation, whereas f_{ave} denotes the average fitness value of the same generation. f' denotes the higher fitness value of a parent pair. The authors of [9] propose values 1.0, 0.5, 1.0, and 0.5 for k_1, k_2, k_3, and k_4 respectively.

Using this approach, the crossover probability decreases as the fitness value of the chromosome approaches f_{max} and is 0.0 for a chromosome with fitness value equal to f_{max}. Also, mutation probability approaches zero as the fitness of a chromosome approaches f_{max}. μ value 0.002 was also used in this GA.

3.3 Adaptive Genetic Algorithm with Modified Probabilities

The values for k_1, k_2, k_3, and k_4 presented in [9] are not obviously appropriate for all applications. Referring to the no free lunch theorem we decided to give value equal to one for each of the k_1, k_2, k_3, and k_4, because the values presented in [9] do not perform so well in this application. This way we may prevent premature convergence, which seemed to be a problem in our application, when 1.0, 0.5, 1.0, and 0.5 were assigned to k_i, $i = 1$, 2, 3, 4, respectively. Since elitist mutation saves the best solution from corruption, we apply the mutation probability of 1.0 to all but the best chromosome. Mutation probability here means that for a specific probability an individual is selected for mutation. This way, although possibly near the optimal solution, we still search efficiently the surroundings for a better one.

3.4 Adaptive Genetic Algorithm with Modified Probabilities and the μ-Term

In the previously presented GAs we used a constant adaptation gain factor μ (see (2) and (3)). Since the characteristics of the input signal vary with time, it is difficult to determine the optimal parameters of the system before the optimization process. Adaptive μ was applied in order to speed up the convergence of genetic algorithm.

The chromosome was modified so that one gene was added to present the μ value. The μ values in the initial population were chosen randomly so that $0 < \mu < 0.009$. Values larger than this can easily make the filter unstable. In crossover the offspring μ value, μ_o, was calculated as follows.

$$\mu_o = r \cdot \mu_{p1} + (1 - r) \cdot \mu_{p2} \tag{12}$$

r stands for a random number between 0 and 1. μ_{p1} and μ_{p2} are the μ values of the first and second parents, respectively. Mutation was applied as follows.

$$\mu_m = \mu + 0.1\mu, r < 0.5$$
$$\mu_m = \mu - 0.1\mu, r \geq 0.5 \tag{13}$$

μ_m presents the value of μ after mutation, and r is a uniformly distributed random number between 0 and 1. The average converged μ value of the 20 runs was 0.0034. This value can be considered as the practical maximum μ value, i.e., values larger than 0.0034 could easily cause the filter to be unstable. The results showed that μ is a sensitive parameter and even small changes can easily make the filter unstable resulting in a low fitness value.

3.5 Adaptive Genetic Algorithm Using Seeding

When analyzing the chromosomes produced by the GA mentioned above, one could see a systematic structure in the converged solutions. All the GAs

Table 2. Example of a converged chromosome. Note the block like structure of the genes

h_1	0 0 0 0 0 0 0 0 0 1 1 1 1 1 1 1 1 1 1 0 0 0 0 0 0 0 0 −1 −1 −1 −1 −1 −1 −1 −1 −1 0 0 0 0
h_2	−1 1 1 1 −1 1 1 1 −1 0 0 0 0 0 0 0 0 0 0 −1 −1 −1 −1 −1 −1 −1 1 0 0 0 0 0 0 0 0 0 −1 1 1 1

Table 3. Performance comparison of the featured GAs. The results are averages of 20 runs

Algorithm	Average best fitness/ standard deviation	Convergence (generations)/ standard deviation	Total run time (min)	Convergence time (min)
Reference GA	3,249/324	837/130	76.4	63.9
AGA	1,472/505	21/11	77.9	1.6
AGA, Modified Probabilities	2,824/557	89/34	76.7	6.8
AGA, Modified Probabilities and μ	3,638/287	329/218	73.9	24.3
AGA, Modified Probabilities, μ, seeding	3,617/329	156/65	73.6	11.5

had a tendency of creating blocks of consecutive equal gene values within a chromosome. An example of a converged chromosome can be seen in Table 2.

This kind of repeating structure was present in nearly all the converged chromosomes. Bearing this in mind, seeding was applied to the GA while generating the initial population. Seeding means applying the information of the structure of the solution to the generation process of the initial set of chromosomes. Seeding was implemented so that the chromosomes in the initial population were formed of blocks of random number of consecutive equal gene values. The block length was set to be 1 at the minimum and the other extreme was a chromosome constituting only of a single gene value.

3.6 Results

The performance of the different GA approaches in MGP filter design is reported in Table 3. The genetic algorithms were given 1,000 generations to converge to a solution, and each genetic algorithm was run 20 times to get some statistical reliability. Elitist mutation scheme was applied in each case, so that the best chromosome was never mutated. A single simulation run of 1,000 generations using initial population of 80 chromosomes and

Table 4. The best converged chromosome with fitness value of 3,985

h_1	1 1 1 0 0 0 0 0 0 0 −1 −1 −1 −1 −1 −1 −1 −1 −1 −1 0 0 0 0 0 1 1 1 1 1 1 1 1 1 1 0 0 0 0
h_2	0 0 0 −1 −1 −1 −1 −1 −1 −1 0 0 0 0 0 0 0 0 0 1 −1 1 1 1 0 0 0 0 0 0 0 0 0 0 0 −1 −1 −1 −1

filter length of 40, took 70–80 min to run on a Dell computer equipped with 2.66 GHz Intel Pentium 4 processor, 512 MB of memory, and MATLAB 6.5.0 software.

There were significant differences in the convergence properties of the algorithms. The adaptive GA proposed in [9] was clearly the fastest to converge, but the algorithm was very likely trapped in a local minimum, and it thus behaved worse in statistical sense than the simple GA used for reference.

Modifying the probabilities of the AGA produced fitness average closer to the reference GA than with original parameters, while simultaneously reducing the number of required generations to one tenth. Adding the adaptive μ term succeeded in achieving higher average fitness, while simultaneously increasing the number of required generations. Applying seeding helped the adaptive μ enhanced AGA to produce best results, indicating highest average fitness with even better convergence characteristics than the AGA with adaptive μ but without seeding. Standard deviations presented in Table 3 indicate that the AGA with modified probabilities, adaptive μ, and seeding produces converges to high fitness solutions rather reliably.

The average time to achieve the best average results was 11.5 min, which is about 18% of the time required by the reference GA to produce results. The gained speedup makes the introduced method a tempting tool for practicing engineers. The average time to go through 1,000 generations is almost the same for all the GAs. Then again, the differences between the average convergence times are notable. The best converged chromosome had fitness value of 3,985 and was achieved using AGA with adaptive μ and seeding. This chromosome is presented in Table 4.

The amplitude response of the filter presented in Table 4 after 600 samples of the evaluation signal can be seen in Fig. 2. This response shows how the GA tries to dampen the harmonic frequencies of the nominal components (i.e., 150, 250,...,750 Hz). The phase delay in samples is shown in Fig. 3. The phase delay is negative in the 50 Hz region, and thus the filter has the desired predictive capabilities. Table 5 shows the calculated total harmonic distortion (THD) values for each of the training signal nominal frequency components before and after filtering. After filtering the THD values as well as the amplitudes of the harmonics are about a tenth of the values before filtering. Figure 3 describes the signals before and after filtering (Fig. 4).

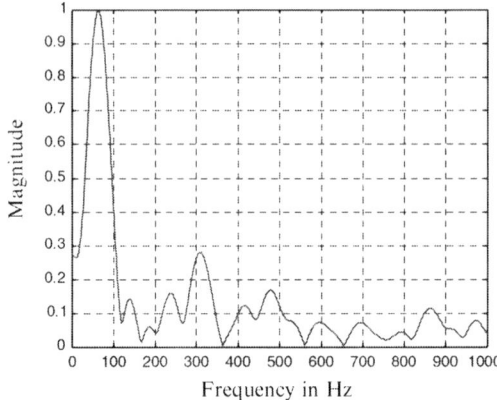

Fig. 2. Instantaneous amplitude responses of the MGP-FIR

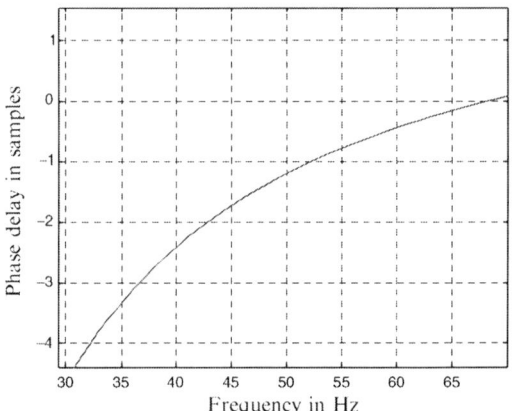

Fig. 3. Phase delay of the MGP-FIR. The phase delay in samples is negative in the 50 Hz region, as it should be in predictive filters

3.7 Two-Population Genetic Algorithm

The ideas presented in Sect. 3.6 concerned mainly modifying the operators of the evolutionary computation scheme. However, it is also possible to increase the performance of the optimization scheme by structuring the solution population into two separate entities. The idea of our two-population genetic algorithm (2PGA) originates from the notion of nature dividing various populations into subpopulations, e.g., a small elite and a large plain, based on their fitness similarities. In our 2PGA scheme the two populations evolve separately in parallel, but they are exchanging chromosomes under certain conditions, i.e., the best chromosome from the plain population is allowed to enter the elite population if its fitness value is high enough. Then

Table 5. Total harmonic distortions (THD) of the best MGP-FIR basis filter

	Input amplitude	Output at 49 Hz	Output at 50 Hz	Output at 51 Hz
Nominal frequency	1	0.87	0.93	0.88
3rd harmonic	0.15	0.018	0.016	0.011
5th harmonic	0.15	0.023	0.025	0.023
7th harmonic	0.15	0.018	0.009	0.008
9th harmonic	0.15	0.012	0.012	0.013
11th harmonic	0.15	0.010	0.008	0.006
13th harmonic	0.15	0.005	0.002	0.007
15th harmonic	0.15	0.005	0.004	0.007
THD	39.7%	4.4%	3.7%	3.6%

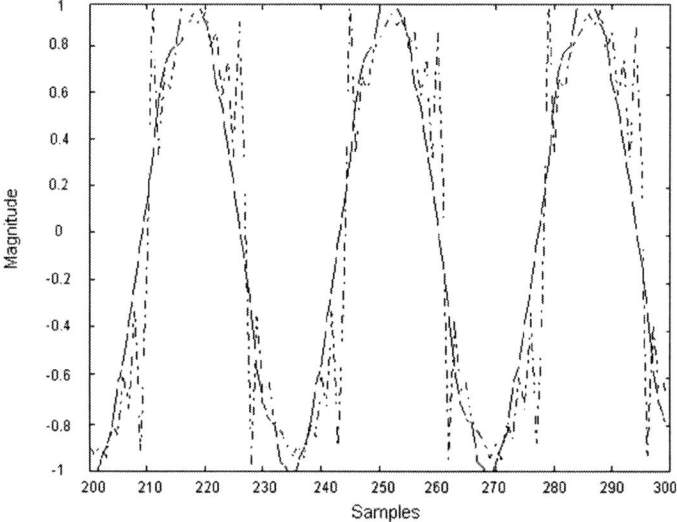

Fig. 4. Test signal defined in (4) (*dash dotted*) is filtered using the best MGP-FIR given in Table 4. Output (*solid line*) is almost sinusoidal

again, the worst chromosome from the elite population is transferred to the plain population to keep the population sizes constant. The evolution in both of the subpopulations is as in a single population GA (SPGA). The principal difference is that in the plain population the mutation probability is higher. The analogy supporting this assumption can be derived from nature, where weaker individuals, in terms of fitness, have to change their behavior or appearance more in order to succeed in competition. Multiple population evolutionary algorithms have usually been experimented in parallel hardware environments, but Gordon et al. proposed in [10] that the mere existence of multiple populations without actual parallel hardware may improve the performance of an evolutionary scheme.

GAs typically have a tendency of finding a good neighborhood fast, but it may take a long time to reach the optimum in that area. This is the reason for using specialized hybrid methods, in which a GA is used for global search, and local search is carried out by some more traditional technique, such as the hill-climbing method. Our 2PGA is an effort to improve the basic GA with little additional computation and no separate algorithms.

The operation of our 2PGA can be divided into seven stages as follows:

1. Generate an initial random population of solutions.
2. Evaluate the fitness (or cost) of the chromosomes in the initial population and divide the population into a small elite population and large plain population.
3. Evaluate the fitnesses of plain and elite populations.
4. Implement reproduction separately in both of the populations.
5. Compose populations for the next generation combining parents, offspring, and possibly migrated chromosomes. If the fitness value of the best chromosome in the plain population supersedes a certain limit value, exchange this chromosome with the worst chromosome in the elite population. The parents not chosen to reproduce in the previous round in the elite population migrate to the plain population.
6. Mutate chromosomes using different mutation probabilities for both of the populations. Elitist mutation that keeps the best solutions in both the populations intact is used.
7. Go to 3 or exit if convergence or run time constraints have been met.

The basic idea of 2PGA algorithm is to conduct global and local search in parallel using two populations. The elite population, having a small mutation probability, searches among the best solutions to find even better solutions, whereas the large plain population, with large mutation probability, searches the whole search space in hope of finding new promising areas of high fitness.

The proposed method is illustrated in Figs. 5–7. Figure 5 describes the division of the initial population into two separate subpopulations in the 2PGA scheme. In this illustrative example, the size of the initial population is only 14 and the elite and the plain populations to be formed contain 4 and 10 chromosomes, respectively. So, the elite population size, es, is 4. This division into subpopulations can be carried out directly after initialization, or alternatively, the initial population can be allowed to converge as a single population GA for some time. We call the point of division as the population division point, n_d.

In addition to the difference in subpopulation sizes, the characteristics of the populations also differ in terms of mutation probabilities; m_p, the mutation probability of the plain population is higher than the mutation probability of the elite population, m_e. If the initial population is not divided directly after initialization, mutation probability for the initial population is described by m. In both plain and elite populations an elitist mutation scheme

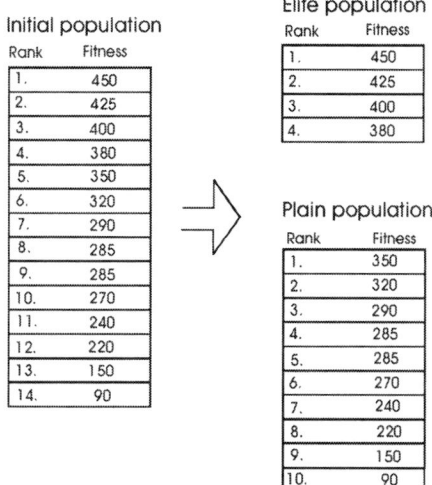

Fig. 5. Example of how the initial population is divided into two subpopulations

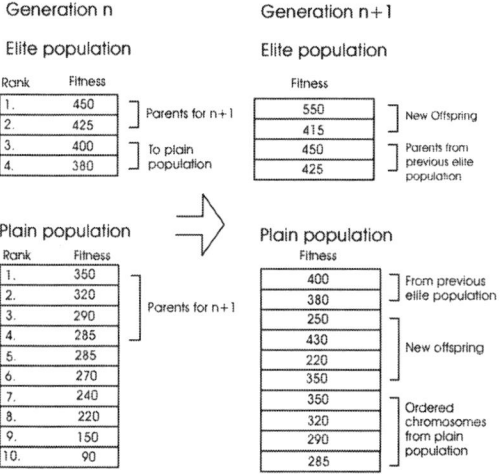

Fig. 6. Example of how the subpopulations of generation $n+1$ are formed from the subpopulations of generation n

is applied so that the best chromosome is never mutated. Thus, two solutions per generation are kept intact in terms of mutation, one for each population.

Figure 6 describes how the populations for the next generation are composed in 2PGA. Within both the elite and plain populations the best half of the chromosomes are selected as parents for the next generation. In Fig. 6 this means that the two best chromosomes from the elite population produce two offspring. The worst two chromosomes in the elite population are transferred

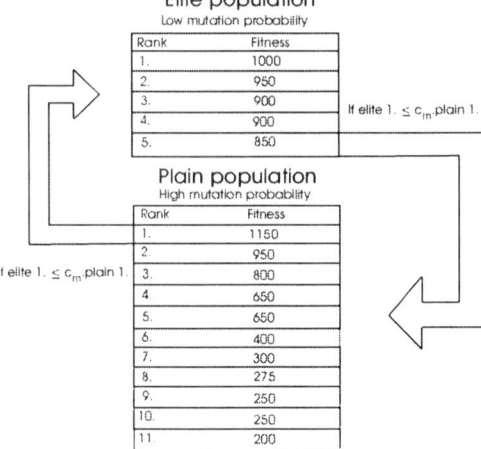

Fig. 7. Example of exchanging chromosomes between the two populations in the case the migration condition is fulfilled

to the plain population. Accordingly, in the plain population the four best chromosomes act as parents for four offspring.

Therefore, the elite population in the next generation is composed of two new offspring and two old parents from the previous generation. The plain population, then again, is composed of the two chromosomes left out from the elite population in the previous generation. In addition, there is also the new offspring of the previous plain population parent chromosomes. To fill the plain population up to the fixed number of chromosomes we add as many chromosomes from the previous plain population as there is space for. The rest of the chromosomes are discarded.

Figure 7 describes the process of how plain chromosomes can enter the elite population. Plain chromosomes need not have access to the elite population during every generation. i_m, the migration interval describes how often the chromosomes are allowed to enter from plain population to the elite population. The migration condition, c_m, describes the condition based on which the migration either does or does not take place. In Fig. 7 c_m equals one, meaning that the best chromosome in the plain population has to be better in terms of fitness value than the best chromosome in the elite population in order migration to take place. Indeed, 1,150 excels 1,000 and thus migration takes place. To keep the population sizes constant, the worst chromosome from the elite population, valued 850, is transferred to the plain population.

The proposed 2PGA scheme can be embedded on every population based evolutionary optimization scheme. It is not designed to compete with highly sophisticated and application specific optimization schemes, such as hybrid evolutionary algorithm methods equipped with gradient-based local search methods. Instead, the proposed scheme is intended to improve the performance

Table 6. The different parameters of the 2PGA scheme

Variable	Description
n_d	Population division point. Describes the point, after which the population is divided into separate subpopulations. Expressed as a percentage of the total number of generations evaluated before the division
i_m	Migration interval. Describes the interval between the points in which the best chromosome from the plain population can be transferred to elite population. Expressed as a number of generation between two migration points
es	Elite size. Describes the size of elite population in percentage of the total population size
c_m	Migration condition. Describes the limit that has to be superseded in order for a plain population chromosome to enter the elite population. Described as a value relative to the best chromosome in the elite population
m	Mutation probability. Describes the mutation probability before the division into separate subpopulations
m_p	Mutation probability. Describes the mutation probability of the plain population after population division
m_e	Mutation probability. Describes the mutation probability of the elite population after population division
f_{ave}	Average fitness
f_{std}	Fitness standard deviation
f_{scaled}	Scaled fitness value. Scaled value 1 corresponds to the average fitness value of the reference SPGA
c_{ave}	Average cost
c_{std}	Cost standard deviation
c_{scaled}	Scaled cost value. Scaled value 1 corresponds to the average cost value of the reference SPGA

of the basic evolutionary algorithm with minimal computational overhead, and thus any modification benefiting a standard evolutionary algorithm will also benefit the proposed scheme. The different parameters of the 2PGA scheme are explained in Table 6.

The implementation of the 2PGA scheme does not require parallel hardware, rather it is suitable for all platforms, both single processor as well as parallel environments. In the 2PGA scheme the two populations are different in size as well as in the mutation probability. The proposed scheme is similar to coevolution schemes, in which separate populations evolve towards a common goal. Also, the 2PGA scheme resembles niching since the individuals can only reproduce with individuals within the same population. However, these individuals can migrate between the populations and this does not require complex computations, as could be the case in niching methods implemented using crowding or fitness sharing.

Table 7. Default parameter values

n_d	i_m	es	c_m	m	m_p	m_e
10%	1	15%	100%	10%	20%	5%

Table 8. Results with different population division points

n_d (%)	f_{ave}	f_{std}	f_{scaled}	t-test
–	3,657	310	1	–
0	3,770	229	1.03	0.0421
10	3,772	231	1.03	0.0384
25	3,810	175	1.04	0.0032
50	3,741	256	1.02	0.1448

Table 9. Results with different elite population sizes

es (%)	f_{ave}	f_{std}	f_{scaled}	t-test
–	3,657	310	1	–
5	3,846	201	1.05	0.1270
25	3,781	214	1.03	0.0071
45	3,780	230	1.03	0.0029

Table 10. Results with different migration intervals

i_m (%)	f_{ave}	f_{std}	f_{scaled}	t-test
–	3,657	310	1	–
2	3,742	261	1.02	0.0005
5	3,802	202	1.04	0.0221
10	3,810	184	1.04	0.0267

Several different parameter values were experimented in the context of the 2PGA scheme, but only one was changed at a time. The default values used for the other parameters are shown in Table 7.

3.8 Result

Results from experimenting with different parameter values are shown in Tables 8–12. The t-test column tells the t-test value between reference value and the result using a specific parameter value. f_{scaled} describes the scaled fitness value, where value 1 corresponds to the reference value.

By examining the results, the following remarks can be made. The total population should be divided into subpopulations before the algorithm has

Table 11. Results with different migration conditions

c_m	f_{ave}	f_{std}	f_{scaled}	t-test
–	3,657	310	1	–
0.99	3,772	263	1.03	0.0487
0.95	3,856	152	1.05	0.0001
0.90	3,814	209	1.04	0.0039

Table 12. Results with different mutation probability schemes

m (%)	m_p (%)	m_e (%)	f_{ave}	f_{std}	f_{scaled}	t-test
40	–	–	3,657	310	1	–
40	40	40	3,637	339	1.00	0.7608
40	80	20	3,857	165	1.06	0.0001
40	20	20	3,534	320	0.97	0.0534

converged, thus eliminating the change of further improvement in fitness values. Experiments suggest that this division should take place before some 25% of the total number of generations is run. Intuitively, everybody cannot be elite, so the size of elite population should be limited to those individuals who posses the highest fitness or lowest cost. Elite size around 5% of the total population is favored by the simulation results.

Migration interval should be long enough for the populations to converge moderately, but not too much. Migration interval of 5–10% of the total number of generations has produced plausible solutions in our experiments. Migration condition sets the limits for the chromosomes from the plain population trying to enter the elite population. Results from the experiments point that the fitness of the plain population chromosome should roughly be at most 1.05 of the cost of the best chromosome in the elite population. This way, relatively moderate chromosomes cannot interfere the search of the local optimum.

Mutation probabilities are very application specific parameters, and only the ratio between the plain and elite population mutation probabilities can be given here. To stress the local search nature of the elite population the mutation probability should be kept rather small, whereas the individuals in the plain population should be more likely to undergo mutations while searching the whole solution space.

The comparison of SPGA and 2PGA results is somewhat problematic. Naturally, the number of function evaluations has to remain the same, as it does in this case. The number of mutations per run, however, rarely is exactly the same. This is likely the case when using for example adaptive mutation probabilities. It is generally not a problem, since the computational burden induced by a mutation operation is generally negligible to that of a solution evaluation. In terms of mutation, what our results show, is that when we find a mutation probability m giving us good results, we can get even better results using two populations, elite and plain, using mutation probabilities of $0.5m$ and $2m$, respectively.

It seems a feasible solution to pick the best parameters from the alternatives presented above, and see how such an algorithm performs. Unfortunately, the relations between the parameters are more complex than that, but picking up the best values for every parameter still produces competitive result. Thus, the interaction of different parameter values remains a research topic for the future.

The computational burden of 2PGA does not differ much from that of the reference SPGA, since both the algorithms use the same amount of crossovers and chromosome evaluations. The need for extra computation arises only during the initialization of the two populations and migration procedure later on. Tests show at maximum a 5% increase in computation time when using 2PGA scheme compared to that of a SPGA. However, the programming overhead is greater than programming a standard evolutionary algorithm.

The results generally show that with reasonable parameter values the 2PGA scheme is capable of producing competitive results compared to its single population counterpart. The different 2PGA parameters discussed in this section are naturally problem dependent but the values shown in Tables 8–12 can be used as a starting point for a search for application specific values. More information regarding the parameters of the 2PGA scheme can be found at [11].

4 Approximation of the Cost Function with Neural Networks

As discussed earlier, MGP-FIRs have been successfully designed using evolutionary computation. The optimization process, however, is still time-consuming, since the fitness of an individual is determined based on the results of applying the candidate filter to a set of test signals. This section introduces a method to speed up the GA-assisted MGP-FIR design process by means of neural networks and fitness function redefinition. Instead of three separate test signals, one for each frequency of 49, 50, and 51 Hz, to determine the fitness of an individual, only one of these test signals, the 50 Hz signal, is actually used and the calculated parameter values are fed to a neural network (NN) [12] to approximate the parameter values of the two other test signals. Also, the previously used fitness function is improved to better respond to the application's requirements.

Neural networks have been used before for fitness function calculations, for example, in evolving color recipes [13] and designing electric motors [14]. The results presented in this section suggest that using neural networks for aiding the fitness function calculations helps to create a competitive algorithm for MGP-FIR basis filter optimization.

4.1 The Reference Genetic Algorithm

The reference genetic algorithm used here is as described in Sect. 3.1. However, there are a few differences: the use of roulette wheel selection instead of pure rank based selection and the redefinition of the fitness function. The new fitness function is expressed as

$$fitness = \frac{1 \cdot 10^6}{a \cdot (ITAE49 + ITAE50 + ITAE51)} \tag{14}$$

where

$$a = \max(h49/10 + NG49, h50/10 + NG50, h51/10 + NG51) \tag{15}$$

The additional *h49*, *h50*, and *h51* terms in (15) stand for the amplification of the third harmonic frequency. Clearly, the addition of this term aims at reducing the amplitude of the third harmonic frequency.

4.2 Neural Network-Assisted Genetic Algorithm

Evaluating a fitness function shown in (14) is time-consuming and in order to speed up the fitness calculations the fitness function can be rewritten as:

$$fitness = \frac{1 \cdot 10^6}{b \cdot \left(IT\hat{A}E49 + ITAE50 + IT\hat{A}E51 \right)}. \tag{16}$$

The *ITAE* term is now calculated only for the 50 Hz test signal and this value is used to approximate the *ITAE* values for 49 and 51 Hz signals using a neural network. This way, we need to filter only one third of the test signals available, namely the 50 Hz test signal. Moreover, *b* is expressed as

$$b = \max(h49/10 + N\hat{G}49, h50/10 + NG50, h51/10 + N\hat{G}51). \tag{17}$$

In *b*, the third harmonic gains are calculated for all the three test frequencies. However, only the noise gain for the 50 Hz test signal needs to be calculated, since noise gains for 49 and 51 Hz test signals were calculated using g_1 and g_2 approximated by a multilayer perceptron network. This neural network consisted of a single hidden layer and five hidden neurons. Figure 8 shows a box plot of the neural network's performance using different number of hidden neurons. Clearly, by using five hidden neurons the median value (marked by a horizontal line) as well as the variance of the results were better than with other number of hidden neurons. These results were calculated using averages of 25 runs. In Fig. 8 + denotes an extreme value. The bottom and top lines of the box correspond to the 25th and the 75th percentiles, respectively.

As inputs the network takes g_1, g_2, and *ITAE* after the 50 Hz test signal has been processed. The outputs of the network are the approximated values

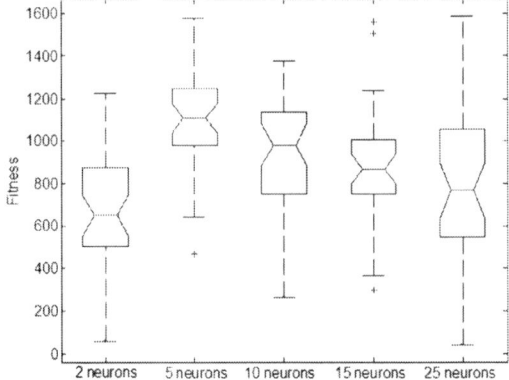

Fig. 8. The effect of the number of hidden neurons on the NN-assisted GA performance

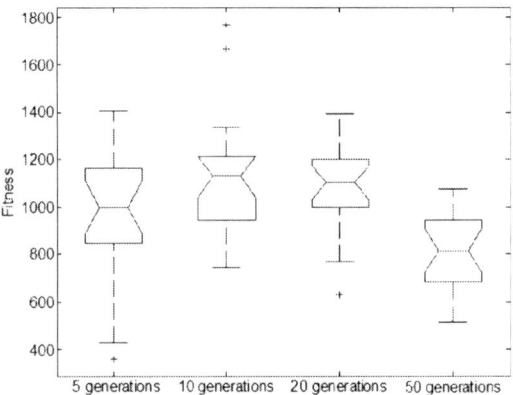

Fig. 9. The effect of the transition point on the NN-assisted GA performance

of g_1, g_2, and *ITAE* for the 49 and 51 Hz test signals. Thus, the neural network approach aims at reducing the computational time required to evaluate individual's fitness by a theoretical two thirds.

Both the standard GA as well as the neural network enhanced GA operate equally up to the 10th generation. By this time, the NN-GA has collected five training and three validation samples per generation, i.e., 50 and 30 individuals, respectively. Early stopping rule is used in training of the neural network [12]. Figure 9 shows the effect of transition point, i.e., the point after which the fitness function is calculated using the assistance of the neural network.

The task of choosing the transition point is a trade-off between accuracy and computing time: the longer we collect the training and validation data the more likely the network is to produce accurate results. However, the sooner the neural network assisted fitness calculation is implemented the more generations the GA is capable of going through during the rest of the given time.

Table 13. Average number of generations evaluated during a single 300-s run using a different transition point

Transition point (generations)	Number of generations
5	122
10	112
20	96
50	68
Reference GA (no NN involved)	66

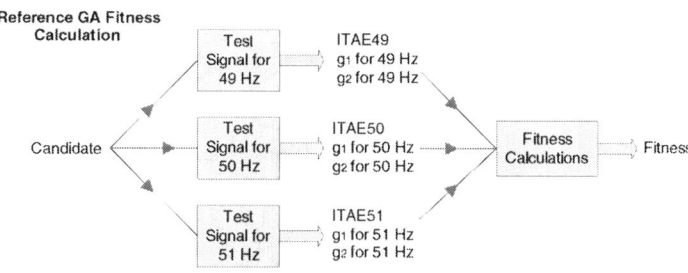

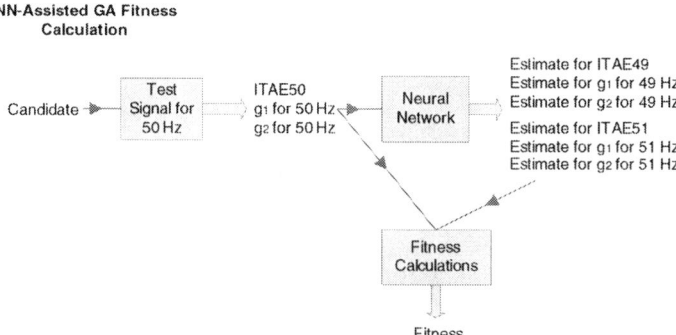

Fig. 10. The principles of the reference GA and the NN-assisted GA fitness calculations

Eventually, 10 generations was chosen to be the transition point. Although the network performs similarly using 10 and 20 generations as the transition point, 10 generations was chosen because the algorithm with this parameter value is capable of evaluating a larger number of generations than using 20 as the transition point. Table 13 summarizes the average number of generations evaluated by the algorithms using different transition point.

Figure 10 illustrates the principles of the fitness calculations of the reference GA and the NN-assisted GA.

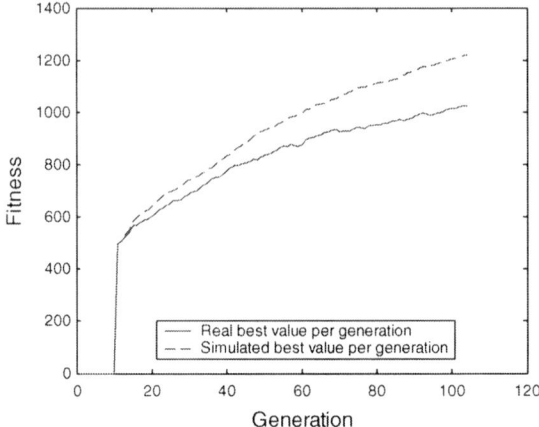

Fig. 11. NN-assisted and real fitness values per generation

4.3 Approximation Capabilities of the Neural Network

Using neural network to model different components of the fitness function is a trade off between speed and accuracy. When calculating the fitness using neural network, we are not concerned how the network eventually maps the true fitness values as long as the fitness-based order of the candidate solutions remains close to the true order. To tackle the inaccuracy of the fitness order based on the simulated values, a roulette wheel selection was used. This kind of selection scheme enables also less-fit individuals to be chosen for the next generation. In theory, it is possible that a low-level simulated fitness would actually be a high-level true-value fitness. In the following the outputs of the neural networks are compared to the true values. The results are calculated based on the averages of 100 runs.

Figure 11 shows the NN-assisted and real fitness values per generation. The simulated value follows closely the real value at the beginning, but eventually the difference increases. This is likely caused by the fact that due to the evolution process the parameter values enter such regions that were not included in the original training set and thus it is difficult for the NN to approximate the rest of the parameter values precisely.

Figures 12–15 show the real and simulation results of the MGP values, i.e., g_1 and g_2, for 49 and 51 Hz signals. Similarly to the overall fitness per generation, the simulated and real MGP values are close to each other in the early generations of the run but separate later on. Again, this can be due to the incapability of the original training set to accurately present the whole parameter space confronted during the optimization process.

Figures 16 and 17 present real and simulated *ITAE* parameters for 49 and 50 Hz signals. The NN seems to approximate the *ITAE* value for 49 Hz signal

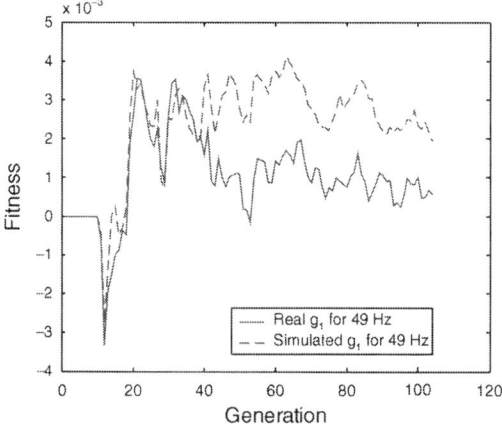

Fig. 12. NN-assisted and real values for g_1 at 49 Hz

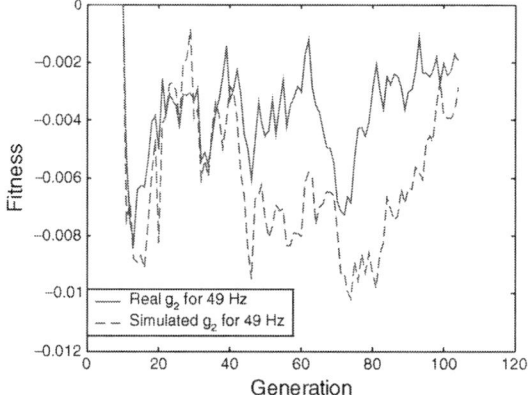

Fig. 13. NN-assisted and real values for g_2 at 49 Hz

well, whereas for the 51 Hz the real and simulated seem to diverge towards the end.

Obviously, based on the results, the approximation accuracy of the neural network decreases as the evolution proceeds. To cope with this problem, the training of the network several times during the evolution with new training sets was experimented. These experiments produced results quite similar to that of the NN-assisted GA trained only once and no dramatic improvement in the performance was observed.

Also, the components of the fitness function were approximated using separate neural networks for MGPs for 49 Hz, MGPs for 51 Hz and the *ITAE* parameters. Using these separate networks for different components of the fitness function produced poor results. However, embedding all the components to the same network seems to bind the approximated values together

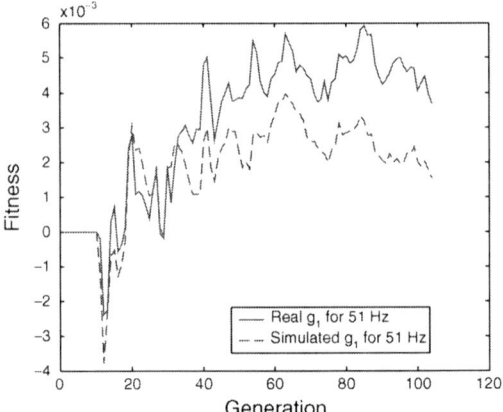

Fig. 14. NN-assisted and real values for g_1 at 51 Hz

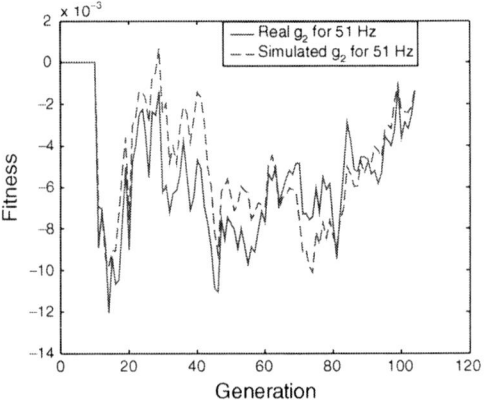

Fig. 15. NN-assisted and real values for g_2 at 51 Hz

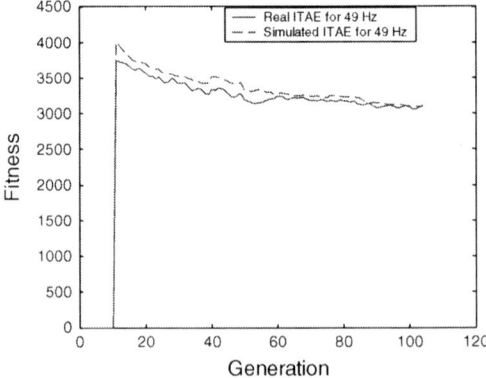

Fig. 16. NN-assisted and real values for $ITAE$ at 49 Hz

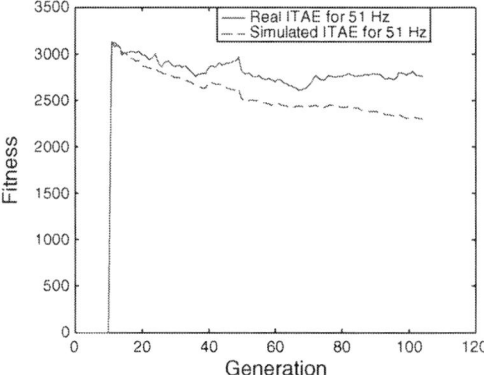

Fig. 17. NN-assisted and real values for $ITAE$ at 51 Hz

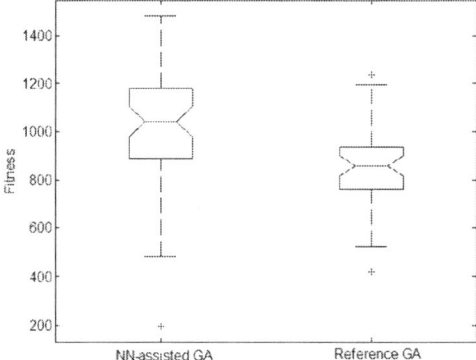

Fig. 18. Box plots for the NN-assisted GA and the reference GA for 50 individual 300-s runs

so that no large approximation errors occur. Approximating the parameter values individually using single NN for each could produce better accuracy, but the advantage is lost in more time-consuming calculations.

4.4 Results

Figures 18 and 19 show the box plots for the averages of 50 individual 300 and 600-s runs. It is clearly visible that the median values are higher in the NN-assisted GA than when using the reference GA. The results of the algorithms should be subjected to a more thorough statistical inspection like the scheme including multiple hypothesis testing and bootstrap resampling [15] to get a more reliable evaluation of the differences between the two algorithms. This kind of scheme, however, requires a lot of data to be collected and in this case a computational time of weeks and it is thus not feasible.

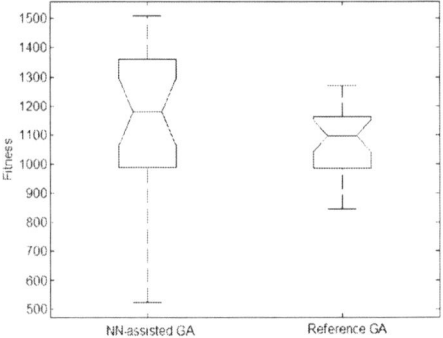

Fig. 19. Box plots for the NN-assisted GA and the reference GA for 50 600-s runs

Table 14. Average THD for a 300-s run

Harmonic	Amplitude	NN-GA	Reference GA
1st	1	1	1
3rd	0.1	0.055	0.047
5th	0.1	0.031	0.054
7th	0.1	0.018	0.021
9th	0.1	0.009	0.024
11th	0.1	0.012	0.016
13th	0.1	0.017	0.015
15th	0.1	0.021	0.015
THD %	26.46	7.26	8.28

Table 15. Average THD for a 600-s run

Harmonic	Amplitude	NN-GA	Reference GA
1st	1	1	1
3rd	0.1	0.059	0.028
5th	0.1	0.030	0.0030
7th	0.1	0.018	0.025
9th	0.1	0.016	0.023
11th	0.1	0.010	0.016
13th	0.1	0.008	0.015
15th	0.1	0.011	0.034
THD %	26.46	7.24	5.99

These MGP filters are intended for suppressing the harmonics in the input signal. Tables 14 and 15 show the performance of filters with median fitness values produced by the different algorithms. In Table 14, THDs are given for filters that after 100 individual 300-s runs have the median fitness values of 1,043 and 858 for the NN-assisted GA and the reference GA, respectively.

In Table 15 THDs are given for filters that after 100 individual 600-s runs have the median fitness values of 1,178 and 1,097 for the NN-assisted GA and the reference GA, respectively.

All the featured filters in Tables 14 and 15 are capable of reducing the THD value of the test signal considerably. In Table 15, the THD value of the reference GA is lower than that of the NN-assisted GA although the fitness value of the former is lower. This is caused by the fact that the THD value is actually not part of the fitness function (17), rather the fitness function consists of other related components, such as noise gain and amplitude of the third harmonic component.

5 Conclusions

This chapter has discussed the design process of MGP-FIR filters with different soft computing techniques. Due to the discrete nature of the filter coefficient set, evolutionary computation methods are the only practical ways to design such filters. A simple genetic algorithm can produce good design performance in most cases but using adaptive parameter values, seeding techniques, and multiple hierarchical solution populations can help to improve the design process. Regardless of how good the optimization algorithm is, a time-consuming objective function evaluation can degrade the performance of the optimization algorithm. This Chapter has also shown a way to approximate parts of an exhaustive objective function with the help of neural networks and thus to speed up the design process. So far, the presented methods have been studied separately, but future research topics will include the fusion of all the presented methods into a single optimization scheme.

The presented filter design problem is a challenging real-word problem, and it clearly shows the need for robust evolutionary optimization schemes. Simple evolutionary algorithms can do a lot, but when the aim is to achieve superior design or operating performance the user can find it useful to experiment with highly tailored and application specific approaches based on different soft computing techniques.

References

1. Vainio O, Ovaska SJ, Pöllä M (2003) Adaptive filtering using multiplicative general parameters for zero-crossing detection. IEEE Transactions on Industrial Electronics 50(6): 1340–1342
2. Ovaska SJ, Vainio O (2003) Evolutionary programming in the design of adaptive filters for power systems harmonies reduction. In: Proceedings of the IEEE International Conference on Systems, Man and Cybernetics 5: 4760–4766
3. Bäck T (1996) Evolutionary Algorithms in Theory and Practice. Oxford University Press, New York

4. Haupt RL, Haupt SE (1998) Practical Genetic Algorithms. Wiley, New York, NY
5. Fogel DB (2000) Evolutionary Computation: Toward a New Philosophy of Machine Intelligence. IEEE Press, Piscataway, NJ
6. Wade G, Roberts A, Williams G (1994) Multiplier-less FIR filter design using a genetic algorithm. In: IEE Proceedings of Vision, Image and Signal Processing 141(3): 175–180
7. Lee A, Ahmadi M, Jullien GA, Miller WC, Lashkari RS (1998) Digital filter design using genetic algorithm. In: Proceedings of the IEEE Symposium on Advances in Digital Filtering and Signal Processing, pp. 34–38
8. Wolpert DH, Macready WG (1997) No free lunch theorems for optimization. IEEE Transactions on Evolutionary Computation 1(1): 67–82
9. Srinivas M, Patnaik LM (1994) Adaptive probabilities of crossover and mutation in genetic algorithms. IEEE Transactions on Systems, Man and Cybernetics 24(4): 656–667
10. Gordon VS, Whitley D (1993) Serial and parallel genetic algorithms as function optimizers. In Proceedings of the Fifth International Conference on Genetic Algorithms, pp. 177–183
11. Martikainen J, Ovaska SJ (2006) Hierarchical two-population genetic algorithm. International Journal of Computational Intelligence Research 2(4): 367–381
12. Haykin S (1998) Neural Networks: A Comprehensive Foundation. 2nd edition. Prentice Hall, Upper Saddle River, NJ
13. Mizutani E, Takagi H, Auslander DM, Jang JSR (2000) Evolving color recipes. IEEE Transactions on Systems, Man, and Cybernetics, Part C 30(4): 537–550
14. Dongjin B, Dowan K, Hyun-Kyo J, Song-Yop H, Chang SK (1997) Determination of induction motor parameters by using neural network based on FEM results. IEEE Transactions on Magnetics 33(2): 1924–1927
15. Shilane D, Martikainen J, Dudoit S, Ovaska SJ (2006) A general framework for statistical performance comparison of evolutionary computation algorithms. In: Proceedings of the IASTED International Conference on Artificial Intelligence and Applications, pp. 7–12

Applying Genetic Algorithms to Optimize the Cost of Multiple Sourcing Supply Chain Systems – An Industry Case Study

Kesheng Wang[1] and Y. Wang[2]

[1] Knowledge Discovery Laboratory, Department of Production and Quality Engineering, Norwegian University of Science and Technology, N-7491 Trondheim, Norway
[2] Nottingham Business School, Nottingham Trent University, Nottingham NG1 4BU, UK

Summary. This chapter is to present a successful industry case applying Genetic Algorithms (GAs). The case has applied GAs for the purpose of optimizing the total cost of a multiple sourcing supply chain system. The system is characterized by a multiple sourcing model with stochastic demand. A mathematical model is adopted to describe the stochastic inventory with the many-to-many demand-supplier network problem and it simultaneously constitutes the inventory control and transportation parameters as well as price uncertainty factors. Genetic Algorithms are applied to derive optimal solutions through the optimization process. A numerical example and its solution demonstrate the detail solution procedure based on Genetic Algorithms and shows that GAs are able to solve such a difficult problem efficiently and easily. Further research is also discussed.

1 Introduction

An abundance of optimization approaches have been applied to solve engineering optimization problems [16]. The algorithms applied are either heuristics or exact procedures based mainly on modifications of dynamic programming and nonlinear programming. Most of these approaches are strongly problem oriented. This means that since they are designed for solving one type of optimization problems, they cannot be easily adapted for solving other problems. In recent years, many studies on optimization problems in the domain of supply chain management apply a universal optimization method based on meta-heuristics (Chi et al., 2007) [2, 8, 9, 14, 15]. These meta-heuristics can be easily adapted to solve a wide range of optimization problems since they hardly depend on the specific nature of the problems being solved. The meta-heuristics are based on artificial reasoning rather than on classical mathematical programming. Their important advantage is that they do not require

K. Wang and Y. Wang: *Applying Genetic Algorithms to Optimize the Cost of Multiple Sourcing Supply Chain Systems – An Industry Case Study*, Studies in Computational Intelligence (SCI) **92**, 355–372 (2008)
`www.springerlink.com`

any information about objective function besides its values corresponding to the points visited in the solution space. All meta-heuristics use the idea of randomness, which is controlled in most cases, when performing a search, but they also use past knowledge in order to direct the search. Such search techniques are known as randomized searches [3].

Genetic Algorithms (GAs) are one of the most widely used meta-heuristics. They were inspired by optimization procedure that exists in nature, the biological phenomenon of evolution. GAs maintain a population of different solution allowing them to mate, produce offspring, mutate, and fight for survival. The principle of survival of the fittest ensures the populations derive towards optimization.

Supply chain management is an integrative approach for planning and control of materials and information flows with suppliers and customers as well as between different functions within a company. This bridges the inventory management focusing on operations management and the analysis of relationships from industrial organization. Nevertheless, most quantitative analysis on supply chain management issues is dominated by the framework of multi-stage serial system or distribution system where relationships between single sourcing are considered. The situation of multiple sourcing supply chain system has received less attention.

Genetic Algorithms (GAs) method has been successfully used to find the optimal solutions for the multiple sourcing supply chain systems. Kim et al. [10] considered a supply chain network consisting of a manufacturer and its suppliers. The case was presented as a multiple-item single period problem within a many-to-one multiple sourcing system. The model gave an optimal solution method for supplier selection. Owing to the nature of the problem described, inventories at the supply source, as well as the distribution cost, were not considered. Ganeshan et al. [4] studied a continuous review, one warehouse, and multiple retailer distribution system. The multiple-supplier aspect was incorporated by splitting the warehouse order into n equal portions in order to reduce effective lead times. Minner et al. [13] considered a periodic review system where, as an alternative to rationing a depot shortage, outstanding orders could be speeded up with a certain probability. Ghodsypour and O'Brien [5] developed a mixed integer non-linear programming model to solve the multiple sourcing problems, which took into account the total cost of logistics, including net price, shortage, transportation and ordering cost. Yokoyama [19] developed a model for a multiple sourcing inventory and distribution system. His model focused on finding the target inventory and transportation quantity that minimizing the total cost of the system by using a random local search method combined with GAs. The model was constructed to provide the advantage in total costs saving to the suppliers. Altiparmak et al. [1] proposed a new solution procedure based on GAs to find the set of Pareto-optimal solutions for multiple-sourcing and multiple-objective supply chain network design problem and a mixed-integer non-linear programming model for the optimization of supply chain network was presented.

In this chapter, a mathematical model is adopted from Han and Damrongwongsiri [6] and it describes the stochastic multiple-period inventory with the many-to-many demand-supplier network problem. Genetic Algorithms are applied to derive optimal solutions through the optimization process. This approach is implemented to solve a real industry case. Our aim is to find optimal inventory level for warehouses and optimal distribution plan concerning various markets' situations. The objective is to arrive at the minimum total supply chain cost. The chapter is organized as follows. Section 2 describes a real industrial case within a multiple sourcing supply chain system and presents a stochastic non-linear programming model for the optimization of the supply chain system. Section 3 indicates the operational procedure of GAs combining the optimization problem considered. Results of the industrial case study, which are obtained by GeneHunter software, are presented in Sect. 4. Some conclusions and suggestions about further research are given in Sect. 5.

2 Problem Description

The problem considered in this chapter is derived from an industry case. The company produces medical products which are to supply the Nordic region markets. The products concerned here are discrete standard products, which are easily attained in traditional warehouses. After the products leave factories, they will be stored at different warehouses. According to the requirement of each market, the products will be further transported form the warehouses to the markets. We assume that there are three warehouses which are located in Falun and Jonkoping (Sweden), and Rovaniemi (Finland). These warehouses will be the supply sources. The main markets in Norway are located in Narvik, Trondheim, Bergen and Oslo respectively, seeing Fig. 1. Each line

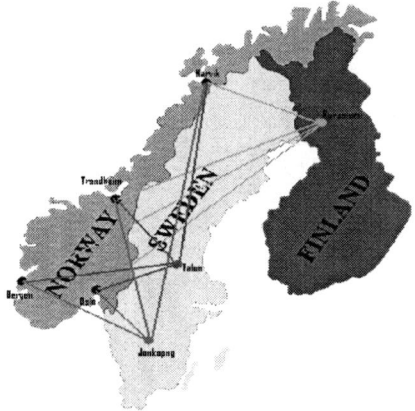

Fig. 1. Warehouses and markets location

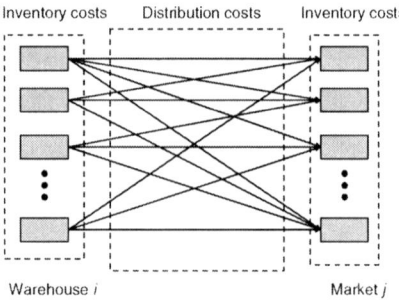

Inventory costs   Distribution costs   Inventory costs

Warehouse i Market j

Fig. 2. A multiple sourcing supply chain system model

connects warehouses and markets through different distribution routs. The goal of the company is to determine the target inventories and the allocation quantities from each warehouse to each market in order to minimize the expected inventory and distribution cost.

Let us derive a generic model of multiple sourcing supply chain networks shown in Fig. 2. The generic model can be simplified to model the industrial case. The problem belongs to single-product and multiple sourcing supply chain optimizations, which may be composed by the following three components:

- Inventory levels and cost at the warehouses.
- Inventory levels and cost at each market.
- Allocation of warehouse-to-market distribution quantities and cost in a certain period.

The optimal performance of the inventory-distribution system is directly related to the solutions from those three components. The supply chain optimization problem can be formulated as a stochastic non-linear mathematical model. The optimization objective functions and some relative constraints are expressed as follows:

$$F = \min \left\{ \sum_{i \in I} \sum_{j \in J} \left[[a_{ij}(t) + b_i(t)] \, x_{ij}(t) + \left[\frac{x_{ij}(t)}{Q_i} \right] VC_{ij}(t) \right] \right. \tag{1}$$

$$\left. + \sum_{t \in T} \left[\sum_{i \in I} {}_wINV_i(t, S_i) + \sum_{i \in I} {}_mINV_j(t) \right] \right\} \tag{2}$$

Subject to :
$$\sum_{i \in I} x_{ij}(t) \leq S_i, \quad \forall i \in I, t \in T \tag{3}$$

$$\sum_{j \in J} x_{ij}(t) \leq M_j, \qquad \forall j \in J, t \in T \tag{4}$$

$$x_{ij}(t) \geq 0, \quad S_i \geq 0 \tag{5}$$

The variables in the objective function can be indicated as follows:

1. *Indices*: i is an index for warehouses ($i \in I$). j is an index for markets ($j \in J$).
2. *Model variables*: $x_{ij}(t)$ is the quantity of product shipped from warehouse i to market j. S_i represents order level at warehouse i, and it is the controllable decision variable of the inventory policy. M_j is the capacity of market j. $_wINV_i$ and $_mINV_j$ represent the inventory cost of warehouse i and the inventory cost of market j, respectively.
3. *Model parameters*: $a_{ij}(t)$ is the unit shipping cost form each warehouse to each market. $b_i(t)$ is the unit price of the item at warehouse i during period t. The unit shipping cost and unit price are assumed to be constant within a finite period t. In addition, each warehouse specifies its cargo capacity Q_i, which describes the maximum number of products that can be fitted into one cargo. $VC_{ij}(t)$ is the distribution cost per cargo from warehouse i to market j. Since we use a synchronized scheduling period t for all warehouses and all markets, the warehouses and the markets will be replenished at the same scheduled time.

2.1 Warehouse Inventory Control Model

A control policy will be used here to indicate the decision-making process. In an inventory control policy, the inventory will be replenished according to the inventory order level S_i in a certain order-period t. The objective is to find optimal order level of each warehouse, which minimizes the total inventory costs at the warehouses. The model assumes an instantaneous supply pattern, in which the warehouse inventory is replenished at the beginning of the period, and then immediately available to the markets. With those conditions, the inventory cost at the warehouses of (2) becomes:

$$_wINV_i(t, S_i) = {}_wC_{ci} \int_0^{S_i} (S_i - {}_wd_i)f({}_wd_i, t)d_wd_i \qquad (6)$$

$$+ {}_wC_{si} \int_{S_i}^{\infty} ({}_wd_i - S_i)f({}_wd_i, t)d_wd_i + {}_wC_{ri}$$

The warehouse inventory cost is composed of three parts: (1). carrying cost, (2) shortage cost and (3) replenishment cost. Warehouse inventory leads to the carrying cost, which is caused by the difference between the ordered amount and the actual demand (first item). Unsatisfied demands lead to shortage costs generated in the warehouse (second item). The third term represents the replenishment cost, which occurs in each order-period. $_wC_{ci}$, $_wC_{si}$, and $_wC_{ri}$ are carrying cost per quantity unit, shortage cost per unit and replenishment cost for each replenishment of the warehouse. $f({}_wd_i, t)$ is the continuous probability density of the demand $_wd_i$ and the demand $_wd_i$ in each order-period could be represented by the probability distribution, such as uniform distribution (7), normal distribution or exponential distribution.

2.2 Market Inventory Control Model Combining with Distribution Model

The market inventory model is a probabilistic system, in which the demand of each market in the period t is uncertain. Here, the reasonable inventory level at each market is examined in order to minimize total cost of all markets, which are the results from the number of units shipped from warehouse i to market j, $x_{ij}(t)$. The variable amount of products distributed will yield a surplus or shortage at the market, which leads to the variable cost at each market. Therefore, to find optimal amounts of products distributed is the task of these two models. The inventory cost at the markets of (2) becomes:

$$
_mINV_j(t) = {}_mC_{cj} \int_0^{\sum_{i\in I} x_{ij}(t)} \left(\sum_{i\in I} x_{ij}(t) - \frac{_md_j}{2}\right) f(_md_j, t) d_m d_j
$$

$$
+ {}_mC_{cj} \int_{\sum_{i\in I} x_{ij}(t)}^{\infty} \frac{\left(\sum_{i\in I} x_{ij}(t)\right)^2}{2_md_j} f(_md_j, t) d_m d_j \tag{7}
$$

$$
+ {}_mC_{sj} \int_{\sum_{i\in I} x_{ij}(t)}^{\infty} \frac{\left(_md_j - \sum_{i\in I} x_{ij}(t)\right)^2}{2_md_j} f(_md_j, t) d_m d_j + {}_mC_{rj}
$$

It is assumed that the market inventory will be replenished at the beginning of the period. The first and the second terms present the average carrying cost. Such costs occur when the distributed products' volumes are more than the demand of the market. The third term states the shortage cost, when the market demand can not be satisfied by the inventory volumes. The last term represents the replenishment cost, which occurs in each distribution-period. $_mC_{cj}$, $_mC_{sj}$ and $_mC_{rj}$ are carrying cost per quantity unit, shortage cost per unit and replenishment cost for each replenishment of the market j, respectively. $f(_wd_i, t)$ is the continuous probability density of the customer demand received at the market j for the scheduling period t.

The distribution cost model has been presented in (1). The cost is composed of shipping cost of the products, the products cost and shipping cost of cargoes. The values of model parameters will be different in each warehouse due to the difference locations of the warehouses.

3 Genetic Algorithms

3.1 GA Procedures

The concept of GAs was developed by Holland and his colleagues in the 1960s and 1970s [7]. A GA is a stochastic optimization method based on the mechanisms of natural selection and evolution. In GAs, searches are performed based on a population of chromosomes representing solutions to the problem. A population starts from random values and then evolves through succeeding generations. During each generation a new population is generated by propagating a good solution to replace a bad one and by combining or mutating existing solutions to construct new solutions. GAs have been theoretically and empirically proven robust for identifying solutions to combinatorial optimization problems [17]. This is due to GAs ability to conduct an efficient parallel exploration of the search space, while only requiring minimum information on the function to be optimized.

Generally, the formulation of Genetic Algorithms can be summarized by the following steps:

1. The development of a fitness function,
2. The representation of the problem,
3. The selection of new generation, and
4. The GAs operation.

In this section, representation, selection and genetic operations which were used in GAs for multiple sourcing supply chain optimizations will be focused. A detailed description of a fitness function will be described in Sect. 4.

3.2 Representation

Representation is one of the important issues that affect the performance of GAs. It happens at the beginning and ending phases of GAs operation. The encoding and decoding mechanism is to convert the parameter value of a possible function into binary strings which represents chromosomes, and vice versa [17]. Here the fixed-length binary strings to represent chromosomes are used. The two kinds of decision variables (S_i and x_{ij}) are represented by binary individuals as the following:

Representation 1

The decision variable in warehouse inventory control model is the inventory order level, S_i. To set up the GA searching process properly the decision variable, S_i, is encoded into a binary string as a chromosome, g_i. The length of the bit stream of a chromosome, g_i, can be determined by the range of the inventory order level at warehouse i, S_i. The required bits of a variable are calculated in the following expression. n_i is the number of the required bits of variable S_i. The domain of variable S_i is $[p_i, q_i]$ and the required precision is "λ decimal point". The precision requirement implies that the range of domain of each variable should be divided into at least $(q_i - p_i) \times 10^\lambda$ size ranges. In the calculation it is supposed that the precision is set as no decimal point ($\lambda = 0$) and the number n can be obtained using the following equation:

$$2^{n_i-1} < (q_i - p_i) \times 10^\lambda \leq 2^{n_i} - 1$$

For example, the range of a variable is between 240,001 and 0. After calculating, n is equal to 18. The total length of a chromosome is 18 bits, which can be represented as follows:

$$g_i : \quad 000101000101001110$$

In addition, the initial population might be generated randomly with the population size k:

$$g_1 : \quad 001011010110101010$$
$$g_2 : \quad 000110101100101010$$

$$\dots$$

$$g_k : \quad 011010010100010101$$

To calculate fitness value, we need to decode the binary string to the real number. The mapping from a binary string to a real number for variable S_i, is calculated by using the equation as follows:

$$S_i = p_i + decimal(substring_i) \times \frac{q_i - p_i}{2^{n_i} - 1}$$

where $decimal(substring)$ is the decimal value of substring for decision variable S_i.

Representation 2

The decision variable in market inventory control model and distribution model is the distribution quantities from warehouse i to market j, x_{ij}. x_{ij} can be represented as the binary string, e.g. $x_{12} \rightarrow 001100100011101001$. The total cost consists of inventory cost form all markets and distribution cost from each distribution line. Therefore, we need represent all of the chromosomes string (the binary string, x_{ij}) into one string (individual) shown in Fig. 3.

An example individual of X_k is:

001100100011101001 . . .001100100110101000 . . .000101101010101001.

The population size k can be set by a user, e.g. $k = 50$ in this case. Each individual can be identified to represent the total number of units shipped form warehouses to markets. After finding the best individual, the binary strings are divided into the chromosomes according to the numbers of distribution lines and then they can be decoded. The decoded result will be the optimal distribution quantities of each distribution line.

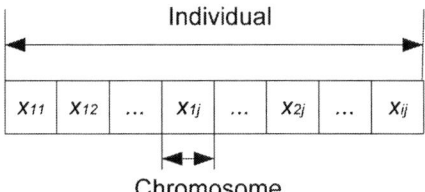

Fig. 3. The structure of the individual X_k

3.3 GA Operators

Crossover

Crossover is the most essential operator in GAs and used to combine multiple parents and create offspring. The crossover points are randomly determined and set in each corresponding chromosome of the selected parents. The chromosomes are split at their crossover point, and further used to interchange and combine to form the child chromosome [18]. The crossover operator tends to preserve good gene segments of both parents. Parent 1 and parent 2 are selected from mating pool for doing crossover operation. Crossover point is defined for these two chromosomes and can be generated randomly. One-point crossover strategy was used in the case study. Offspring is then generated by concatenating the left sub-chromosome of one parent and right sub-chromosome of other parent. The crossover rate P_c may be selected between 0.7 and 1.0.

Mutation

Mutation is used to avoid becoming trapped at a local area by exchanging bit of the chromosome. Mutation works by randomly choosing a gene and changing it from 0 to 1 or from 1 to 0 with very small probability, P_m, (usually between 0.01 and 0.1%) [20]. In the "current" population after the crossover operation, it generates a random number r. If $r < P_m$, then the bit is mutated.

3.4 Selection Methods

After GA operations, the problem is how to perform selection, that is, how to chose the individuals in the next generation population pool. The purpose of selection is to find their offspring that have even better fitness. Selection must be balanced with variation from crossover and mutation: too strong selection may result in convergence to a local optimal; too weak selection will result in too slow evolution. Numerous selections methods have been proposed in GA literature, such as Fitness-proportionate selection with "Roulette wheel" and "Stochastic universal" sampling, Sigma scaling, Elitist strategy, Rank selection, Tournament selection, Steady-state selection and Boltzmann selection. In this chapter, Elitist strategy selection method is used [11]. It is an addition to many selection methods that forces the GAs to retain some number of the best individuals to each generation. Such individuals can be lost if they are not selected to reproduce or if they are destroyed by crossover or mutation. Many papers have shown that Elitist strategy method can significantly improve the GA performance.

4 Numerical Simulation

As we mentioned above, three warehouses locate in Falun and Jonkoping, Sweden, and Rovaniemi, Finland, and the main markets in Norway are located in Narvik, Trondheim, Bergen and Oslo respectively. To simplify the representations of the warehouses and markets, we use some codes to replace them. W_1, W_2 and W_3 represent three warehouses in Falun, Jonkoping and Rovaniemi, M_1, M_2, M_3 and M_4 represent four markets in Narvik, Trondheim, Bergen and Oslo, respectively. Since the capacity of each warehouse is relatively stable, the allocation from one source that offers the lowest price to market region might not be feasible. Therefore, this study will explore and reveal the strategy to achieve the minimal cost of the entire supply chain.

4.1 GeneHunter [17]

One of the optimization tools that recently appeared in the market is Gene-Hunter from Ward Systems. GeneHunter is a genetic algorithm product designed to solve optimization problems, for example, finding the best schedules, financial indicators, mixes, model variables, locations, parameter settings and portfolios, etc. This software is more powerful than traditional optimization ones. It includes both an Excel spreadsheet add-in and a programmer's tool kit. The Microsoft® Excel add-in allows the user to solve an optimization problem from an Excel spreadsheet. The programmer's tool kit, on the other hand, has a Dynamic Link Library (DLL) of GA functions that can be called from programming languages such as Microsoft® Visual Basic or C++. Among other optimization problems, GeneHunter can be used to solve problems in the following areas: Financial optimization, Resource allocation, Scheduling, Teaching method optimization, Genetic algorithm studies.

To solve the problem, GeneHunter was used in the project. With the Excel spreadsheet it is possible to set a lot of parameters in the Genetic Algorithms operation, but the default option is applicable in most cases. Figure 4 shows the options window in GeneHunter with the specification in solving the problem.

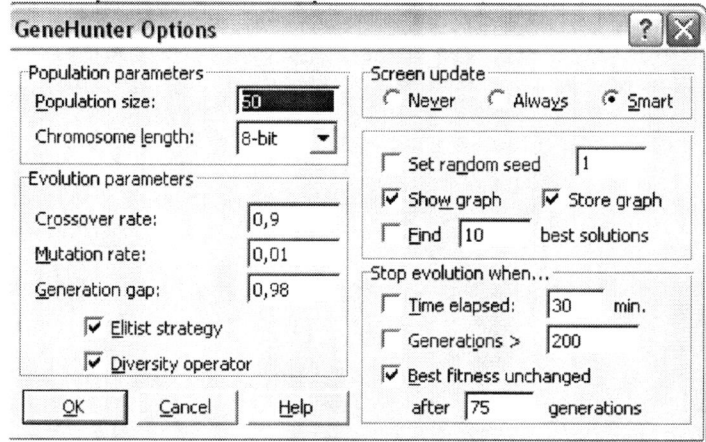

Fig. 4. GeneHunter Options window

Table 1. Warehouse inventory input parameters

Inventory	Product Price ($)	Unit Carrying Cost ($)	Unit Shortage Cost ($)	Replenishing Cost Per Period ($)	Demand	Standard Deviation
W1	50.00	15.00	30.00	1000.00	100000	16667
W2	55.00	16.50	33.00	1200.00	80000	13334
W3	45.00	13.50	27.00	800.00	120000	20000

The population size is 50, the crossover rate is 0.9 and the mutation rate is 0.01. The selection approach is to use the elitist strategy. The GAs will stop evolution when the best fitness unchanged after 75 generations.

4.2 Warehouse Inventory Control System

In this inventory stage, the objective function in (6) can be directly used as a fitness function in a GA. The demand history of all warehouses is given. According to some basic forecasting method, the initial input for the model is obtained. The demand at each warehouse is generated randomly following a normal distribution with the mean and standard deviation. The input parameters are shown in Table 1. The planning horizon $T = 1$, and the order level S_i should be continuous integer values. The optimal order level for each warehouse and the relative cost of warehouse W1 are shown in Table 2. The optimal order levels for the warehouses are $S_1 = 109,105$ (W1), $S_2 = 89,138$ (W2), $S_3 = 114,307$ (W3), and the total average inventory cost is 887061,65$. By using software GeneHunter, the best solution is found after 15 s running when the best solution do not change for 75 generations, seeing Fig. 5.

Table 2. The solution of the inventory problem for warehouse W1

Warehouse W1							
Period	Historical Demand	Unit Carrying (+)/Shortage (−)	Units in Inventory after Replenishing	Replenishing Period	Inventory Cost	Replenishing Cost	Total Cost
1	87919	21186	109105	2	317789,62	1000	318789,62
2	110646	−1541	109105	3	46229,77	1000	47229,77
3	117469	−8364	109105	4	250934,77	1000	251934,77
4	101578	7527	109105	5	112897,62	1000	113897,62
5	67452	41653	109105	6	624787,62	1000	625787,62
6	114304	−5199	109105	7	155984,77	1000	156984,77
7	92379	16726	109105	8	250882,62	1000	251882,62
8	91196	17909	109105	9	268627,62	1000	269627,62
9	119195	−10090	109105	10	302714,77	1000	303714,77
10	10642	2653	109105	11	39787,62	1000	40787,62
11	83394	25711	109105	12	385657,62	1000	386657,62
12	107271	1834	109105	13	27502,62	1000	28502,62
13	104765	4340	109105	14	65092,62	1000	66092,62
14	79359	29746	109105	15	446182,62	1000	447182,62
15	93278	15827	109105	16	237397,62	1000	238397,62
16	1099127	−22	109105	17	674,77	1000	1674,77
17	86209	22896	109105	18	343432,62	1000	344432,62
18	117040	−7935	109105	19	238064,77	1000	239064,77
19	92228	16877	109105	20	253147,62	1000	254147,62
20	96018	13087	109105	21	196297,62	1000	197297,62
21	97107	11998	109105	22	179962,62	1000	180962,62
22	109024	81	109105	23	1207,62	1000	2207,62
23	133200	−24095	109105	24	722864,77	1000	723864,77
24	84717	24388	109105	25	365812,62	1000	366812,62
25	135712	−26607	109105	26	798224,77	1000	799224,77
26	88247	20858	109105	27	312862,62	1000	313862,62
27	135306	−26201	109105	28	78044,77	1000	787044,77
28	92181	16924	109105	29	253852,62	1000	254852,62
29	128812	−19707	109105	30	591224,77	1000	592224,77
30	89073	20032	109105	31	300472,62	1000	301472,62
					Average Cost		296887,50

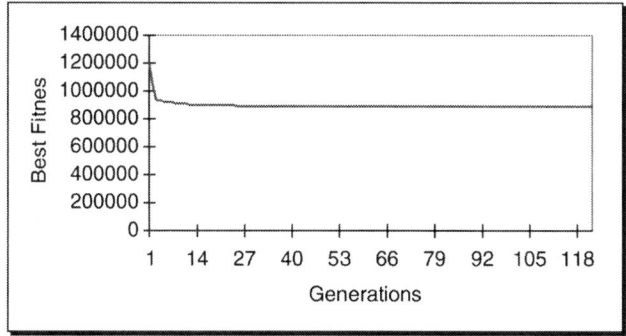

Fig. 5. GA running situation for warehouse control

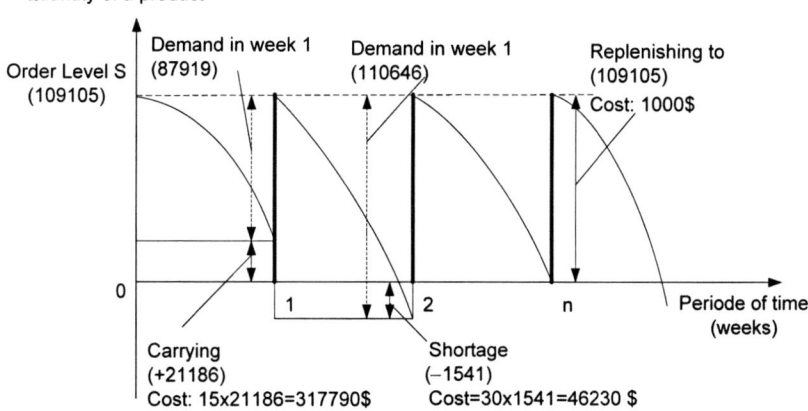

Fig. 6. Probabilistic warehouse inventory control policy (t, S) in W_1

An example of the cost calculation of the inventory control system $(t. S)$ in warehouse 1 is shown in Fig. 6, which includes carrying cost in week 1 (317790\$), shortage cost in week 2 (46230\$) and replenishment cost (1000\$). The solution from the GA process will provide the optimized inventory order level S_1 (109105).

4.3 Distribution Control and Market Inventory Control System

As described earlier, this problem involves both the transportation cost and the market inventory cost. The objective functions (1) together with (6) can be used as fitness function at this stage. This problem is solved for each period t. In practice this means that t disappears from the model. The decision variables are the amount distributed from each warehouse to each market. The calculation of the transportation cost is straight forward. Note that the number representing cargoes shipped has to be an integer value because it is not possible to ship 1.4 cargoes.

Since we are looking at only one time t, we have to use the expected value for the inventory cost as given earlier. To make the calculation relatively simple, the uniform distribution for the uncertain demand could be used. This means that the demand at each market is uniformly distributed between to values. With uniform distribution the formula for market inventory cost becomes:

$$
{}_m INV_j(t) = \frac{3_m c_{cj} \left(\sum_{i \in I} x_{ij}(t) \right)}{4(b-a)} + \frac{{}_m c_{cj} \left(\sum_{i \in I} x_{ij}(t) \right)^2 \left(\ln b - \ln \sum_{i \in I} x_{ij}(t) \right)}{2(b-a)} +
$$

$$
\frac{{}_m c_{sj}}{b-a} \left[\frac{b^2}{4} - b \sum_{i \in I} x_{ij}(t) + \frac{3 \left(\sum_{i \in I} x_{ij}(t) \right)^2}{4} + \frac{\left(\sum_{i \in I} x_{ij}(t) \right)^2 \left(\ln b - \ln \sum_{i \in I} x_{ij}(t) \right)}{2} \right]
$$

$$
+ {}_m c_{rj} \tag{8}
$$

When the uniform distribution is used to describe the demand, it also gives the upper limit for each market. The maximum demand at each market is the maximum amount one can order to that market. This was added as a constraint in the GeneHunter. Another constraint is that the sum ordered from each warehouse cannot be higher than the S_i. The maximum demand is chosen as an upper range for the variables.

In this case study, there are four markets, M_1, M_2, M_3 and M_4, which are located in Norway. The demands at each market are generated randomly according to the known normal distribution (μ, the mean value and σ, the standard deviation), which is same as warehouse inventory. The demand of each market is uncorrelated to each other. The optimal number of units distributed will yield a surplus or shortage at the market place constrained by the objective of minimizing the total supply chain cost. Each warehouse offers a different product price for the market regions mainly because of the difference in warehouse operation and shipping cost. Input parameters are showed in Table 3. The prices of the products at the market are given. The carrying cost, shortage cost and replenishment cost are different due to the market condition, currency exchange, labor cost and taxes. Table 4 shows the input parameters for each market.

Using the GA to calculate the optimal solutions for the units distributed from each distribution line, the best fitness value reveals the minimum total costs, which include market inventory cost and distribution cost. After running GeneHunter, the numbers of units shipped from each line, x_{ij}, the distribution cost and inventory cost are obtained. The distribution situation is showed in Table 5. After running the GAs, the following results are obtained: total inventory cost is 12855268,79$; total distribution cost is 6583752,97$ and total cost is 19439020,76$. The solutions are obtained after 1,250 trails and the generation process takes 45 s. A non-optimal policy to this problem is to just look at

Table 3. Input parameters for each distribution line

Distribution	Unit price $	Shipping cost per unit	Cargo cost per cargo $	Unit fit in one cargo $	Distribution Line	Distribution Quantity
W1 to M1	50,00	0,10	1700,00	25600	X11	26388
W1 to M2	50,00	0,10	1900,00	25600	X12	4918
W1 to M3	50,00	0,10	1840,00	25600	X13	12655
W1 to M4	50,00	0,10	1725,00	25600	X14	1434
W2 to M1	47,00	0,09	1900,00	25200	X21	13788
W2 to M2	47,00	0,09	1750,00	25200	X22	22004
W2 to M3	47,00	0,09	1640,00	25200	X23	18292
W2 to M4	47,00	0,09	1710,00	25200	X24	34448
W3 to M1	52,00	0,11	1640,00	25300	X31	0
W3 to M2	52,00	0,11	1820,00	25300	X32	0
W3 to M3	52,00	0,11	1750,00	25300	X33	0
W3 to M4	52,00	0,11	1810,00	25300	X34	0

Table 4. Input parameters for each market

Market	Product price $	Unit carring cost $	Unit shortage cost $	Replenishing cost $	Demand	Standard deviation
M1	250,00	75,00	225,00	1000,00	90000	15000
M2	270,00	81,00	243,00	1200,00	60000	10000
M3	225,00	67,50	202,50	800,00	70000	11667
M4	255,00	76,50	229,50	1000,00	80000	13333

Table 5. Distribution situations

Distribution Line	Distribution Quantity
X11	979
X12	6571
X13	6187
X14	2061
X21	35058
X22	18289
X23	16504
X24	18688
X31	4018
X32	1910
X33	8155
X34	14932

the expected demand in each market, and then order an equal amount of products from each warehouse. With this model, a total cost is 95485450.64\$. The GA method computes the trade-off of all parameters and derives the optimal allocation of inventory and distribution plan, which leads the reduction of the total cost to be more than 80%.

The use of GAs to manage the multiple sourcing inventory distribution supply chain system provide a flexible tool to handle markets changes and uncertain. By modifying the input parameters to our model, we will be able to quickly obtain a modified solution to adjust the market change. The calculation from GA provides the optimal number of unit distributed from each warehouse to each market zone. Further, the results from GA demonstrate the minimum total supply chain cost resulting from inventory and distribution cost.

5 Conclusions

This chapter studies multiple sourcing supply chain problems with stochastic demands. The mathematical models have been used as the objective functions of the optimization process. First, the order level at each warehouse based on statistical data which gather from historical record is considered. Then the distribution system is derived according to optimal number of units shipped from each warehouse to each market. A practical industry case with one product and multiple sourcing supply chain has been solved using Genetic Algorithms. The corresponding objective functions can be directly used as fitness functions of the GAs. In the warehouse stage, the minimum inventory cost and the corresponding order level S_i for each warehouse i can be obtained using the GA. In distribution control stage, the GA provides the minimum total supply chain costs including both distribution costs and inventory costs in markets. By adjusting the input parameters, a practitioner will be able to determine the appropriate inventory allocation and manage the distribution process with minimum costs. GeneHunter is used as the development tool for the GAs.

The practical case shows that the GAs have been a popular universal tool for solving supply chain optimization problems. This is due to the following benefits: (1) They can be easily implemented and adapted in various problems; (2) They usually converge rapidly on solutions of good quality; (3) They can easily handle constrains of an optimization problem; (4) They produce variety of good solutions simultaneously, which are important in the decision-making process.

Further research might extend the mathematical model, which should be more appropriate for the different inventory and distribution systems. Moreover, the GA process can be improved to treat multi-objective problems in supply chain management.

References

1. Altiparmak, F., Gen, M., Lin, L. and Paksoy, T., (2006), A genetic algorithm approach for multi-objective optimization of supply chain networks, *Computer & Industrial Engineering*, Vol. 51, pp. 197–216.
2. Chan, C. H., Cheng, C. B. and Huang, S. W., (2006), Formulating ordering policies in a supply chain by genetic algorithm, *International Journal of Modelling and Simulation*, Vol. 26, No. 2, pp. 129–135.
3. Droste, S., Jansen, T. and Wegener, I., (2002), Optimization with randomized search heuristics – The (A)NFL theorem, realistic scenarios, and difficult function, *Theoretical Computer Science*, Vol. 287, No. 1, pp. 131–144.
4. Ganeshan, R., Tyworth, J. E. and Guo, Y., (1999), Dual sourced supply chains: The discount supplier option, *Transportation Research* Part, E 35, pp. 11–23.
5. Ghodsypour, S. H. and O'Brien, C., (2001), The total cost of logistics in supplier selection, under conditions of multiple sourcing, multiple criteria and capacity constraint, *International Journal of Production Economics*, Vol. 73, pp. 15–27.
6. Han, C. and Damrongwongsiri, M., (2005), Stochastic modelling of a two-echelon multiple sourcing supply chain system with genetic algorithm, *Journal of manufacturing Technology Management*, Vol. 16, No. 1, pp. 87–108.
7. Holland, J. H., (1975), Adaptation in natural and artificial systems, Ann Arbor: University of Michigan Press.
8. Huang, M., Ding, J., Ip, W. H., Yung, K. L., Liu, Z. and Wang, X., (2006), The research on the optimal control strategy of a serial supply chain based on GA, Lecture Notes in Computer Science (including subseries Lecture Notes in Artificial Intelligence and Lecture Notes in Bioinformatics), Vol. 4221 LNCS – I, Advances in Natural Computation – Second International Conference, ICNC 2006, Proceedings, pp. 657–665.
9. Jeong, B., Jung, H. and Park, N., (2002), A computerized causal forecasting system using genetic algorithms in supply chain management, *Journal of Systems and Software*, Vol. 60, No. 3, pp. 223–237.
10. Kim, B., Leung, J. M. Y., Park, K. T., Zhang, G. and Lee, S, (2002), Configuring a manufacturing firm's supply network with multiple suppliers, IIE Transactions, Vol. 34, pp. 663–677.
11. Konak, A., Coit, D. W. and Smith, A. E., (2006), Multi-objective optimization using genetic algorithms: A tutorial, *Reliability Engineering & System Safety*, No. 91, pp. 992–1007.
12. Maiti, M. K. and Maiti, M., (2007), Two-stage inventory model with lot-size dependent fuzzy lead-time under possibility constraints via genetic algorithm, *European Journal of Operational Research*, Vol. 179, pp. 352–371.
13. Minner, S., Diks, E. B. and de Kok, A. G., (1999), Two-echelon inventory system with supply lead time flexibility, Faculty of Economics and Management, Ottovon-Guericke-University Magdeburg. IIE-Transactions.
14. Naso, D., Surico, M., Turchiano, B. and Kaymak, U., (2007), Genetic algorithms for supply-chain scheduling: A case study in the distribution of ready-mixed concrete, *European Journal of Operational Research*, Vol. 177, No. 3, pp. 2069–2099.
15. O'Donnell, T., Maguire, L. and Humphreys, P., (2006), Minimizing the bullwhip effect in a supply chain using genetic algorithms, *International Journal of Production Research*, Vol. 44, No. 8, pp. 1523–1543.

16. Ravindran, A., Ragsdell, K.M. and Reklaitis, G.V., (2006), Engineering optimization: methods and applications, Wiley, Hoboken, NJ.
17. Wang, K., (2005), Applied Computational Intelligence in Intelligent Manufacturing Systems, Advanced Knowledge International Pty Ltd, Australia.
18. Xing, L., Chen, Y. and Cai, H., (2006), An intelligent genetic algorithm designed for global optimization of multi-minima functions, *Applied Mathematics and Computation*, Vol. 178, pp. 355–371.
19. Yokoyama, M., (2002), Integrated optimization of inventory-distribution systems by random local search and a genetic algorithm, Computer & Industrial Engineering, Vol. 42, No. 2–3, pp. 175–188.
20. Zhang, W., Lu, Z. and Zhu, G., (2006), Optimization of process route by *Genetic Algorithms, Robotics and Computer-Integrated Manufacturing*, Vol. 22, pp. 180–188.

Printed in the United States
116960LV00002B/97-105/P